T0174489

NO STONE UNTURNED

NO STONE UNTURNED

A HISTORY OF FARMING, LANDSCAPE AND ENVIRONMENT IN THE SCOTTISH HIGHLANDS AND ISLANDS

• • •

ROBERT A. DODGSHON

EDINBURGH
University Press

For
Zara, Caitlin, Adam and Toby

© Robert A. Dodgshon, 2015

Edinburgh University Press Ltd
The Tun – Holyrood Road
12 (2f) Jackson's Entry
Edinburgh EH8 8PJ
www.euppublishing.com

Typeset in 10/12.5 Sabon by
Servis Filmsetting Ltd, Stockport, Cheshire,
and printed and bound in Great Britain by
CPI Group (UK) Ltd, Croydon CR0 4YY

A CIP record for this book is available from the British Library

ISBN 978 1 4744 0074 9 (hardback)
ISBN 978 1 4744 0075 6 (webready PDF)
ISBN 978 1 4744 0351 1 (epub)

The right of Robert A. Dodgshon to be identified as author of this work has been asserted in accordance with the Copyright, Designs and Patents Act 1988 and the Copyright and Related Rights Regulations 2003 (SI No. 2498).

CONTENTS

Acknowledgements vi
List of Figures viii
List of Abbreviations x

1. Writing the History of Highland Farming, Landscape and
   Environment 1

2. The Prehistoric Footprint 8

3. Prehistory into History 43

4. The Late Medieval and Early Modern Landscape: Stasis or Change? 80

5. On the Eve of the Change: a Look through the Surveyor's Eye 118

6. The Highland Toun through Time: an Interpretation 153

7. Landscapes of Change, 1750–c.1815: The Broadening Estate 192

8. Landscapes of Sheep, Deer and Crofts: Change after c.1815 227

9. The Years of Change: an Overview 270

Select Bibliography 282
Index 288

ACKNOWLEDGEMENTS

The aim of this book is to bring together a number of themes that I have worked on over the years regarding farming, landscape and environment in the Highlands and Islands, and to combine my data and thinking with that of other scholars so as to produce a rounded, long-term view or history. Work on the background material for the text has been spread over quite a period of time, with the gathering of relevant documentary sources being backed up by increasingly focused site visits across the Highlands and Islands. As with my previous books, I am especially grateful to my former university at Aberystwyth for providing both research leave and grants for the work involved prior to my retirement, both documentary work in Edinburgh as well as the field work. Some of the specific material dealing with livestock farming on the Breadalbane estate was funded as part of my involvement with the NTS Ben Lawers project. Again, I am extremely grateful for their support. Over the years, I have consulted documentary material in a number of archives, from the NAS and NLS in Edinburgh to family muniments in the Highlands: the help of all those who dealt with my requests and queries has been greatly appreciated. Most of the maps were drawn by Ian Gulley and Antony Smith, the cartographers in my former department, the Institute of Geography and Earth Sciences at Aberystwyth University. Sadly, Ian Gulley died during the later stages of the book's final preparation. As someone who has produced maps for a number of my publications over the past twenty years, I would like to take this opportunity to record my deep appreciation of all his fine work as a cartographer for me. Lastly, I am extremely grateful to John Watson and Ellie Bush of Edinburgh University Press for helping to steer my book through to publication.

This is the first book that I have written surrounded, on more than a few occasions, by my grandchildren. It is appropriate, therefore, that I have dedicated it to them.

Robert A. Dodgshon
Aberystwyth and Stratton Audley

FIGURES

2.1   Neolithic dwelling at Scord of Brouster, Shetland   14
2.2   The Prehistoric Footprint   14
2.3   Varieties of prehistoric dwelling   15
2.4   Neolithic dwelling at Skara Brae, Orkney   16
2.5   Bronze/Iron Age hut circle at Dola, near Lairg, Sutherland   19
2.6   Late prehistoric round houses   26
2.7   Broch at Dun Carloway, Lewis   27
2.8   Part of the 'village' that surrounded the broch at Gurness, Orkney 27
2.9   Radial piers of wheelhouse at Jarlshof, Shetland   28
2.10 Segmentation of space around wheelhouses/Atlantic
      roundhouses, North Uist   30
3.1   Forms of Pictish settlement   45
3.2   Pictish 'figure-of-eight' dwellings, Gurness, Orkney   46
3.3   Duns in the South-west Highlands and Islands   52
3.4   Viking longhouse, Jarlshof, Shetland   58
3.5   Viking and Late Norse dwellings, after Bigelow   59
3.6   Soil mound, Sanday, Orkney   62
4.1   Tiree, touns lying waste, 1662   84
4.2   Former settlement Illeray, North Uist   86
4.3   Land lost through sand blows, Tiree, 1768   87
4.4   Ramasaig, Skye, 1810   94
4.5   Ardvoile and Ballemore, Lochtayside, 1769   95
4.6   Daugh of Deskie, Strathavon, 1761   98
4.7   Daugh of Taminlienin, Glenlivet, 1761   100
4.8   Abstract representation of an infield–outfield system   108
5.1   Carie, Rannoch, 1756   122
5.2   Infields and 'sheeling' grounds in Assynt, 1769   125

5.3   Balewilline, Tiree, 1768–9   126
5.4   Cultivation rigs on South Harris   129
5.5   Houses at Laxobigging, Nesting, Shetland, 1827, after Thomson   131
5.6   Runrig at Funzie, Fetlar, Shetland, 1829, after Thomson   131
5.7   Cultivation rigs at Europie, Lewis   137
5.8   Shielings on northern slopes of Ben Lawers   138
5.9   Sheiling sites, Duirinish, Skye   139
6.1   Rosal, Strathnaver, 1811   154
6.2   Premodern enclosures on Coll   156
6.3   Forms of post-medieval dwellings   162
6.4   Blackhouse, Arnol, Lewis   163
6.5   Blackhouse interior, Arnol, Lewis   163
6.6   Dwelling at Illeray, North Uist   166
6.7   Cultivation rigs on slopes overlooking Loch Odhairn, Lewis   168
6.8   Former toun at Borrafiach, Waternish, Skye   169
6.9   Enclosures at Borrafiach, Waternish, Skye   170
6.10   Enclosures and settlement, north Kingsburgh, Trotternish, Skye 171
6.11   House sites and enclosures at Glenarigolach, Gairloch   173
6.12   Enclosures at Righcopag, south of Loch Naver, Sutherland   174
6.13   Pre-Clearance enclosures at Righcopag, Sutherland   175
6.14   How the farming toun evolved: an interpretation   179
7.1   Pre-crofting arable around Grimsay, North Uist, 1799   204
7.2   Settled islet beside Grimsay, North Uist   205
8.1   Deer forests in the Scottish Highlands, 1884   237
8 2   Crofts at Illeray, North Uist, 1799   242
8.3   Ladder crofts at Borghastan, Lewis   243
8.4   Crofts and pre-crofting rigs at Arnol, Lewis   244
8.5   Touns, clearances and crofts, Park district, Lewis   246
8.6   Land use on Lewis, 1884   248
8.7   Ormiscaig, Gairloch, 1845 and 1875   250
8.8   Crofts at Badluarach and Scoraig, Little Loch Broom   251
8.9   Peat cutting, Lewis   255
8.10   Heather burning in the Grampians   259

ABBREVIATIONS

ARCHIVES AND MSS COLLECTIONS

AP	Argyll Papers
CDC	Clan Donald Centre, Armadale
DC	Dunvegan Castle
FE	Forfeited Estates
HULL	Hull University Library
IC	Inveraray Castle
MDP	Macleod of Dunvegan Papers
MP	Lord Macdonald Papers
NAS	National Archives of Scotland (now National Records of Scotland), Edinburgh
NLS	National Library of Scotland, Edinburgh
GD16	Earls of Airlie Papers
GD44	Gordon Castle Muniments
GD46	Seaforth Papers
GD50	John McGregor Collections
GD64	Campbell of Jura Papers
GD80	MacPherson of Cluny Papers
GD92	Macdonald of Sanda Papers
GD112	Breadalbane Muniments
GD128	Fraser Mackintosh Papers
GD170	Campbell of Barcaldine Muniments
GD174	MacLaine of Lochbuie
GD201	Clanranald Papers
GD221	Macdonald of Sleat Papers
GD305	Cromartie Papers

GD312   Gordon Family Papers, Marquis of Huntly
GD403   Mackenzie Papers
SP   Sutherland Papers

PUBLICATIONS

APS   Acts of Parliament of Scotland
BPP   British Parliamentary Papers
JHG   *Journal of Historical Geography*
NSA   *New Statistical Account*
OSA   *Old Statistical Account*
PSAS   *Proceedings of the Society of Antiquaries of Scotland*
SGM   *Scottish Geographical Magazine*
SHR   *Scottish Historical Review*
SS   *Scottish Studies*
TIBG   *Transactions of the Institute of British Geographers*

AREAL MEASURES

Acreages down to and including Chapter 7 are in Scots acres, which equalled 1.26 of the imperial acre, though some later surveys do not make it absolutely clear that Scots acres are being used. Except for Blackadder's figures cited on page 239, acreages in Chapter 8 (following the 1824 Weights and Measures Act standardising measures) are given in imperial acres.

WRITING THE HISTORY OF HIGHLAND FARMING, LANDSCAPE AND ENVIRONMENT

This book looks at the long-term history of farming communities in the Highlands and Islands and the impact they had on landscape and the environment. Until comparatively recently, the prospect for such a history was limited by the fact that while a great deal had been written about the changes of recent centuries, enabling us to understand the major discontinuities represented by the Clearances and spread of crofting, there was a dearth of studies for the centuries that stretched back from this point. In part, this neglect had to do with the evidence to hand. An abundance of both cartographic and written estate surveys become freely available from the mid-eighteenth century onwards that have been used to cast a revealing light as much on what existed immediately prior to the Clearances and crofting as on what followed. By comparison, the documentary sources that are available for the centuries stretching back prior to the mid-eighteenth century are fewer in number, more fragmented in character and more limited in content. Faced with this thinning of material earlier than the mid-eighteenth century, it was hardly surprising that our first position on what the traditional or pre-Clearance landscape of the region looked like should have relied heavily on what mid-eighteenth-century surveys could tell us. Of course, given the richness of the surviving field archaeology, field surveys and excavations always had the potential to offset these documentary limitations. Indeed, the point at which field archaeology and excavation can become the prime source of understanding for farming and its landscape history is probably reached at a shallower point in the history of Highland landscapes than elsewhere in Britain. Until recently, though, approaches to field archaeology in the region, as elsewhere, tended to be site based, ignoring the wider landscape context.

Yet in part, this neglect of work prior to the mid-eighteenth century was also bound up with a particular reading of this history. Some early commentators solved the problems of the countryside's pre-eighteenth-century history by assuming that what we could see in the detail of its mid-eighteenth-century landscapes answered for earlier centuries. It was even argued that the institutional forms that underpinned the eighteenth-century farming landscape, notably infield–outfield cropping and the runrig organisation of landholding, had a lineage that extended back into late prehistory. In fact, for Estyn Evans, these institutional forms were seen as present throughout the Atlantic ends of Europe (the Scottish Highlands, Ireland, Wales, Cornwall, Brittany, western Spain). Far from being exclusively Celtic, he saw them as having roots that took them back to the very origins of farming in western Britain and Ireland during the Neolithic.[1] Such assumptions of archaism are actually deep rooted in the historiography of the Scottish Highlands and Islands. The tone was set by the government itself with even the 1752 Annexing Act for settling the Highlands referring to the 'barbarism' of the region and the need to 'civilize the inhabitants'.[2]

That the region may have seen little progress was a theme taken up by writers such as John Walker. On the basis of two separate tours (1764 and 1771),[3] Walker produced a report on the Hebrides for the Board of Commissioners for the Annexed Estates. In style, he was an excellent field observer, attentive to detail and context, and engaging with mundane issues in a matter-of-fact way. His description of agriculture on Lewis captures this approach. In a paragraph that ranges over the use of the caschrom and spade, harrows drawn by men, ploughs that were only lately known on the island, wool being pulled rather than shorn, barley being pulled up at the roots and grain dried by graddaning, he did not condemn such practices but he clearly saw the island as a cultural backwater, a museum of the living past.[4] Later, he made his position clear, seeing such practices as of great antiquity, elements of husbandry that had survived from the very dawn of agriculture in the region.[5] Nevertheless, he saw the continuing utility of hand-tool cultivation to the extent of advocating the spade husbandry as part of his solution for the smallholder. As Jonsson put it in his analysis of Walker's thinking, he 'took the side of prehistory against Lowland agriculture'.[6]

Variants of such a perspective recur through other reports or diaries on the region. In 1773, having travelled north from London on his *Tour*, Dr Samuel Johnson used a leisurely moment as he and Boswell entered the Highlands to record his thoughts on the nature of Highland societies such as those he was about to observe. He reasoned:

> Mountainous countries commonly contain the original, at least the oldest race of inhabitants, for they are not easily conquered, because they must be entered by narrow ways, exposed to every power of mischief from those that occupy the heights . . . As mountains are long before they are conquered, they are likewise long before they are civilized.[7]

Another variant of such a reading can be found in the travelogue of A. Campbell, written in 1802. In fact, he had hardly reached Strathyre before he was noting the 'happy inmates of these huts', 'their rude implements of husbandry; the dwarfish appearance of their cattle', and concluding that 'every article about their dwellings, is characteristic of a people as yet but in the unpolished state of infant society'.[8] The first Improvers began writing about the region at about the same time. Their ideological positioning is also made clear in much of what they had to say. Many of the techniques and husbandry practices being used by the Highlander, and especially the Hebridean, were seen by them as having descended in a pristine, unmodified way from late prehistory if not from the very beginnings of agriculture.

When, in the nineteenth century, scientists and scholars first began writing extensively about the region, this portrayal of it as deeply conservative, 'a window on the Iron Age' to adapt Jackson's phrase,[9] even became rooted in the academic debate. Among the early field scientists who researched the region, the reading of Highland landscapes by geologists may have been especially influential in helping to turn what had previously been mere assumption into what appeared to have the solidity of fact or science about it. At the core of their influence was the interpretation of Highland geology popularised by Murchison. His thesis, a thesis that sparked off what became known as the Highlands controversy, was that not only did the region have the oldest rocks in Britain but that there was a progressive shift to older and older rocks as one moved westwards, the Lewisian gneiss of the Outer Hebrides being both the oldest and most westerly rocks.[10] Clearly, at this point, there would have existed an isomorphism between what people assumed for the appearance of the cultural landscape of the region and what was being argued for its physical landscape. In an age where scholars moved freely between different fields of interest, we can be fairly certain that those who were writing about the cultural forms of the region were also aware of what was being said about its physical forms and that they each, somehow, made the other more plausible. The geologist, Geikie, an early supporter of Murchison, even wrote in his *Landscape in History* that the physical character of an area and the character of its people were interlinked.[11] When, later in the same text, he wrote that the Highlands in the centuries before the Forty-five were seen as being at an early stage of social development,[12] he must have been acutely conscious of the analogy that he was drawing between the physical and cultural forms of the region.

Such a reading was reinforced rather than corrected when specialist ethnographers first began to observe the region. For Carmichael, the runrig townships that survived into the late nineteenth century were rooted in an ancient tribalism.[13] Likewise, for Curwen, the region was a 'cultural backwater' even when seen in the early twentieth century and manifested an 'extraordinary culture-lag', its people, with their handtools of husbandry and blackhouses, being in 'the same state as the Celtic people of the pre-Roman Iron age in Wessex'.[14]

There is a direct lineage here between such thinking and Evans's presumption about the prehistoric roots of core institutions such as infield–outfield. Yet such assumptions had ramifications. By assuming a deep-rooted cultural continuity, they effectively suppressed the region's experience of change and the need to research that experience. Instead, the farming landscapes which we first see emerging into the light of the eighteenth century were seen as answering for earlier landscapes because they contained all the cultural information that we needed to know about what had gone before, with the immediate pre-Clearance landscapes and their prehistoric precursors belonging to the same genus. Of course, a prehistoric origin for some institutional forms is always a possibility but the point being made is that their antiquity was an assumed, rather than demonstrated, antiquity. Working from such an assumption made light of our limited knowledge. We knew little but what we knew was, nevertheless, deemed sufficient.

Over the past decade or so, the dim light in which these pre-eighteenth-century landscapes were seemingly cast has given way to a more powerful illumination that is being energised in different ways. First, archaeologists have made reconstructing the landscape that surrounded sites a vital part of their work. Agencies such as the RCAHMS have started to produce excellent regional and local surveys of archaeological landscapes.[15] Likewise, focused research programmes by a number of university archaeology departments, particularly in the Western[16] and Northern Isles,[17] have incorporated a strong landscape element to their work, one that has added greatly to our understanding of the deep history of Highland landscapes. Recording work by the county archaeology services has also generated a significant increase in the material available for landscape analysis, creating high-quality databases about past landscapes. Approaching the problem from a different angle, work by specialist groups such as Medieval or Later Medieval Settlement Research (MOLRS) Group has brought more focus to such themes.[18] Work at individual sites has also helped to move the debate forward. Building on the early work at Lix,[19] Rosal,[20] and The Udal,[21] new studies have added to the range of in-depth analyses of particular settlement sites.[22] Together, these various strands of approach have started to outline a clearer context within which to discuss the field evidence for the different elements of the farming landscape from prehistory down into the eighteenth century.

Second, others have extended the debate by exploring the available documentary data.[23] Though the pre-eighteenth-century material is not abundant, when used in conjunction with pre-Clearance or pre-crofting surveys, it enables us to gain some sense of the processes shaping landscapes in the centuries before the Clearances. When combined with the insights being generated by the field evidence, it puts us in a much better position to analyse these landscapes, rather than just to describe them, and to ask whether they have a history of stasis or change behind them.

Providing a third strand of approach, one that again has produced a range of fresh insights over the past decade or so, has been a more direct interest in how Highland society had an impact on the natural landscape of the region, changing its soils, clearing its woods and so on. With so much apparent wilderness, it is easy to assume that the farming landscape of the region is a tightly prescribed one, occupying only a small proportion of the total surface, very much the junior rather than dominant partner in landscape formation. Yet, while the imprint of human settlement was always confined, its impact on the natural landscape and its ecologies reached out far beyond the core of settlement. Well-known examples, such as the 'skinning' of land in north-west Lewis for peat or turf, can be matched by other examples.[24] In Orkney, human impacts from at least the Viking age onwards not only produced deepened top soils in some areas but has also led to an accumulation of soils under farmsteads and their outbuildings in a number of settlements, with successive generations of the latter literally rising in the landscape on a growing bed of midden.[25] On the Scottish mainland, myths about the late and dramatic clearance of the Great Wood of Caledon have now given way to a more informed history based on a more phased process of clearance.[26] Similarly, where the wood had retreated either before the axe or climatic degradation, we now have a better understanding of how stock grazing, heath-management practices, such as burning, and the cutting of turf and peat all played a part in creating semi-natural habitats, or cultural landscapes, out of the wilderness.[27]

When brought together, these various strands of research have opened up the possibility of writing a more comprehensive history of Highland farming and its impacts. It is not yet the basis for a definitive history, one in which there is a settled understanding for the main issues, and there are still gaps. There is, however, enough for us to lay out a preliminary framework for such a history. This book sets out to present such a text. Its core approach will be to draw out the institutional basis of the farming landscape, that is, the cultural, social and economic forms around which it was organised and how they changed or did not change over time. The prime sub-themes will include how well we can distinguish between phases of stasis and phases of change and how communities interacted with their environment and the variegated ecologies of the region. As a complex, multidimensional, multiphased artefact, one with layers of meaning, writing about the farming landscape needs an open approach as regards sources, one that tries to blend excavated data, data for upstanding surface remains, place names, documentary sources and cartographic material. The challenge in using these disparate sources lies in the way each presents landscape from a slightly different perspective. Where or when we are dependent on a single source, we need to be sensitive over how it slants our perspective. Thus, using eighteenth-century cartographic sources to look at settlement patterns or field layout can impose a sense of contemporaneity or sameness to what we see. By comparison, a survey of upstanding remains on a pre-Clearance site can reveal that many comprise structures and dykes of different styles and vintages. One gives a picture of unity and stasis, the other of compounded change. Likewise, no matter how systematic our approach, the reconstruction of farming landscapes through field evidence cannot fully offset the absence of documentary data about the tenure of land.

The substantive discussion of these issues is divided into eight chapters, covering different chronological phases or aspects of landscape change from prehistory down to *c.*1914. While accepting that trying to cover how the countryside changed from prehistory down to *c.*1914 is a challenge, it is also clear to me that fundamental questions about the role of continuity/discontinuity in the Highland landscape need that kind of overview on what has endured and what has not. To break down chronologically such a discussion into a series of disconnected studies, without the means of cross-relating them, risks a repeat of the Indian story about the blind men and the elephant. It tells how one has hold of its trunk and says, 'I think I know what an elephant looks like': another a tusk, saying 'I know what it looks like': another its tail, and so on.

Notes

1. E. E. Evans, 'The Atlantic Ends of Europe', *Advancement of Science*, 15 (1958), pp. 54–64.
2. APS, 25 George II *c.*41.
3. M. M. McKay (ed.), *The Rev. Dr. John Walker's Report on the Hebrides of 1764 and 1771* (Edinburgh, 1980).
4. Ibid., pp. 42–3.
5. J. Walker, *An Economical History of the Hebrides and Highlands of Scotland* (Ediuburgh, 1808), i, p. 120.
6. F. A. Jonsson, *Enlightenment's Frontier: The Scottish Highlands and the Origins of Environmentalism* (New Haven, 2013), p. 36.
7. S. Johnson, *A Journey to the Western Islands of Scotland*, ed. By M. Lascelles (New Haven, 1971), p. 43.
8. A. Campbell, *A Journey from Edinburgh through parts of North Britain* (London, 1802), p. 148.
9. K. H. Jackson, *The Oldest Irish Tradition: A Window on the Iron Age* (Cambridge, 1964).
10. D. R. Oldroyd, *The Highlands Controversy. Constructing Geological Knowledge through Fieldwork in Nineteenth-Century Britain* (Chicago, 1990).
11. A. Geikie, *Landscape in History and Other Essays* (London, 1905), pp. 1–2.
12. Ibid., p. 119.
13. A. Carmichael, 'Grazing and agrestic customs of the outer Hebrides', *Celtic Review*, 10 (1914–16), pp. 40–54, 144–8, 254–62, 358–75.
14. E. C. Curwen, 'The Hebrides: A cultural backwater', *Antiquity*, 12 (1938), pp. 261–89.

15. RCAHMS, *North East Perthshire* (Edinburgh, 1990); RCAHMS, *Strath of Kildonan. An Archaeological Survey* (Edinburgh, 1993); RCAHMS, Waternish (Edinburgh, 1993); RCAHMS, *Well Sheltered* (Edinburgh, 2001); RCAHMS, *Historic Landscape of the Cairngorms* (Edinburgh, 2001); RCAHMS, Eigg (Edinburgh, 2003); RCAHMS, *Mar Lodge Estate Grampian* (Edinburgh, 1995); RCAHMS and Historic Scotland, *But the Walls Remained* (Edinburgh, 2002).

16. For example, K. Branigan and P. Foster, *From Barra to Berneray* (Sheffield, 2000); Parker Pearson, M. (ed.), *From Machair to Mountains* (Oxford, 2012).

17. For example, J. W, Hedges, *Bu, Gurness and the Brochs of Orkney*, (Oxford, 1985), 3 vols; B. E. Crawford, and B. Ballin Smith, 1999, *The Biggings, Papa Stour Shetland. The history and archaeology of a royal Norwegian farm*, Society of Antiquaries of Scotland Monograph Series no. 15 (Edinburgh, 1999).

18. For example, S. Govan (ed.), *Medieval or Later Rural Settlement in Scotland: 10 Years On* (Edinburgh, 2003).

19. H. Fairhurst, 'The deserted settlement at Lix, West Perthshire', *PSAS*, 101 (1968–69), pp. 160–99.

20. H. Fairhurst, 'The surveys for the Sutherland Clearances of 1813–1820', *SS*, 8 (1964), pp. 1–18.

21. I. Crawford, 'Contribution to a history of domestic settlement in North Uist', *SS*, ix (1965), 34–65.

22. O. LeLong, and J. Wood, 'A township through time: excavation and survey at the deserted settlement of Easter Raitts, Badenoch, 1995–1999', in J. A. Atkinson, I. Banks and G. MacGregor (eds), *Townships to Farmsteads. Rural Settlement Studies in Scotland, England and Wales* (Oxford, 2000), pp. 40–9; D. H. Caldwell, R. McWee and N. A. Ruckley, in Atkinson, Banks, and MacGregor (eds), *Townships to Farmsteads. Rural Settlement Studies in Scotland, England and Wales* (Oxford, 2000), pp. 58–68; J. A. Atkinson, 'Settlement Form and Evolution in the Central Highlands of Scotland, ca. 1100–1900', *International Journal of Historical Archaeology*, 14 (2010), pp. 316–34.

23. R. A. Dodgshon, *From Chiefs to Landlords* (Edinburgh, 1998), pp. 51–63.

24. F. Darling, *West Highlands Survey* (Oxford, 1955), pp. 275–6.

25. D. A. Davidson, D. D. Harkness, and I. A. Simpson, 'The Formation of Farm Mounds on the Island of Sanday, Orkney', *Geoarchaeology: An International Journal*, 1 (1986), pp. 45–60; I. A Simpson, 'Relict properties of anthropogenic deep top soils as indicators of infield management in Marwick, West Mainland, Orkney', *Journal of Archaeological Science*, 24 (1997), pp. 365–80.

26. T. C. Smout, 'Highland land use before 1800', in T. C. Smout (ed.), *Scottish Woodland History* (Dalkeith, 1997), pp. 5–22; T. C. Smout, A. R. Macdonald and F. Watson, *A History of the Native Woodlands of Scotland, 1500–1920* (Edinburgh, 2005).

27. R. A. Dodgshon and G. A. Olsson, 'Heather moorland in the Scottish Highlands: the history of a cultural landscape, 1600–1800', *JHG*, 32 (2006), pp. 21–37.

THE PREHISTORIC FOOTPRINT

Of No Fixed Abode

Mesolithic hunter-gatherers were the first people known to have settled the Highlands and Islands. They were established in parts of the Hebrides by the seventh millennium BC, though a case has been made for an even earlier settlement.[1] By nature, their occupation sites were ephemeral, leaving few traces. Potential sources of field evidence, such as post holes for temporary shelters or windbreaks, cooking hearths, lithic assemblages, bone dumps or debris, and shell middens, are not easily located, especially where subsequent environmental change has submerged them under a rising sea level, or covered them with peat, silt or blown sand. Cave sites, such as those at Oban, or the midden sites on Oronsay represent straightforward loci of search but other sites have been found only through the chance exposure of buried sites. Techniques such as systematic field walking and the mapping of lithic scatter, however, have now enabled other sites to be recovered.

Viewed geographically, more Mesolithic sites are to be found on or near the coast than in the main body of the Highlands. This reflects the greater chance of locating coastal sites but we must not ignore the fact that proximity to coastal or estuarine ecologies was attractive to hunter-gatherer communities because of their high and diverse energy yield. Survival for such communities depended on putting together a year-round system of support based more or less on the direct consumption of resources, given that few storable items of food (for example, nuts) were available.

This meant bringing together a set of ecologies within their annual cycle of resource foraging, fishing or hunting that would answer for their minimum needs during each season. Within such a scheme of things, the gathering of seeds, tubers, fruit and nuts, the hunting of small and large animals, wildfowling, fishing and the collecting of shell food would all have had a particular value in particular seasons. In most cases, these seasonal strategies of food procurement would have involved shifts between two or more settlements during the year. While some sites may have been occupied all year round, others were no more than seasonal camps, designed to take advantage of local opportunities for collecting, fishing, wildfowling, and so on, for only for a few months of the year. In the southern Hebrides, for instance, where a great deal of recent archaeological work has been carried out, it has been argued that local Mesolithic communities were 'generalist foragers, without substantial, let alone permanent, base camps . . . the epitome of what Mithen et al. has [*sic*] termed *Thoughtful Foragers*',[2] their subsistence drawn out across a 'local-ised pattern of settlement and movement'.[3] Patently, this would have led to a diffuse if not undetectable impact on landscape.

When we ask what this impact amounted to, that created by their occupa-tion was the least substantial of all prehistoric settlement amounting, at most, to shelters, windbreaks, cooking hearths, working floors for stone or flint, bone and skins, and food refuse or middens. Where we are dealing with sea-sonal rather than main camps, what existed on the ground would have been even more ephemeral, with all but the camp's refuse, broken or discarded tools, arrowheads, scrapers, and so on. While we can hypothesise about the potential differences that may have existed between base camps occupied for long periods of the year and those occupied for only short periods with more specialised functions, such differences are difficult to validate in the field. The problems are well rehearsed by Wickham-Jones's report on the Mesolithic site at Kinloch beside Loch Scresort on the east side of Rum. Overall, the island had sufficient resources to maintain a small band of hunter-gatherers all year and, archaeologically, the site excavated yielded balanced assemblages but, as the excavators made clear, the evidence can be interpreted in different ways. The fine-scale differentiation of lithic material within the overall site, with concentrations of microliths to the south and scrapers to the north, could be read as denoting contemporaneous parts of a single site, perhaps expressing different areas of male and female work, but they could also be read as sea-sonal sites occupied at different times, with the microliths being the debris of a summer camp and the scrapers that of a winter camp.[4] We get a different slant on the same problem from sites in the southern Hebrides. There is a case for seeing the close proximity of Islay, Jura, Colonsay and Oronsay as offer-ing local Mesolithic communities a cluster of easily accessible ecologies which, when integrated, ensured their all-year-round survival. Their surviving traces present an interesting problem, however. Those sites dated to before 4800 BC and after 3400 BC have been found on Jura and Islay but not on Oronsay. Conversely, sites have been found on Oronsay datable to the years between 4800 and 3400 BC, but not on Jura and Islay. This 'chronological gap', as Mithen called it, raises the question of whether Mesolithic communities were able to subsist entirely on the limited ecologies of Oronsay during it or whether we can explain away the gap by presuming that our archaeological distribution of sites is incomplete.[5] Logically, the latter might seem the more likely explanation given the specialised subsistence ecology of Oronsay, with ready access to shell food, fish, marine mammals and seabirds but not much else. On the face of it, Oronsay middens might appear the result of 'many short-term visits by foragers based principally on the larger islands of Colonsay, Jura and Islay'.[6] Analysis of human bones deposited in some of the island's middens[7] concluded, however, that they belonged to people whose diet was entirely based on marine foods. In other words, those communities who lived on the island between 4800 and 3400 BC may have been resident all year round.[8]

Potentially, most Mesolithic settlements would have comprised a small cluster of shelters, windbreaks, cooking hearths, working areas for tasks such as the dismemberment of carcasses and the preparation of hides or for the making of flints and other stone tools, together with the inevitable debris of broken tools, shells, discarded bones, and so on. The temporary nature of such camps and the extended territory over which their occupants hunted or foraged from them meant that, while they had an impact on landscape, it was subdued. Hunting strategies may have added to these traces. Some studies have reported signs of disturbances to the woodland cover contemporary with Mesolithic communities, with reduced tree pollen, increases in open-ground herbs and raised charcoal levels. Among the anthropogenic explanations for such disturbances is the possibility that they may have used fire setting, the deliberate firing of woodland, so as to create an area of younger, more herbaceous growth that was attractive to foraging herbivores such as deer. The problem with this explanation is that much of the birch–hazel woodland that existed across parts of northern and western Scotland at this point would already have been more open in character anyway, obviating the need for fire setting.[9] Furthermore, even if fire setting was practised, establishing an association between disturbances observable in the palaeobotanical record and Mesolithic communities would be far from straightforward, simply because other factors, notably climate shifts, were also at work at this point.[10] Using data for both coastal and inland areas around Oban, Macklin et al. were able to show that raised levels of charcoal were associated with drier spells and lower levels with wetter spells during the Mesolithic, suggesting an 'underlying climatic control of fire frequency'.[11]

The First Farmed Landscapes

If we rely on the presence of cereal-type pollen, the earliest signs of the farming communities in the region are dated to the late fifth millennium BC, though signs of cereal-type pollen prior to *c.*4000 BC are easily confused with wild grass grains from the *Hordeum* group and are therefore not always a reliable indicator of early farming.[12] Less uncertainty exists over the increasing amount of evidence for cereal-type pollen that can be found after *c*.4000 BC. We should keep in mind, however, that current debate over the Mesolithic–Neolithic transition is far from settled. Some see the shift involved as marked by a clear break in lifestyle between hunter-gatherers and the earliest farmers, one that is likely to have been underpinned by new communities moving into the region. In a study which relied on the later phases of Mesolithic at Oronsay, Schulting and Richards used bone isotopes to characterise the sharpness of the break in dietary terms, with Mesolithic communities relying more on a marine-based diet (fish, shell food, and so on) and Neolithic communities having more of a land-based diet (cereals, meat, and so on).[13] For others, it was a more drawn-out process of change. Hunter-gatherer communities survived alongside the first farmers in parts of the region, the two forms hybridising so that, in time, there may have been hunter-gather communities who manipulated animal movements and farming communities who also hunted or foraged for food.[14] Whilst we can see these differences as differences over how we interpret the transition in the region as a whole, some see them as capturing *regional* differences, with areas to the west – contra Schulting and Richards – experiencing a more drawn-out transition as new peoples or new ideas diffused northwards along the easier pathways of the western seaboard but areas to the north, such as Orkney, and on parts of the mainland experiencing a sharper transition, one marked by the appearance of fully fledged farming communities, replete with fixed settlements, fields, tombs and ceremonial monuments.[15]

Being a more intensive form of energy capture, one that invested labour in the working of the land and enabled communities to adjust to more localised territories, farming marked a fundamental moment for landscape. That said, the woodland clearances documented by the palaeobotanical evidence after 4000 BC were generally small scale and temporary, so their initial impact on landscape must not be overstated.[16] A lot depended on the ecologies that faced the first farmers. In the north, both in Shetland and Orkney, the earliest farmers had created permanent clearings and field systems by the mid- to late Neolithic. There were reasons for this. Woodland habitats there were more open to begin with and, once cleared, trees did not regenerate easily. Work on sediments associated with Eilean Domhnuill, a Neolithic site set down on an islet in Loch Olabhat, North Uist, highlights another aspect. Cores sampled from the site show tree pollen as starting to decline from 4000 BC onwards. Around the time when Eilean Domhnuill was first occupied, or 3700 BC, the rate at which total arboreal pollen declined increased sharply. This decline was accompanied by a reduction in the level of total organic matter in the sediment profiles sampled and by the start of what proved to be a long-lasting phase of soil erosion.[17] Mills et al. concluded that, by the end of the Neolithic, the land surrounding Eilean Domhnuill had lost its tree cover, and its surface or soil cover had become destabilised. This environmental degradation was not attributed to the impact of settlement at Eilean Domhnuill alone. In fact, it was concluded that it may have been only a seasonally occupied site, possibly one with a ritual signifi-cance.[18] Other sites lay around it, so much so that Neolithic activity on North Uist needs to be seen as 'surprisingly dense, leaving little room for any surviv-ing wildscape as would have existed around the fringes of settled land on the Scottish mainland'.[19] In such a cultural landscape, we can conclude that envi-ronmental degradation would have been widespread. The clearing of woodland would have been one reason for it. The cutting of turf for fuel and for use as a basic raw material for house construction also played a part, however.[20]

For comparison, some have concluded that mainland areas in western Scotland saw no significant change in the level of woodland disturbance until the start of the Bronze Age,[21] though others have detected a 'marked' increase in cereal-type pollen at c.3500 BC, at least in coastal areas if not inland.[22] Such local differences are apparent in parts of the Hebrides, such as on Skye, but we always need to keep in mind the difficulties of reading what may have been very small-scale clearances from the pollen record. Nor should we overlook the fact that some signs of disturbance may signal the impact of pastoral farming rather than arable. Work in Glen Affric, for instance, concluded that sites which show no signs of charcoal and, therefore of burning, yet which show signs of a reduction in tall herbs, are likely to have been associated with these early grazing systems. In fact, there is a case for arguing that grazing may have been a widespread factor in the disturbance of woodland habitats across large parts of the northern and western highlands during the Neolithic, with the concentration of herds playing a part in opening up ground cover and grazing out young growth and tall herbs.[23]

Davies and Tipping's work in Glen Affric reached another interesting con-clusion. As well as confirming the small-scale and localised nature of woodland clearances for arable, they stressed the variegated nature of the habitats worked by Neolithic communities.[24] At this stage, farming was solely about outputs, not about balancing outputs with inputs, so that the challenge for Neolithic farmers was to capitalise on the pockets of natural fertility among these varie-gated habitats. This is why early attempts at cropping were initially based on the small-scale, temporary clearance of woodland, each new clearance provid-ing access to a fresh source of fertility. Some of this activity would have involved the clear felling of timber but some would have been directed at simply opening up the tree canopy and clearing the ground cover by fire, enabling vegetation and fertility to recover once cultivation was moved on. These were farmers for whom the adage 'the forest creates, the farmer removes' applied. Those in Glen Affric provide us with a glimpse of another solution, for they displayed a prefer-ence for cultivating the open, shrubby vegetation of alluvial soils, opportunisti-cally relying on seasonal flooding to refresh what they depleted.[25] In fact, it was farming strategy that was to endure in the Highlands.

In the more exposed northern and western parts of the region, early farming communities faced a different challenge. Because of their greater wetness and exposure, woodland habitats there were generally more open and conditions more marginal so that, as mentioned above, trees struggled to recover once cleared.[26] In these circumstances, clearing woodland to access stored fertility was not a sustainable strategy. Even during the Neolithic, some appear to have worked their way around the problem by cropping former middens, small anthropogenically derived sites that formed rich concentrations of plant nutrients. It was not a case, however, of former middens being spread across arable as an additive but of their cultivation *in situ* as small but highly productive garden plots.[27] The use of such sites surely confirms that early Neolithic cultivators were already conscious of the need to target nutrient stores, whether natural (that is, forest land) or artificial (that is, middens).

By its nature, farming enabled communities to adjust to a more intense, localised form of energy capture and to a more sedentary existence. As the Neolithic unfolded, signs of this settled existence emerged in the form of fixed settlements, clearance cairns and small enclosed fields, together with monumental ritual sites and funerary monuments. Among these different impacts, the emergence of fixed settlement had the greatest significance for landscape. Surviving examples are not spread evenly, occurring in relative abundance across the Northern Isles and to a reasonable extent in the Hebrides but less so on the mainland, especially on the eastern side of the Highlands. This may be about survival rather than presence for, when it came to the building materials, more use appears to have been made of timber and earth in the east, whether for domestic, ceremonial and funerary purposes, whereas more use was made of stone and turf in the far north and in the Hebrides. Ultimately, the spread of Neolithic cultures and ideas led everywhere to 'a settled lifestyle, widespread division of land and the use of new resources'.[28] Two sites illustrate this new, more intense relationship to land. One of the earliest sites of occupation is the Neolithic settlement and field system at the Scord of Brouster, Shetland. It comprises a small group of oval-shaped huts (Figure 2.1), two definite and a third probable, surrounded by five or six associated fields, some with signs of former ard (scratch plough) cultivation (Figure 2.2a).

Its initial occupation belongs to the mid-fourth millennium BC, so it ranks among the earlier settlements for the Northern Isles. Other early settlements elsewhere on Shetland, such as Stanydale and Gruting (Figures 2.3a and 2.3b), were also based around small oval huts, though a few were distinctive in having more regular, rectangular plans. By comparison, those settlements built later, or *c.*3000 BC+, such as at major sites like Skara Brae and Barnhouse, both in Orkney, were larger and more elaborate in their construction. The later huts built at the Skara Brae site were recessed deep in sand, and interconnected one with another (Figure 2.4).

On the mainland of Orkney, Barnhouse comprised a small cluster of circular huts, with even the smaller hut circles have distinctive designed

Figure 2.1 Neolithic dwelling at Scord of Brouster, Shetland.

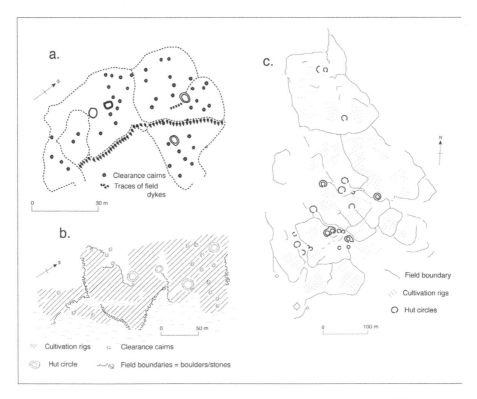

Figure 2.2 The Prehistoric Footprint: a. Neolithic settlement and fields at Scord of Brouster, Shetland, based on Calder, 'Stone Age sites', p. 371; b. Bronze/Iron Age hut circles and fields at Kilphedir, Strath of Kildonan, Sutherland, based on hut circle 1, Fairhurst and Taylor, 'Hut-circle Settlement at Kilphedir', p. 69; c. Late Iron Age hut circles and fields at Drumturn Burn, Perthshire, based on RCAHMS, *North-East Perth*, p. 46.

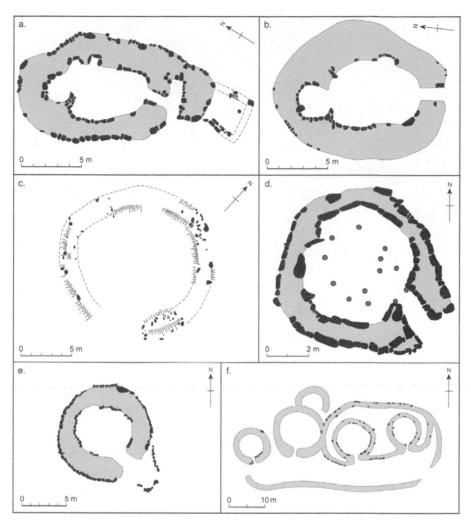

Figure 2.3 Varieties of prehistoric dwelling: a. Neolithic dwelling, Stanydale, after Calder, 'Report . . . Stone Age Sites', p. 341; b. Neolithic dwelling, Gruting, Shetland, after ibid., p. 345; c. Bronze/Iron Age dwelling, Kilphedir, Sutherland, after Fairhurst and Taylor, 'Hut-circle Settlement at Kilphedir', p. 72; d. Bronze/Iron Age hut circle at Coile a Ghasgain, Skye, after Armit, *Archaeology of Skye*, p. 104; e. Bronze Age hut circle at Cùl a'Bhaile, Jura, after Stevenson, 'Cùl a'Bhaile', p. 133; f. Hut circles at Alyth Burn, Perthshire, after RCAHMS, *North-East Perth*, p. 30.

features such as drainage. Apart from its scale, the prime feature of the large hut circle that dominates the group is that, while its outer bounding enclosure is circular, the hut site itself has a rectilinear layout of walls and inner space that, in effect, squares the circle.[29]

Figure 2.4 Neolithic dwelling at Skara Brae, Orkney.

A number of Neolithic sites have also been excavated along the western seaboard, on the mainland and in the Hebrides.[30] That at Eilean Domhnuill in Loch Olabhat, North Uist, has already been mentioned. It was set down on a small island site, linked to the mainland by an artificial causeway. The settlement itself only ever comprised a single dwelling, with turf or earthen walls raised over a simple stone footing. Over time, the dwelling was successively demolished and rebuilt, creating 'an almost tell-like accumulation of superimposed floors and hearth mounds'.[31] The fact that the site was subjected to frequent flooding, notably between *c.*2800 and 2700 BC,[32] was probably a factor in this successive rebuilding. That excavated at Machrie Moor, Arran, is one of the more complex sites, with hut, fields, clearance cairns and pits scattered across ground that was subsequently affected by peat development. Its most interesting features are its fields which were laid out as a series of long, rectangular fields, 200 metres in length and 50 metres in width, running down slope.[33] Just how spread settlement was, even by the Neolithic, is surely underlined by the traces of early fields on St Kilda. That at Gleann Mór, as well as the Tobar Childa system, have been dated to some time between the Neolithic and early Bronze Age, with ample surviving field evidence for the stone tools used to work the soil, including hoe blades or mattock heads.[34]

This more intense possession of space during the Neolithic was carried over into the henge monuments and earthen enclosures that emerged as ritual sites and the funerary monuments used to bury their dead. Henge monuments,

such as those at Callanish (Lewis) and the Ring of Brodgar (Orkney), display
an enhanced concern among early farming communities with the cosmology
of their world and with its ceremonialised performance through ritual. Such
rituals, especially those focused on the interplay between cosmology and
the farming calendar, sat comfortably in a world in which farming, though
capable of producing more subsistence, faced greater risks because of the narrower ecology on which it rested. The megalithic tombs, which also emerged
over the Neolithic period, can be seen, too, within a meaningful landscape
context. Islands such as Rousay shed light on this landscape context. As well
as being sited in prominent places, their scatter of chambered cairns or burial
monuments appears linked with modern-day blocks of arable, an apparent
territorialisation that suggests they were family tombs for ordinary farming
communities, not those of chiefly status.[35] We are given the impression of a
modular landscape, with each community or extended family having its own
block of arable overseen by its ancestral tomb on the higher ground inland.
Different things may have been intended by having such a visible, monumental
cairn but its prominence as a sort of megalithic charter of ownership of the
land now needed for survival was possibly one of them. Such equivalences
between family sites contrasts with what emerged around the heart of Orkney.
Initially, the ceremonial complex that developed around the Ring of Brodgar
and the Stones of Stenness appears to have been a site used, and managed, by
farming communities in the area, one that enabled them to engage with their
perceived cosmology of the world through a real-world enactment of its ritual.
The discovery of the more substantial housing at Barnhouse and the large
ceremonial building now located between the Ring of Brodgar and Stones of
Stenness suggests that, in time, control over belief and its rituals may have
passed into the hands of a more specialised priesthood.

The Broadening Impact

Quite apart from witnessing the appearance of the first metal-using cultures,
the Bronze Age, *c.*2200 BC+, saw a number of innovations that changed how
communities interacted with landscape. Some of these innovations have been
bundled together under the title of a secondary products revolution, including
the greater use of animal products, such as wool and possibly milk, together
with the greater use of animals for traction in pulling ploughs and carts.
Positioned at the end of a long chain of diffusion, it is difficult to assess when
a region such as the Highlands and Islands first acquired such innovations,
though one could argue that easy movement along the coast may have worked
in favour of their earlier adoption across the Western and Northern Isles. What
is clear, however, is that the Bronze Age overall experienced an expansion of
settlement and farming. This happened against a backcloth of environmental
changes. Even by the time Bronze Age communities had emerged, the warmer

conditions which formed the so-called post-glacial or Holocene climatic optimum (5000–3000 BC) had started to give way to a phase of cooler, wetter conditions. Initially, this shift towards cooler, wetter conditions was not sufficient to inhibit the continued expansion of settlement and farming that appears to have started during the closing phases of the Neolithic. By the late Bronze Age, however, the worsening climate had started to bring about two significant environmental changes. First, they brought to an end the northward expansion of woodland, an expansion that had seen hazel and birch colonise parts of the Northern and Western Isles, and oak woodland in parts of the latter. As conditions became cooler, wetter and windier over the third and second millennium BC, tree growth became more difficult in these exposed areas, with woodland either dying back or struggling to regrow after clearance. Pine was also affected. During the first half of the third millennium BC, encouraged by a phase of climatic amelioration, it followed hazel and birch northwards into parts of Sutherland and Caithness. By the start of the second millennium BC, though, and the onset of wetter conditions, this expansion was reversed, with pine retreating south to core areas such as Abernethy and Rothiemurchus Forest and the catchment of the Beauly River, though anthropogenic impacts may have had some role to play in its decline at some sites.[36] Second, the cooler, wetter condition ultimately led to waterlogging and the start of peat formation. Its development, including the emergence of blanket peats, was not something that happened synchronously across the region. Different peat hollows and different areas of blanket peat had different histories.[37] We know that, by the close of the Bronze Age, some farm ground was being slowly overwhelmed by peat, forcing the communities off the higher ground down on to the better-drained valley slopes, though other factors, such as the impact of the Hekla 3 (2310 BC) and Hekla 4 (c.950 BC) eruptions in Iceland may have had a contributory role through the impact of its tephra outfall on crops and stock. Where we glimpse the remains of hut sites and fields walls that lie partially covered by peat, it is not difficult to imagine a once more extensive landscape lying wholly beneath it. Along parts of the west coast, but especially in parts of the Western Isles, these problems facing Bronze Age communities were compounded by sea level changes. The Holocene saw both sea level recovery and an uplift of land surfaces. The fact that these took place at different rates meant that, in low-lying areas like those that fringed the western side of the Long Island, there were times when the land gained at the expense of the sea and, at other times, when it lost. Overall, the net effect was an encroachment of the sea from the Bronze Age onwards, with extensive areas of machair lost.[38]

Bronze Age settlement is typically evidenced by their hut circles, together with a scatter of small clearance cairns, small irregularly shaped or curvilinear field clusters and large irregular enclosures (Figures 2.2b and 2.2c). The paucity of dated sites, however, coupled with the fact that, as a settlement form, hut circles continued in use down into the Iron Age, mean that it is not

Figure 2.5 Bronze/Iron Age hut circle at Dola, near Lairg, Sutherland.

always clear whether such sites are of Bronze Age or Iron Age origin. Yet, even with such a qualification, it is clear that Bronze Age settlements were widely distributed across the region. Those in Sutherland have been the most comprehensively studied. More numerous in the east than the west, a RCAHMS survey first published in 1911 noted over 250 separate sites where Bronze Age hut circles, either singly or in groups, were to be found, most lying just beyond the zone of modern-day intensive farming and site disturbance, but some surviving within (Figure 2.5).

Later work by Fairhurst and Taylor increased this figure to over five hundred groups, embracing over two thousand separate hut circles.[39] Many can be found strung out along the sides of the straths that run inland from the east and north coasts, such as the Strath Fleet, Strath of Kildonan, Strath Brora and Strathnaver, as well as on the lower, wetter ground around their headwaters.

Detailed mapping of Bronze Age settlement in Sutherland has now been carried out in two areas: around Dunbeath and in the Strath of Kildonan. What comes across from the survey of the former is the extent to which environmental change from the late Bronze Age must have subjected local field economies to environmental stress with sites along the valley slopes of Houstry Burn and Dunbeath Water now being covered with peat as well as turf.[40] The RCAHMS has also provided us with a detailed survey of those to be found in the Strath of Kildonan (Sutherland) where hut circles, clearance cairns, burnt mounds and associated field systems occur widely, stretching inland along the

River Helmsdale and its tributaries as far as Loch Badanloch. Settlement in the strath actually began during the Neolithic but this stage is represented only by burial cairns, standing stones and cup-and-ring marks. The earliest occupation sites are provided by hut circles of early to mid-Bronze Age date. Broadly circular in shape, their internal diameters varied mostly between 5 metres and 12 metres (Figures 2.3c–e). The earliest sites are those comprising small clusters of between two and three huts, though some contain as many as six. They occur in association with clearance cairns, some linked together by banks. We can contrast these with those thought to be of later date that comprise larger clusters of hut circles associated with clearly developed field systems, delineated by banks, walls and lynchets.[41] This distinction between, on the one hand, sites with hut circles and clearance cairns and, on the other, those involving more developed hut circles 'with trailing banks of stones and boulders marking out what appear to be minute fields which are devoid of the clearance cairns' was also recognised by Fairhurst and Taylor but they saw them as contemporaneous with each other, dating both to a shallow phase of occupation in the mid-first millennium BC.[42] That the fields were used for arable is confirmed by the presence of narrow rigs similar to so-called cord-rigs.[43]

We generally do not find the same abundance of field data for Bronze Age settlement in the Western Highlands and Islands as we do for the eastern side of the region. There are areas, however, where there is good field evidence. The fairly numerous farm mounds of the Uists, for instance, provide us with a valuable context for the study of Bronze Age settlement. Like others in the far west and north, they were sites on which settlement debris was accumulative, with different layers of occupation succeeding each other so that, over time, the site became a tell-like structure. Those in the Uists are invariably sited on the machair.[44] In fact, sand blows played a part in their formation by, on the one hand, helping to seal one occupation layer from another and, on the other, slowly deflating the surrounding surface, so that the mound became more upstanding. That at the Udal in North Uist was the first to be excavated, with occupation layers from the Neolithic down to the eighteenth century AD, being uncovered, including layers of Bronze Age settlement.[45] Latterly, a substantive excavation report on other farm mounds in the Uists has highlighted the value of seeing the long-term history of settlements through a single site. Two of its conclusions stand out. First, it demonstrated that the character of dwellings and other buildings experienced ongoing changes and adjustments in style so that we cannot speak of a Bronze Age dwelling or an Iron Age dwelling as if it was a fixed, immutable form. Early Bronze Age dwellings at some sites, for instance, were timber based, with turf also being used, whereas those of the late Bronze Age were the 'typical' stone-based hut circle.[46] Second, by their very nature, the analysis of such sites provides a clearer guide to breaks in the continuity of settlement. Those farm mounds studied given some support for those who have proposed a break in settlement between the Early and Late Bronze Age.[47]

Elsewhere, traces of Bronze Age settlement are less easily disentangled from morphologically similar Iron Age forms. As with the hut circles at An Sithean, in the north-west of Islay, only excavation can resolve questions of chronology.[48] An Sithean was associated with a network of fields or enclosures. Like many other Bronze Age sites, it was affected by environmental degradation, with progressive waterlogging and peat formation forcing its eventual abandonment, though part was eventually reoccupied during the medieval or post-medieval periods.[49] The abandonment of Bronze Age sites as climate worsened during the second millennium BC was, in fact, a problem across the region, with many former hut sites and field systems, as at North Dell on Lewis, now buried under peat.[50] Elsewhere, the report for the excavations at Kebister on the Shetland mainland concluded that the growth of peat across the upper slopes of the township led to 'some contraction of settlements onto the more fertile lower slopes during the second millennium BC',[51] the excavators reiterating Fojut's observation that the spread of peat in Shetland during the second millennium BC forced a general concentration of settlement on the coast.[52]

There are also good reasons for believing that, bound up with the intensification of settlement, the Bronze Age also saw the emergence of settlement that worked a larger area of arable when compared to their Neolithic forebears. This is confirmed by the marked increase around the start of the Bronze Age in cereal-type pollen found at a number of sites,[53] as well by the archaeological record. Thus, work on the farm mound at Baleshare (North Uist) suggests that the Bronze Age community there now cropped an arable area of *c.*3 hectares, exclusively for barley, and maintained it in continuous cultivation.[54] Of course, any intensification of output had to address the question of how to maintain nutrient levels. Data from the excavation of farm mounds in the Uists show that Bronze Age communities responded by adopting a more active manuring of arable. Instead of simply taking advantage of former midden sites, and cultivating them *in situ*, they now appear to have actively spread their midden refuse across arable. By the Late Bronze Age, some were also adding organic manures such as peat and turf.[55] In the Northern Isles, sites such as Old Scatness show that communities still relied for their arable on the opportunism afforded by their old midden sites but now 'fresh' kitchen waste and hearth ash was being added.[56] As with its Neolithic antecedents, Bronze Age farming in the Northern Isles displayed a high dependence on domestic waste and the 'absence' of organic manures,[57] though signs of some peat and night soil being used are present on some sites.[58] As in the Western Isles, the greater effort now being made to add fertiliser, whether domestic waste alone or waste plus organic manures, left its legacy in an extension of arable and a deepening of their topsoils.[59] The 2 to 3 hectares of arable cropped by the typical Bronze Age farm, though, was still modest compared to that of the average early modern farming toun.

Stock was probably at least as important to the Bronze Age farm economy as arable. Bone dumps around sites of settlement and the stock pens provide obvious clues but asking how, on what scale and where pasture was used is more demanding. Prior to the climatic worsening that had set by the mid- to late Bronze Age, many high-ground pastures must have been much more productive than they are today. The survey of the Strath of Kildalton noted some isolated hut circles as being possible shieling sites, implying that they were dependent on a parent site elsewhere, perhaps in the main body of the Strath. Yet, in the context of the Early Bronze Age, some upland pastures as high as 300 to 400 metres would have been able to support specialised pastoral communities settled in such habitats all year round.

Filling out the Landscape

From the start of the Iron Age, at about the mid- to first millennium BC, we can observe a number of changes in landscape development. First, whereas Bronze Age settlement was open, Iron Age settlement was more enclosed and, in some cases, defensive in character.[60] Second, the spread of iron, a metal that was harder and could be more easily procured, meant that farming communities made greater use of it for farms tools. Coupled with other changes during the mid- to first millennium BC, notably the replacement of the saddle quern by the greater output of the rotary quern, the adoption of new manuring techniques and, by the late Iron Age, the adoption of oats as a field crop, arable farming acquired a greater capacity for further extending and intensifying farm output. Third, driven by an expansion of numbers, there was a filling out of settlement on the lower ground. Most niches occupied by eighteenth-century farming touns were probably utilised in some form by Iron Age communities.[61]

We can organise the review of Iron Age settlement under a simple geographical distinction: the Central and Eastern Highlands, and an Atlantic province embracing the Hebrides and Northern Isles.

Settlement Forms: the Central and Eastern Highlands

Three types of settlement existed in the Central and Eastern Highlands. First, there were drystone-based hut circles found on the higher ground around the eastern and southern edge of the Highlands. Harris's study of late prehistoric hut circles in south-east Perthshire surveyed around seven hundred hut sites, all on high ground and occurring in small groups or clusters.[62] They covered a range of typologies, from single- to double-walled sites, from relatively small huts (internal diameter, 5 metres) to much larger examples (internal diameter, 18 metres), and from huts linked or conjoined to each other to others that were free-standing (Figure 2.3f).

Chronologically, we are dealing with a type of site that was in use from the second millennium BC down to the early first millennium AD.[63] A survey of similar sites was carried out by the RCAHMS in north-east Perthshire, including the Glen Shee and Strathardle area. Hut circles not only represented the most commonly occurring prehistoric monument in the area, with 845 examples in all, but, in the words of the survey, they 'form one of the densest concentrations of prehistoric settlement remains known in Scotland',[64] with other local concentrations nearby in the Glen Isla and Pitlochry area. As in southeast Perthshire, collectively, they represent a diversity of structures but single-walled hut circles formed the vast majority, many set down on platforms.

The second type of Iron Age settlement found in the Central Highlands, one that also occurs widely in the South-west Highlands and Hebrides, is the crannog. Crannogs were settlements built on an artificial islet in inland lochs, usually with an artificial causeway linking them to the shore. Those in the main body of the Highlands, such as on either side of Loch Tay and along Loch Awe, were built out timber.[65] In comparison, those found in the Western Isles tended to make more use of stone both for the settlement itself and for the foundations though some, like the North Tolsta crannog, were timber based.[66] Excavated examples in the main body of the Highlands, such as Oakbank on the north side of Loch Tay, have yielded dates that suggested it was occupied for only a four-hundred-year period from the late Bronze Age down into the early Iron Age[67] but evidence from other sites suggests that many were occupied throughout the Iron Age and more than a few may have continued in occupation well into the first millennium AD.[68] Despite their defensive siting, crannogs have been portrayed as simply farmsteads, albeit farmsteads detached from the farmland that they worked. The way in which they were spaced along the shoreline of lochs such as Loch Tay supports such a conclusion. An initial survey located seventeen examples on either side of the loch, a number later revised to eighteen, each one appearing to command a pocket of onshore arable.[69]

The defended enclosures or forts found across the southern and eastern parts of the Highlands provide us with a third type of settlement, one that hints at a much more hierarchical society. The precise distinction between what is a defended enclosure and what is a fort is not always clear.[70] Both occupied naturally defensive positions such as rock outcrops or hilltops across a range of altitudes. In scale, they enclosed an area that ranged from 250 square metres to more than 5 hectares. Forming a distinctive sub-type within the broad class of fortified enclosures are the timber-laced and vitrified forts found widely across the Northern, Eastern and Southern Highlands and dating from the early first millennium BC.[71] Their name derived from the way in which the stone ramparts had been fused by the burning of the timbers with which they were interlaced, a process that may have started out as an accidental process but which, once the strength of its consequences were apparent, may have been carried out deliberately.[72]

Settlement Forms: the Atlantic Province

Our understanding of settlement in the Atlantic province of the Highlands and Islands during the Iron Age has made considerable advances over the past decade. In 1990, Armit was able to write that areas such as 'the Western Isles exist on the periphery of archaeological awareness: a forbidding tangle of brochs, duns, galleried duns, island duns, wheelhouses etc.'.[73] He saw this neglect as due partly to the 'very richness of the prehistoric landscapes'.[74] Since then, some of the tangle has been unravelled, not least by Armit himself. The early emphasis was on the typological uniqueness of each form, their differences and the need to interpret their origins separately. This approach is best exemplified by MacKie's early work on brochs,[75] in which he not only argued for their initial appearance within the Northern Isles but also attributed their development to the arrival of new migrants from outside the region. He subsequently modified his original position but he still sees Shetland as playing a key role in their initial development and continues to see them as reflecting influences from outside the region.[76] Others have adopted a different approach. Instead of seeing the various forms that made up Iron Age settlement in the Atlantic Province – the roundhouses, brochs, cellular houses, wheelhouses, aisled wheelhouses, duns and galleried duns – as discrete types of structure that need to be distinguished from each other on the basis of style, origin and chronology and whose more innovative elements are best interpreted as exogenous, they suggest that we should see them within a single context of debate as structures that shared overlapping, not successional, chronologies, all drawing on an indigenous pool of building know-how.[77]

The starting point for this rethink is the work of Armit. Within the broad generic grouping of all roundhouse forms, he drew the distinction between simple and complex forms.[78] The 'simple atlantic roundhouses' were those that lacked the defining traits of the broch, even if large and substantial in character, while the 'complex atlantic roundhouses' were those that incorporated some or all of these traits.[79] His proposals have not been accepted without question but they have undoubtedly given fresh momentum to the debate.

The earliest of the roundhouse structures were the simple roundhouses. These began to appear in the Northern Isles by *c.*700 BC. Prior to this point, settlement in the far north consisted of small cellular hut clusters, some enclosed and some open. In comparison to the latter, the first roundhouses were a different, more monumental form of settlement. They were more massive in size, both in terms of their overall diameter and in having thicker walls. This is shown well at Howe, by Stromness, Orkney where the full sequence of Iron Age settlement – as well as earlier forms – is played out on the same site. By about the eighth century BC, the enclosed 'nucleated' settlement of the late Bronze Age had given way to a large stone-built roundhouse with walls that were 4 metres thick.[80] By their scale and upstanding nature, even simple roundhouses made new demands. As Armit has observed, their height in an exposed environment demanded insulation, an insulation now provided by thicker walls. Likewise, their larger diameter also placed a heavier demand on timber for roofing,[81] at a time when timber was becoming scarce in the Northern and Western Isles. Though massive in construction, the early roundhouses cannot be classed as defensive, or as a response to increasingly unsettled conditions. The reason for their development, and possibly the reason for their purposeful visibility, may lie instead in changes over landholding. Environmental changes may have reduced the amount of good land available and, in the process, generated more intense competition for land, a competition that persuaded those who had control to publicise their possession in a conspicuous way.[82]

The more complex roundhouses first emerged during the second and third centuries BC.[83] They appear in different forms but the most notable was the broch tower. As with other late prehistoric roundhouses, these not only appear in number but also seem to partition areas suitable for arable. In construction, they formed a tower built from two concentric walls with passages, intermural staircases and cells inserted into the space between them (Figure 2.6a and b).

Their usable space was organised into at least two levels, a ground level, possibly a store, and a second level where the main living quarters were based. Many broch towers are now badly ruined so that we can only guess at their original height but not all were necessarily as high as multistorey brochs such as Mousa or Dun Carloway (Figure 2.7) when in use.

In fact, when defined simply as all tower-like structures, brochs were surprisingly varied. Even on Orkney, where some of the best-preserved examples have survived, they 'do not conform to a single plan',[84] with large variations in the thicknesses of their walls and in the overall diameter of the broch: the former varied from 2.75 metres to 5.2 metres while the internal diameter of brochs, their floorspace, ranged from 7.3 metres to 13.7 metres.[85] Adding to the complexity of roundhouses by the second century BC is the fact that some in the far north, mainly on Orkney but including some in Caithness, acquired a tightly drawn outer enclosure around them, within which was developed a cellular system of huts (Figure 2.6a), the whole complex, broch plus cellular huts, becoming, as Hedges put it, a 'keep with a village'.[86] At Howe, this phase also involved the construction of a new, more developed broch.[87] Yet, though best developed at broch sites, some cellular hut clusters were also developed beside roundhouses that had no hint of aspiration to being a broch. If we include these in the sample, there were over twenty such 'villages' on Orkney. For Hedges, their growth in Orkney was a natural development, for small clusters of late Bronze or early Iron Age cellular huts had persisted in use alongside roundhouses. All that now happened was that the two forms were brought together as if part of a single social system.[88] The village at Gurness is one of the few such sites to have been excavated. It comprised 'a gatehouse, a

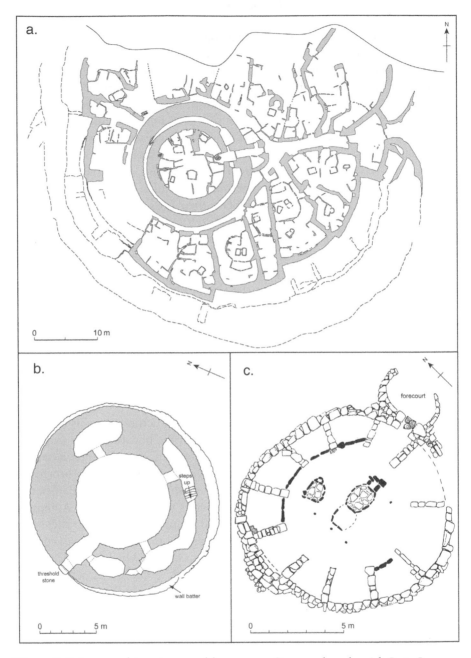

Figure 2.6 Late prehistoric round houses: a. Gurness broch with Late Iron Age 'village', based on Hedges, 'Surveying the Foundations', p. 29; b. Broch at Carloway, Lewis, based on C. Tabraham, 'Excavations at Dun Carloway broch', *PSAS*, 108 (1976–77), p. 157; c. Wheelhouse at A'Cheardach Mhor, South Uist, based on Armit, *The Later Prehistory*, p. 61.

Figure 2.7 Broch at Dun Carloway, Lewis.

Figure 2.8 Part of the 'village' that surrounded the broch at Gurness, Orkney.

Figure 2.9 Radial piers of wheelhouse at Jarlshof, Shetland.

passage to and round the broch, this giving access to a number of radial conjoined outbuildings, their backs forming a curtain wall which lined the inner
side of the Great Ditch'.[89] Altogether, the different hut cells at Gurness filled
the space between the broch and the outer surrounding wall of the site (Figure
2.8). When we combine the families that lived in the broch itself with those of
the village, it formed a substantial settlement.

Wheelhouses consisted of a circular structure set *into* the ground and
accessed via a short passage. Internally, they were arranged into a series of cellular spaces laid out around their inner edge, each cell demarcated by a series
of radial piers, disposed like the spokes of a wheel, that gave way to a central
open space at the core of the wheelhouse (Figures. 2.6c and 2.9).

This central floor space at the core was covered by a timber roof that projected above ground but the perimeter cells were corbelled and buried beneath
a capping of soil or sand, a roofing arrangement that was probably a response
to the increasing scarcity of timber.[90] The number of cells and the size of the
wheelhouse were closely correlated. Smaller wheelhouses of 7 to 8 metres
diameter had eight or nine cells while the larger wheelhouses, such as Sollas
(North Uist), had an overall span of 9 to 11 metres and contained eleven to
thirteen cells.[91] Whereas the larger, more complex roundhouses were built to
be visible, the wheelhouse seems to have been designed more to conceal itself,
or at least shelter itself, more effectively. Except for the timber roofing over
its core, they were recessed into the ground but their inhabitants were not

averse to setting it down in the midden soils of earlier settlements or their own midden. As Parker Pearson and Sharples put it, their inhabitants lived 'encased in their middens and dug into the very soil which gave them their livelihood'.[92] Once seen as a form of settlement that replaced the brochs, they are now seen as contemporary with the brochs and, therefore, as in use during the closing centuries of the first millennium BC[93] and continuing in use down to the first to second centuries AD.

In the case of the so-called broch villages of the far north, the 'village' component may have continued in use after the broch itself was abandoned. In time, probably by the third and fourth centuries AD, they were replaced by small hut clusters, some of a simple cellular type, such as those at Gurness,[94] but others of a type known as figure-of-eight huts, as at Buckquoy.[95] In character, cellular houses formed 'a detached structure in which a central chamber is surrounded by rounded or rectilinear cells, the whole entered by a single entrance'.[96] Often, as at Howe (Orkney) and Dun Bharabhat (Lewis), they were built over part of the broch or wheelhouse site, using stone from earlier structures.[97] As a slight, almost regressive form of settlement structure when compared to the roundhouses which they replaced, it has taken time for these cellular huts to become accepted for what they are, that is, 'standard settlement units of the period'.[98] Their reduction in size relative to earlier forms has been interpreted as marking a discontinuity, one denoting new influences or people,[99] but others prefer to see them as part of an ongoing 'unified tradition of architecture' that continued into the post-broch period without any external input.[100]

The distribution of Iron Age settlement in the Atlantic province has attracted much attention. Whatever its preferences, it was affected by two ongoing environmental changes. The first was the climatic degradation that had set in by the end of the second millennium BC which served to force settlement away from the wetter, marginal sites.[101] The second was that which occurred along the highly dynamic Atlantic-facing coasts of the Hebrides. Quite apart from what was lost through changes in sea levels, some of these Atlantic-facing coasts experienced a considerable loss of land from sand blows wherever the machair was disturbed by cultivation or by storms, with seaward areas being extensively eroded and inland areas being buried by blown sand. Over time, sites, such as the wheelhouse at Sollas (North Uist), were 'periodically inundated' by sand.[102] In fact, most of the large-scale archaeological programmes which have surveyed islands such as the Uists and Lewis over recent years have drawn attention to the land loss caused by the seaward erosion that was actively taking place by the first millennium BC.[103] In the Uists, it was a doubled-edged process, with inland sites lost through waterlogging and peat formation, and coastal sites, such as those around Baleshare, lost through coastal retreat.[104]

Despite these environmental pressures, what is striking about Iron Age settlement in the Atlantic Province is its abundance and its spread, with most sites with any suitability for cultivation being occupied in some form. We can

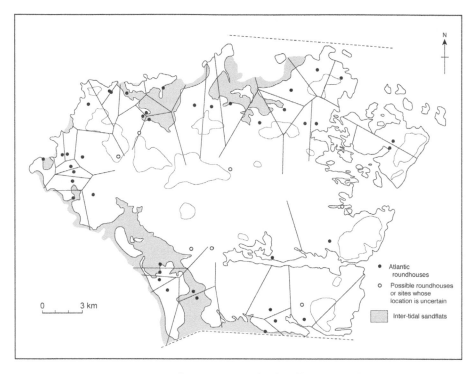

Figure 2.10 Segmentation of space around wheelhouses/Atlantic roundhouses, North Uist, using Thiessen polygons, based on Armit, *The Later Prehistory*, p. 118.

see this from three particular surveys. The survey by Armit et al. of the Bhaltos Peninsula on the northern coast of Lewis disclosed over fifty sites covering the prehistoric, early historic and medieval periods. Though many sites remain to be dated, those that have been dated to the Iron Age suggest that they represent a fair proportion of its main settlement, including three roundhouses and two wheelhouses. The Bhaltos site embraces an area of machair erosion on its eastern side where surviving archaeological remains have actually assisted in binding the sand, helping to resist local erosion. As a consequence, many of the archaeological sites 'survive as islands in a deflated machair plain, reduced to isolated points of survival'.[105] The broken, eroded nature of the machair suggests that those occupying the wheelhouses at Bhaltos may not have had access to the same amount of resource as those living in the earlier roundhouses. North Uist was equally well settled during the Iron Age. Its coastal areas had already seen a great deal of machair erosion over the centuries but it was still closely settled in terms of roundhouses and wheelhouses, with their distribution showing a broad correlation with areas of modern-day arable (Figure 2.10).

By comparison, in the south of the Long Island, detailed surveys of the islands lying beyond South Uist, such as Barra, Vatersay and Mingulay, have supported a different conclusion. There, roundhouses and wheelhouses occupied different types of site. The former were in prominent, sometimes exposed positions that were subsequently abandoned, while the wheelhouses not only occupied more low-ground sites but were to be found in the same areas as the later, early modern touns.[106] What is especially striking about the evidence for each of these three areas is the way in which some roundhouses appear to cluster in close proximity to one another, such as on the south side of the Borve valley or above Gortien on Barra, where five or six are found within a square kilometre.[107] Whereas Branigan has made a case for those round Gortien and below Ben Verrisey as being occupied at the same time,[108] Campbell thought those at Vallay, North Uist, were not.[109]

We should also note the significant inter-island differences as regards Iron Age settlement. There have long been perceived differences between the Northern Isles and Hebrides as regards the broch, with the more complex forms of the Northern Isles leading some to see them as developing there, with those of the Western Isles being later, degraded copies. Others completely inverted this logic, seeing their initial traits as first evolving in the Western Isles before spreading northwards and becoming more sophisticated forms in the process. Recent thinking suggests that trying to distinguish between the Northern and Western Isles in terms of what is immature/mature as regards broch structures may not be the best approach, for both areas have a mix of structures that could be described as antecedents. Thus, far from having only degraded versions, the Western Isles contain examples of both simple and complex roundhouses, with some of the latter anticipating some of the features of the broch. On South Uist, there are even sites that might be seen as antecedent to simple roundhouses, based on turf or stone and turf walls, and thought to date from the late Bronze/early Iron Age, or early first millennium BC. More significant is the presence of complex roundhouses that contained some of the structural features of the broch, such as galleried walls and intramural staircases. Some of the perceived differences between settlement in the Western Isles and that of the South-west Highlands and Islands have also come under closer scrutiny. In particular, the fact that some duns in the latter appear to incorporate so-called broch traits, such as double walls, intermural cells and staircases, opens up the possibility that such features may even have spread into the Western Isles from there. Yet, as Nieke observed, most duns in Argyll are relatively simple drystone-wall affairs, dating to the first millennium AD, making them post-broch in use.[110] Further, those duns possessing a more complex structure tend to occur on islands such as Tiree and Coll,[111] suggesting that the flow of influence was probably in the reverse direction. To complicate matters, when we look at islands such as Barra and Skye, islands on which brochs as well as duns are present, it has long been recognised that the two settlement forms occupy complementary distributions suggesting that they enjoyed a degree of contemporaneity.[112]

Asking what settlement patterns can tell us in the Outer Isles, Armit felt that there was no standard answer. Mapping the fifty-one roundhouses that have been located on North Uist led him to suggest that most areas suitable for settlement were represented by at least one roundhouse so that, overall, we appear to have a full settlement pattern in terms of the opportunities available. He compared the geography of roundhouses with eighteenth-century touns and concluded that the bulk of those who held and worked the land, and not just the dominant elite, must have been represented by those who inhabited the roundhouses. A large structure, such as the wheelhouse at Sollas, could have represented a sizeable family group, one involving multiple generations and extended kinship affiliations, even allowing for the fact that its hut cells were not all used as living space.[113] In South Uist, though, the situation was different. There, as noted earlier, fewer roundhouses existed so that those who occupied them may not have represented the total population. The settlement pattern of South Uist may express a different social order from that depicted for us by roundhouses on North Uist, with a relatively small elite occupying the roundhouses, including the brochs, and controlling the occupation of land and those who lived in a yet-to-be-identified form of settlement.[114] A detailed survey of South Uist revealed a total of 225 settlements along its machair belt, in addition to many settlements sited on the peat land. Of the 225 settlements on the former, the 'vast majority of these probably date broadly to the Iron Age'.[115] Among them are likely to have been not just the roundhouses but, arguably, the settlements of those with less standing.

The Iron Age Farming Landscape

Most Iron Age sites were farmsteads in one form or another, set down beside the land that gave them their subsistence. Speaking about this in relation to the siting of Atlantic roundhouses, Armit saw it as expressing 'the self-sufficiency of the community, its control of its small pocket of land and its permanence in the face of a hostile environment'.[116] Though machair strips such as that of South Uist form an obvious exception, the variegated nature of the Highland environment generally, the way soils most suited for arable lie in detached pockets, made this association between particular settlements and specific blocks of arable a prominent feature.

The successive reuse of these detached pockets has meant that traces of how the earliest communities worked the land were repeatedly worked over by successive generations. What may have existed in prehistoric times, by way of field walls, dykes, ditches and earthen banks, have subsequently been removed, modified or allowed to decay. The higher ground or marginal areas tend to offer the most abundant evidence of hut circles and the most extensive evidence of the small, irregular networks of accreted fields, clearances cairns and stock enclosures that accompany them in regions such as eastern Sutherland and north-east Perthshire. Discussing the extensive patterns of settlement that appeared at the head of Dunbeath Water during the first millennium BC, Morrison drew attention to the way in which many late prehistoric huts and roundhouses had 'associated field plots' with 'straggling stone boundaries'.[117] Likewise, the RCAHMS survey of archaeological landscapes in north-east Perth records fine examples of how late prehistoric communities created working landscapes around their settlements systems, such as that around the headwaters of Drumturn Burn,[118] with their network of fields and clearance cairns (Figure 2.2c). Traces of late prehistoric field systems can also be found in parts of the Hebrides, such as those recorded at An Sithean on Islay.

Turning to how land was used, most excavated Iron Age sites have shown that both arable and pasture were important to the subsistence economy. Part of the evidence for how pasture was used is derived from the bone assemblages found at excavated sites. Both in the Northern and Western Isles, cattle, sheep and small numbers of pigs were present, with horses appearing during the Iron Age. A feature of the bone data, one prominent in the assemblages for Hebridean sites such as Dun Vulan (South Uist) and Baleshare (North Uist), is the bias towards young calves and lambs, most being under a year old.[119] This early slaughter of animals has been ascribed by some to the scarcity of winter fodder. Unless one presumes very high levels of stocking, though, one would have expected the available machair at these sites to answer for outdoor winter grazing. Addressing comparable Norse data for the Northern Isles, Bigelow has suggested that the imbalance reflected the deliberate early slaughter of calves so as to aid the production of milk and, therefore, of cheese and butter.[120] If this is also relevant to the Iron Age data, it would favour the indoor wintering of stock, at least overnight, evidence for which has been found at the wheel-houses of Allasdale and Clettravel.[121] Another clue as to the evolving nature of stock husbandry is provided by the evidence for the Iron Age use of shielings in the Hebrides, such as around Loch Aineort and Loch Baghasdail in South Uist.[122] The function of shielings was multipurpose. Whatever role they served as places where butter and cheese could be made, they kept stock away from growing crops at a critical time of year, and saved vital winter grazings.

Initially, Iron Age communities appear to have been dependent largely on barley or bere as their main field crop. This was the case in the Western and in the Northern Isles, the type sown being mostly the hulled six-row barley. Excavations at Dun Vulan (South Uist) also yielded signs of wheat, oats and rye being present but, at this stage, it was concluded that these were 'likely to represent weeds of the barley rather than crops in their own right'.[123] Excavations at sites such as Tofts Ness, Sanday (Orkney) demonstrated that arable was exploited through the cropping of hulled six-row barley from the Neolithic down to the early Iron Age when the site was abandoned.[124] This

was probably the case with arable throughout the Northern Isles. It was a relatively high-yielding crop and provided communities with enough output for a
storable surplus, one that could be carried over to start of the next season.[125]
Only during the mid-Iron Age was it supplemented by the introduction and
field sowing of oats, both *Avena sativa* and *Avena strigosa*.[126]

Work on a number of sites in the Northern Isles enables us to fit these crops
into a scheme of husbandry that enhances our reading of the Iron Age landscape. Sites such as Tofts Ness (Sanday) and Old Scatness suggest that, by the
late Bronze Age, the core of the settlement's field economy had become concentrated on a small area of arable, no more than a hectare or two at most (cf.
p. 21), that was intensively managed and manured to produce high yields of
barley.[127] The manurial inputs used were largely ash and turf.[128] Significantly,
no animal manure was used at this point, though some of the ash may have
been derived from animal manure burnt as fuel. The emergence of such intensively worked arable has been ascribed to the climatic worsening during the
Bronze Age. Communities were seen as responding by investing more in less,
offsetting the risks by concentrating their efforts and manure on a particular plot and on a high-yielding crop. Dockrill described these small plots of
intensively prepared arable as infield.[129] Their intensive treatment is certainly
suggestive but there are important ways in which they do not easily equate
with such infields, being more confined in area than later infields, making little
or no use of animal manure, and lacking the landholding significance of later
medieval infields.

The evidence for the Northern Isles suggests at least three interrelated
changes transformed this pattern of arable during the mid- to late Iron Age.
First, there are signs that an areal expansion of arable occurred. The increased
output is seen as accounting for the shift from the saddle to the rotary quern
that occurred during the mid- to late Iron Age, the latter's greater capacity
enabling more grain to be processed. Second, the introduction of oats, *Avena
sativa* and *A. strigosa*, as a field-sown crop, enabled communities to cultivate
more marginal land, though it has been suggested that, in the Western Isles, it
was not widely sown as a field crop until the medieval period.[130] As a crop, *A.
strigosa* or black oats, coped better with poorer, more difficult soils than barley
did. It also coped better than barley with wetter conditions but it brought its
own risks. During phases of climatic degradation, its longer growing season
exposed it to an even greater risk of storms during spring and autumn.[131] We
need only look at how, during the early phases of the Little Ice Age in the fourteenth and fifteenth centuries AD, the cropping of oats was actually abandoned
in Iceland, with barley being the only grain crop sown owing to its shorter
growing season. The adoption of oats during the mid- to late Iron Age allowed
communities to cope with more marginal ecologies but only by accepting a
greater risk of crop failure.

Third, by the late Iron Age, use was also made of animal dung as a field manure. The shift in manuring technique underpinned the marked expansion of arable and the adoption of oats. A growth of dairying and the indoor wintering of stock would have helped to expand the extra supply needed. In its effects, animal dung was far more effective and more lasting as a fertiliser than either domestic waste or turf. There was a synergy here about the changes being made. Not only did animal dung have a more lasting effect as a fertiliser but, as a crop, oats proved better at being able to prolong the cropping of manured land. It may well be that this was the point at which Highland farmers introduced what became a standard cropping regime for the region, with a crop of barley receiving all the available manure followed by one or two years of oats. With arable now being intensively manured using the dung accumulated over winter from the byre, we can start to speak of something that more resembled the infields to be found in the pre-Clearance touns of the eighteenth century. Even as expanded arable, though, their size compared only with the smallest of infields present in later pre-Clearance touns, and some vital characteristics were still missing, but a core feature – the use of winter manure – was now present. To capture this basic similarity, we can label these Late Iron Age arable systems as *proto* infields.

The Later Prehistoric: What Endured?

Disagreement over how we strike the balance between continuity and discontinuity has long been a part of the debate over the prehistoric landscape. No small part of this disagreement arises from how we actually frame the question. At its most generalised level, we can use to it to ask how cultural change came about. For a time, the balance of opinion tended to read all the more striking changes, such as farming or the appearance of the broch, as best explained by the migration of new peoples into the region. There is still a strong case for seeing the spread of farming as being about new peoples, traits or ideas spreading into the region but subsequent changes, such as the 'sudden' appearance of the roundhouse of the Middle Iron Age, probably have more claim to being indigenous developments. As Parker Pearson put it, once farming had been established, the region may have been experienced a long drawn-out process of indigenous change in which 'a landscape of networks linking ancestral tombs, ceremonial monuments and other significant places was replaced by a landscape of houses, settlements and fields within a new tradition of rootedness and place commencing in the Middle Bronze Age around 1500 bc' and then 'culminating in the concept of the substantial house in the Iron Age'.[132]

We can also direct the question of continuity/discontinuity at the occupation of particular sites. Did different groups at different times select the same sites for their arable and settlement? The answer to the first part of this question is yes. The repeated use of the same land reflects the realities of the region's physical environment, with the fragmentation of arable into a well-defined mosaic of opportunities, a mosaic that helped to steer the choice of successive communities.[133] Nevertheless, the choice was helped by the lower opportunity costs offered by existing arable because less clearing of roots, stones and so on would have been involved. As Cavers and Hudson have recently put it, this successive duplication of choice turned patches of better land into 'palimpsests of activity spanning several millennia'.[134] As regards the question of whether there was continuity in the siting of settlement, we can answer in two ways. First, the farm mounds of the Uists, as well as accumulated sites like Jarlshof, Old Scatness and Howe in the Northern Isles, are striking testimony to the way successive communities built over what had gone before. Even at such sites, however, it was a qualified continuity, for the excavation of these multiperiod sites has revealed that there were significant breaks in their occupation as well as ongoing adjustment in building styles, materials and orientation. Thus, speaking about the excavations at Toft Ness and Pool on Sanday (Orkney), Hunter talked about 'the lack of continuity' at each site, but also how 'the settlement gaps in one site' were 'filled by occupation in the other, and vice versa, over the period from the fourth millennium BC to the thirteenth century AD'.[135] He saw the early second millennium BC and the end of the later Norse settlement in the twelfth- thirteenth centuries AD as laying claim to the most significant moments of settlement discontinuity.[136]

As Hunter's conclusion helps bear out, the end of the Iron Age is not defined by a marked cultural disjuncture in the Highlands and Islands. The one shades into the other. Inevitably, this raises questions over what was carried over as regards landscape. Needless to say, the extent to which work by Armit and by Parker Pearson et al. has revealed the Late Iron Age landscape as a closely settled one increases the possibility of there being some form of continuity in terms of what we see on the ground. While, however, it makes a straightforward case for duplication of site choice, it does not, in itself, make a case for institutional continuity, that is, in how farms and fields were organised. To establish the latter requires a fuller grasp not just of how the Iron Age landscape was organised in terms of its institutional underpinnings (that is, landholding, field layout) but also of the institutional underpinnings to the Medieval and Early Modern landscape, as well as a consideration of how sociopolitical changes of the historic period may have changed these institutional underpinnings. These are matters that will be taken up in later chapters.

Notes

1. K. J. Edwards and S. Mithen, 'The colonisation of the Hebridean islands of western Scotland: evidence from the palynological and archaeological records', *World Archaeology*, 26 (1995), pp. 348–65.
2. S. Mithen, A. Pirie, S. Smith and K. Wicks, 'The Mesolithic–Neolithic transition

in western Scotland: a review and new evidence from Tiree', *Proceedings of the British Academy*, 144 (2007), p. 514.

3. S. Mithen (ed.), *Hunter-gatherer Landscape Archaeology: The Southern Hebrides Mesolithic Project 1988–98* (Cambridge, 1988), vol. 2, p. 603.

4. C. Wickham-Jones, *Rhum: Mesolithic and later sites at Kinloch. Excavations 1984–6* (Edinburgh, 1990), pp. 166–7.

5. Mithen, *Hunter-gatherer*, vol. 2, pp. 298–304.

6. Ibid., p. 298.

7. R. J. Schulting and M. P. Richards, 'The wet, the wild and the domesticated: the Mesolithic–Neolithic transition on the west coast of Scotland', *European Journal of Archaeology*, 5 (2002), pp. 147–89.

8. S. Mithen, *To the Islands: An Archaeologist's Relentless Quest to Find the Prehistoric Hunter-Gatherers of the Hebrides* (Uig, 2010), pp. 218–24.

9. R. Tipping, 'The form and fate of Scotland's woodlands', *PSAS*, 124 (1994), p. 18.

10. Ibid., pp. 15–18.

11. M. Macklin, C. Bonsall, F. M. Davies and M. R. Robinson, 'Human-environmental interactions during the Holocene: new data and interpretations from the Oban area, Argyll, Scotland', *Holocene*, 10 (2000), p. 113; R. Tipping et al., 'Prehistoric *Pinus* woodland dynamics in an upland landscape in northern Scotland: the roles of climate change and human impact', *Vegetation History and Archaeobotany*, 17 (2008), pp. 39–40. For a more supportive view of an anthropogenic role, see K. J. Edwards, P. G. Langdon and H. Sugden, 'Separating climate and possible human impacts in the early Holocene: biotic response around the time of the 8200 Cal yr. BP event', *Journal. of Quaternary Science*, 22 (2007), pp. 77–84.

12. Macklin et al., 'Human–environmental interactions', p. 113; G. Noble, *Neolithic Scotland: Timber, Stone, Earth and Fire* (Edinburgh, 2006), p. 32.

13. Schulting and Richards, 'The wet, the wild and the domesticated', pp. 147–89; P. J. Ashmore, *Neolithic and Bronze Age Scotland* (London, 1996), pp. 24–5; Noble, *Neolithic Scotland*, p. 21.

14. Tipping, 'Scotland's woodlands', p. 14; Mithen et al., 'Mesolithic–Neolithic transition', p. 536.

15. Noble, *Neolithic Scotland*, pp. 24–7, 40; Mithen et al., 'Mesolithic–Neolithic transition', pp. 514–15.

16. Ibid., pp. 511–41; Tipping, 'Scotland's woodlands', pp. 1–54; Macklin et al., 'Human–environmental interactions', p. 113; A. L. Davies and R. Tipping, 'Sensing small-scale human activity in the palaeoecological record: fine spatial resolution pollen analyses from Glen Affric, northern Scotland, *The Holocene*, 14 (2004), pp. 233–45.

17. C. M. Mills, I. Armit, K. P. Edwards, P. Grinter and Y. Mulder, 2004, 'Neolithic land-use and environmental degradation: a study from the Western Isles of Scotland', *Antiquity*, 78 (2004), pp. 888, 894.

18. Ibid., p. 889; I. Armit, 'The Drowners: Permanence and transience in the Hebridean Neolithic', in I. Armit, E. Murphy, E. Nelis, and D. Simpson (eds), *Neolithic Settlement in Ireland and Western Britain* (Oxford, 2013), p. 99.
19. Ibid., p. 99.
20. Ibid., p. 93.
21. Mithen et al., 'Mesolithic–Neolithic transition', pp. 511–41.
22. Macklin et al., 'Human–environmental interactions', pp. 113, 120.
23. Davies and Tipping, 'Sensing small-scale human activity', pp. 233–45.
24. Ibid., p. 241.
25. Ibid., p. 241.
26. Tipping, 'Scotland's woodlands', p. 23.
27. E. B. A. Guttmann, S. J. Dockrill and I. A. Simpson, 'Arable agriculture in Prehistory: new evidence from the Northern Isles, *PSAS*, 134 (2004), pp. 53–64.
28. Noble, *Neolithic Scotland*, p. 22.
29. Ashmore, *Neolithic and Bronze Age Scotland*, pp. 49–50; C. Richards, 'Monuments as Landscape: Creating the Centre of the World in Late Neolithic Orkney', *World Archaeology*, 28 (1996), pp.191–2.
30. For example, ibid., pp. 39–40.
31. Mills et al., 'Neolithic land-use', p. 888.
32. Ibid., p. 889.
33. Noble, *Neolithic Scotland*, p. 37.
34. A. Fleming and M. Edmonds, 'St Kilda: quarries, fields and prehistoric agriculture', *PSAS*, 229 (1999), pp. 119–59; A. Fleming, *St Kilda and the Wider World* (Macclesfield , 2005), pp. 50–2, 62.
35. C. R. Renfrew, *Approaches to Social Archaeology* (Edinburgh, 1984), pp. 182–7.
36. Tipping, 'Scotland's woodlands', p. 26; Tipping et al., 'Prehistoric *Pinus*', pp. 251–67.
37. Tipping, 'Scotland's woodlands', p. 55.
38. A. Dawson, *So Foul and Fair a Day. A History of Scotland's Weather and Climate* (Edinburgh, 2009), p. 80.
39. H. Fairhurst and D. B. Taylor, 'A Hut-circle Settlement at Kilphedir, Sutherland', *PSAS, 103* (1970–71), p. 67.
40. A. Morrison, 1996, *Dunbeath. A Cultural Landscape*, Department of Archaeology, University of Glasgow, Occasional paper no. 3 (1996), pp. 35–41.
41. RCAHMS, *Strath of Kildonan*, 6–10; D. C. Cowley, 'Identifying marginality in the first and second millennia BC in the Strath of Kildonan, Sutherland', in C. M. Mills and G. Coles (eds), *Life on the Edge. Human Settlement and Marginality*, Association for Environmental Archaeology no. 13 (Oxford, 1998), pp. 165–71.
42. Fairhurst and Taylor, 'Hut-circle Settlement at Kilphedir', p. 68.
43. Cowley, 'Identifying marginality', pp. 165–71.
44. Parker Pearson and Sharples, 'South Uist in the Iron Age', p. 16; Armit, 'Brochs and Beyond', p. 61.
45. I. Crawford and I. Switsur, 'Sandscaping and C14: the Udal, N. Uist', *Antiquity*,

51 (1977), pp. 124–36; I. Crawford, 'The present state of the settlement history in the West Highlands and Islands', in A. O'Connor and D. V. Clarke (eds), *From the Stone Age to the 'Forty-Five* (Edinburgh, 1983), pp. 350–67.

46. J. Barber, *Bronze Age Farm Mounds and Iron Age Farm Mounds of the Outer Hebrides*, Scottish Archaeological Internet Reports (www.sair.org.uk), 3 (2003), pp. 229–30.

47. Ibid., p.28.

48. J. Barber and M. M. Brown, 'An Sithean, Islay', *PSAS*, (1984), pp. 161–88.

49. Ibid., pp. 184, 186.

50. Barber, *Farm Mounds*, p. 22.

51. O. Owen and C. Lowe, *Kebister. The Four-Thousand-Year-Old Story of One Shetland Township*, Society of Antiquaries of Scotland Monograph Ser., no. 14 (Edinburgh, 1999), p. 253.

52. Ibid., p. 253. See also, N. Fojut, *A Guide to Prehistoric and Viking Shetland* (Lerwick, 1994), p. 323.

53. Macklin et al., 'Human-environmental interactions', p. 117.

54. Barber, *Farm Mounds*, p. 235.

55. Ibid., p. 235.

56. E. B. Guttmann, I. A. Simpson, N. Nielsen and S. J. Dockrill, (2008), 'Anthrosols in Iron Age Shetland. Implications for arable and economic activity', *Geoarchaeology: An International Journal*, 23 (2008), p. 801.

57. E. B. Guttmann, 'Midden cultivation in prehistoric Britain: arable crops in gardens', *World Archaeology*, 37 (2005), p. 231.

58. E. B. Guttmann, I. A. Simpson, D. A. Davidson and S. J. Dockrill, 'The management of arable land from prehistory to the present: case studies from the Northern Isles of Scotland', *Geoarchaeology: An International Journal*, 21 (2006), p. 87.

59. Ibid., p. 87; Barber, *Farm Mounds*, p. 235.

60. Cowie and Shepherd, 'The Bronze Age', in K. J. Edwards and I. B. M. Ralston (eds), *Scotland After the Ice Age: Environment, Archaeology and History* (Edinburgh, 2003), pp. 161, 166.

61. J. R. Hunter, *Fair Isle. The Archaeology of an Island Community* (Edinburgh, 1996), p. 95; I. Armit, 'Land and freedom: implications of Atlantic Scottish settlement patterns for Iron Age land-holding and social organisation', in B. Ballin Smith and I. Banks (eds), *In the Shadow of the Brochs. The Iron Age in Scotland* (Stroud, 2002), p. 22.

62. J. Harris, 'A preliminary survey of hut circles and field systems in SE Perthshire', *PSAS*, 114 (1984), pp. 199–216.

63. Cowie and Shepherd, 'The Bronze Age', p. 158.

64. RCAHMS, *North East Perth*, p. 2.

65. T. N. Dixon, 1982, 'A survey of crannogs in Loch Tay', *PSAS*, 112 (1982), pp. 17–38; J. Henderson, 1998. 'Islets through time: the definition, dating and distribution of Scottish crannogs', *Oxford Journal of. Archaeology*, 27 (1998), pp. 227–44.

66. I. Armit, 'Brochs and Beyond in the Western Isles', in I. Armit (ed.), *Beyond the Brochs: Changing Perspectives on the Later Iron Age in Atlantic Scotland* (Edinburgh, 1990), p. 51.

67. J. Miller, 'The Oakbank crannog', in Ballin Smith and Banks (eds), *In the Shadow of the Brochs*, p. 35.

68. N. R. Nieke, 'Fortifications in Argyll: Retrospect and Future Prospect', in Armit (ed.) *Beyond the Brochs*, p. 134.

69. Dixon, 'A survey of crannogs in Loch Tay', pp. 17–38; Henderson, 'Islets through time', pp. 227–44.

70. I. Ralston, K. Sabine, and I. W. Watt (1983), 'Later prehistoric settlement in North-East Scotland: a preliminary assessment', in J. C. Chapman and H. C. Mytum (eds), *Settlement in North Britain 1000 BC–AD 1000*, BAR, British Series, 118 (1983), p. 156.

71. R. W. Feachem, 'Hillforts of Northern Britain', in A. L. F. Rivet (ed.), *The Iron Age in Northern Britain* (Edinburgh, 1966), pp. 65–77.

72. Ralston et al., 'Later prehistoric settlement', p. 161.

73. Armit, 'Brochs and Beyond', p. 41.

74. Ibid., p. 41.

75. E. W. MacKie, 'The origin and development of the broch and wheelhouse building cultures of the Scottish Iron Age', *Proceedings of the Prehistoric Society*, 31(1965), pp. 93–146; E. W. MacKie, 'Brochs and the Hebridean Iron Age', *Antiquity*, 39 (1965), pp. 266–78.

76. E. W. MacKie, 2008, 'The broch cultures of Atlantic Scotland: origins, high noon and decline. Part 1: 'Early Iron Age beginnings c.700–200 BC', *Oxford Journal of Archaeology*, 27 (2008), p. 272, and Part 2: 'The Middle Iron Age: High noon and decline c.200 BC–AD 550', *Oxford Journal of Archaeology*, 29 (2010), p. 90.

77. Armit, 'Brochs and Beyond', pp. 41, 53, 59; M. Parker Pearson and N. Sharples, 'South Uist in the Iron Age', in M. Parker Pearson and N. Sharples, *Between Land and Sea: Excavations at Dun Vulan, South Uist*, Sheffield Environmental and Archaeological Research Campaign in the Hebrides, vol. 3 (Sheffield, 1999), pp. 1–2, 364; D. W. Harding, *The Hebridean Iron Age: Twenty Years On*, University of Edinburgh, Department of Archaeology, Occasional Papers no. 20, p. 5.

78. I. Armit, 'The Atlantic Scottish Iron Age: five levels of chronology', *PSAS*, 123 (1991), pp. 182–3.

79. I. Armit, 'Epilogue: The Atlantic Scottish Iron Age', in Armit (ed.), *Beyond the Brochs* (Edinburgh, 1990), pp. 199, 201.

80. B. Ballin Smith (ed.), 1994, *Howe: Four Centuries of Orkney Prehistory Excavations 1978–1982*, PSAS Monographs, no. 9.

81. Armit, 'Epilogue', p. 196.

82. I. Armit, 'Archaeological field survey of the Bhaltos (Valtos) peninsula, Lewis', *PSAS*, 124 (1994), pp. 78, 90.

83. M. Parker Pearson and N. Sharples, 'Solving the problem of the brochs', in M. Parker Pearson and N. Sharples, *Between Land and Sea*, p. 359; Ballin Smith (ed.), *Howe*, p. 273.

84. J. W. Hedges, *Bu, Gurness and the brochs of Orkney*, Part III (Oxford, 1985), p. 32.

85. Ibid., p. 5.

86. Ibid., p. 13; J. W. Hedges, 'Surveying the foundations: Life after brochs', in Armit (ed.), *Beyond the Brochs*, p. 31.

87. Ballin Smith, *Howe*, p. 273.

88. Hedges, *Bu, Gurness*, Part III, p. 38.

89. Ibid. Part III, p. 37.

90. Harding, *Hebridean Iron Age*, p. 14.

91. E. Campbell, 'Excavations of a wheelhouse and other Iron Age structures at Sollas, North Uist, by R. J. C. Atkinson in 1957', *PSAS*, 12 (1991), p. 137.

92. Parker Pearson and Sharples, 'South Uist in the Iron Age', p. 16.

93. Armit, 'Brochs and beyond', p. 61.

94. A. Ritchie, 'Excavation of Pictish and Viking-age farmsteads at Buckquoy, Orkney', *PSAS*, 108 (1976–67), p. 182.

95. Ibid., p. 182; Parker Pearson and Sharples, 'South Uist in the Iron Age', p. 3.

96. Ritchie, 'Pictish and Viking-age farmsteads', p. 182.

97. Armit, 'Brochs and beyond', p. 65; D. W. Harding and I. Armit, 'Survey and Excavation in West Lewis', in I. Armit (ed.), *Beyond the Brochs* (Edinburgh, 1990), p. 82; Hedges, *Bu, Gurness*, Part III, p. 41.

98. Armit, 'Brochs and beyond', p. 64.

99. For example, Crawford and Switsur, 'Sandscaping and C14', p. 129.

100. Armit, 'Brochs and beyond', pp. 64–5; Parker Pearson and Sharples, 'South Uist in the Iron Age', p. 15.

101. I. Armit, *The Archaeology of Skye and the Western Isles* (Edinburgh, 1996), p. 133.

102. Campbell, 'Excavations of a wheelhouse', p. 139.

103. For example, Armit, 'Bhaltos (Valtos) peninsula', pp. 72–3, 78.

104. For example, Parker Pearson and Sharples, 'South Uist in the Iron Age', p. 12.

105. Armit, 'Bhaltos (Valtos) peninsula', pp. 69–70.

106. F. K. Foster and R. Krivanek, 'An archaeological survey on the southern Outer Hebridean Islands of Sandray, Pabbay, Mingulay and Berneray', in K. Branigan and P. K. Foster, *From Barra to Berneray*, p. 126.

107. K. Branigan, 'The later prehistory of Barra and Vatersay', in K. Branigan and P. K. Foster, *From Barra to Berneray*, p. 344.

108. Ibid., p. 344.

109. Campbell, 'Excavations of a wheelhouse', p. 166.

110. M. R. Nieke, 'Fortifications in Argyll: Retrospect and Future Prospect', in Armit (ed.), *Beyond the Brochs*, pp. 133–4.

111. Ibid., p. 134.

112. Campbell, 'Excavations of a wheelhouse', p. 166; Parker Pearson and Sharples, 'South Uist in the Iron Age', p. 10.
113. 3. Cf. Campbell, 'Excavations of a wheelhouse', p. 166.
114. Armit, 'Land and freedom', p. 25.
115. Parker Pearson and Sharples, 'South Uist in the Iron Age', p. 14.
116. Armit, *Archaeology of Skye*, p. 130.
117. Morrison, *Dunbeath*, p. 62.
118. RCAHMS, *North East Perth*, pp. 444–9.
119. Parker Pearson and Sharples, 'South Uist in the Iron Age', p. 21.
120. G. F. Bigelow, 'Domestic architecture in Medieval Shetland', *Review of Scottish Culture* 3 (1987), pp. 23–38.
121. Parker Pearson and Sharples, 'South Uist in the Iron Age', p. 23.
122. Ibid., p. 14; see also Armit, *Archaeology of Skye*, p. 217.
123. H. Smith, 'The plant remains', in Parker Pearson and Sharples, *Between Land and Sea*, p. 334.
124. Dockrill, 'Brochs', p. 156.
125. Ibid., p. 158.
126. J. Bond, 'Pictish pigs and Celtic cowboys: food and farming in the Atlantic Iron Age', in Ballin Smith and Banks (eds), *In the Shadow of the Brochs*, p. 183.
127. Dockrill, 'Brochs', p. 157.
128. Ibid., p. 156.
129. Ibid., p. 157.
130. Smith, 'The plant remains', p. 334.
131. R. A. Dodgshon, 'Coping with risk: subsistence crises in the Scottish Highlands and Islands', *Rural History*, 15 (2004), pp. 9–10.
132. Parker Pearson and Sharples, 'South Uist in the Iron Age', p. 1.
133. Hunter, *Fair Isle*, p. 95.
134. G. Cavers and G. Hudson, *Assynt's Hidden Lives. An Archaeological Survey of the Parish* (Assynt, 2010), p. 58.
135. Hunter, *Fair Isle*, p. 95.
136. Ibid., p. 87.

PREHISTORY INTO HISTORY

The period from Late Prehistory down to the twelfth and thirteenth centuries has long been seen as yielding to the historian of the countryside only a few tantalising clues. For the Highlands and Islands it yields even less. Work on the available archaeological, place-name and documentary sources has started to add more substance to what we know about the period generally but, when it comes to how the farming landscape was organised and how it may have changed, there are still clear gaps in our understanding. Even a fundamental question, such as whether the different cultural groups that were settled in, or became settled in, the Highlands and Islands during this period – the Picts, Scots and Norse – organised themselves on the ground cannot yet be answered in a definitive way. If we maintain a focus on how the countryside was occupied and its resources exploited, the gaps in what we know inevitably make for an unsatisfactory discussion, with a growing capacity for debating some topics in detail but a coverage of others that remains inadequate.

Locating the Picts

The one indigenous culture to be found in the region at the start of the early medieval period to which we can give a name was that of the Picts. Disagreements still remain as to when we can first speak about their presence. Archaeologists no longer talk of brochs as 'Pictish towers' but it has become commonplace for Late Iron Age communities across the north and west of the region, and in the east, to be referred to as 'proto-picts' and for some of the post-broch settlement structures that have been excavated in the Northern and Western Isles, and variously dated to between the third and eighth century AD, to be seen as occupied by 'proto-Pictish' communities.[1] Writers such as Ritchie have challenged this early appearance, arguing that, if we rely on the archaeological record, then we should not use the term before about AD 600, the point when so-called Class I Pictish standing stones were first erected,[2] and that any attempt to qualify that position by using a term like 'proto-Pictish' was 'unhelpful'.[3] For some, imposing such a late chronology ignores the fact that the symbols depicted on Class I stones formed a mature system of symbolic representation and were unlikely to have *first* appeared with the stones. In all probability, they had been used in other ways (that is, on wood) before they were inscribed on stone around AD 600.

The Pictish imprint on landscape is discernible through three different types of field evidence: symbol stones; the archaeological evidence for settlement; and *Pit-* (or *Pett*) place-name elements. The Class I and II symbol stones form a distinctive but restricted cultural marker of where the Picts lived, at least at the time when erecting such stones was in fashion. They occur widely along the eastern flanks of the Highlands and adjacent low ground, from central Fife to south-east Sutherland, but with a few tantalising outliers in the Northern Isles, Skye and the Outer Hebrides. The earlier Class I stones are associated with a wide range of symbols, some based on animals, such as boars, snakes or fishes, others on items such as mirrors and combs, and others on geometric designs. In practice, symbols occur in either pairs or threes. Analysis of their meaning as standing stones has led to support for three possible interpretations. The first is that they were erected so as to make a statement about land and its ownership. As such, they would have formed a vital part of the landscape and its territorialisation. The second is that they were commemorative stones erected in honour of a particular person, either of who they were or the office they held.[4] Again, as commemorative stones, their siting would have had territorial connotations. The third is that they were stones celebrating a marriage alliance between lineages or clans, the presence of two symbols being the clans or lineages involved while, if three symbols were present, the third invariably was a comb or mirror, denoting the payment of brideprice.[5] This interpretation supposes a landscape not only inhabited through kin groups but one that was, first and foremost, identified through them, with Class I stones again having a territorial meaning, perhaps by marking out boundaries rather than the core area of settlement. Significantly, Class I stones rarely occur immediately beside settlements bearing *Pit-* place names. In some cases, such as around Rhynie in Aberdeenshire, we find a local cluster of stones but no *Pit-* place names.[6] Of course, the distribution of symbol stones may be incomplete, especially where examples may have been moved from their original locations and reset in churchyards.[7]

The debate over Pictish or 'proto-Pictish' settlement archaeology has focused on a diverse range of sites, including souterrains, figure-of-eight or cellular huts, Pitcarmick dwellings and wags (Figure 3.1).

Souterrains, or underground passages, occur widely across the Highlands and Islands, as well as in adjacent lowland areas such as Angus. Their precise function is open to some doubt despite a number of excavations. Even where

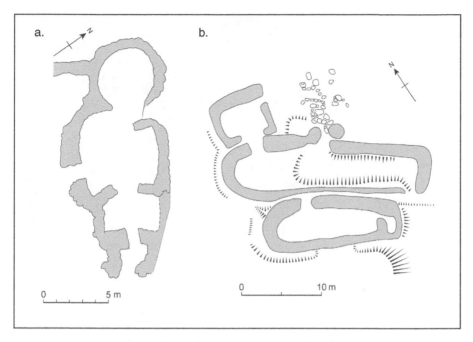

Figure 3.1 Forms of Pictish settlement: a. Figure-of-eight huts at Buckquoy, based on Ritchie, 'Pictish and Viking-age farmsteads', p. 176; b. Pitcarmick West, based on RCAHMS, *North-East Perth*, p. 155.

seen as storage cellars, there is added dispute over what might have been stored in them: grain, dairy produce or, though much less probable, animals? Those in the far north and west were actually built within the living space of settlements, such as in the Late Iron Age roundhouse at Howe,[8] while those in areas such as Angus were free-standing structures established beside farmsteads. Most appear to have been in use across the Late Iron Age and early historic period: those in the Angus area, for example, were deliberately filled in by the second to third centuries AD, as if their role had ceased. Though some have questioned whether souterrains should be treated as 'proto-Pictish', the consensus view is that, at the very least, they are Pictish by geographical association. Their concentration in the area of southern Pictland broadly coincides with an area in which Pictish symbol stones are concentrated, suggesting that the communities who built the one may be connected by descent to those who built the other,[9] and forming part of a 'settlement continuity from Iron Age to early historic times'.[10]

Acquiring more prominence in the Pictish debate over recent years has been a form of settlement found in post-broch contexts known as cellular huts and their associated forms, including a form known as figure-of-eight huts (Figure 3.2).

Figure 3.2 Cellular or figure-of-eight dwellings beside broch at Gurness, Orkney.

During the more systematic excavations of recent decades, the presence of cellular huts built alongside, revetted into or built over the remains of brochs and wheelhouses has been noted across a number of sites as well as in free-standing positions.[11] One of the key excavated sites for such structures is that of Buckquoy.[12] Sited on Birsay in Orkney, it formed a settlement overlooking Birsay Bay. What survived on the site had been seriously reduced by coastal erosion by the time it was excavated. The site produced an example of a figure-of-eight structure and two cellular buildings, with a Norse longhouse subsequently laid out across part of it (Figure 3.1a). Similar examples have also been excavated at sites such as The Udal (N. Uist). At the latter, cellular houses were fashioned from vertical slabs which were made into a 'detached structure in which a central chamber is surrounded by rounded or rectilinear cells, the whole entered by a single entrance'.[13] Similar examples have also been excavated at sites at Loch na Berie and Dun Bharabhat on Lewis and Dun Cuier on Barra in the Western Isles.[14] Dating suggests both figure-of-eight and cellular housing were in use during the middle centuries of the first millennium AD, that is, during the two or three centuries immediately prior to the Norse occupation. Though some attribute their style to Celtic influences filtering into the area from the south-west, others see them as a locally derived or indigenous style that can be labelled as proto-Pictish.

Another type of dwelling whose dating now appears to place them squarely within the Pictish period is that known as Pitcarmick-type dwellings. These are found in the south-east corner of the Highlands, deriving their type name from those around Pitcarmick Loch. When first recognised as a distinct form, Pitcarmick-type dwellings were thought to be simply a distinct sub-type of the much larger concentration of Bronze Age and Iron Age hut circles and settlement structures to be found on the higher ground in this area. In design, they were elongated structures, oval in terms of their overall ground plan, extending to as much as 15 to 25 metres in length (Figure 3.1b). In some examples, one end appears lower or sunken, suggesting that it may have served as a byre for stock and hinting at a system of husbandry that needed to house stock over the winter, presumably to ensure an ongoing supply of milk in winter and manure for arable. The enclosures which are still visibly attached to some examples would fit in with this picture of a mixed field economy. Recent excavations also show them to be distinct as regards their dating as well as their construction. Initially, they were thought to be in occupation during late prehistory, with the possibility of some surviving into the early medieval period.[15] In fact, recent work has shifted their chronology of use to the mid-first millennium AD. This redating brings a further conclusion into focus. The contrast between Pitcarmick-type dwellings and the large Bronze Age to Iron Age hut circles that lie in proximity to them, together with the fact that the field enclosures attached to them are distinct, suggest that in this area at least, what we see in late prehistory and what we see in the early historic period does not support a straightforward continuity across this critical divide.

Though it is far from conclusive, there is also a case for seeing a type of site known as a wag as also being occupied at some point during the Late Iron Age or early medieval period and, therefore, as used by communities that may have had Pictish associations. Their distribution is limited, being found only in Caithness and Sutherland. Their typology as a distinct category of site is also far from clear-cut. Part of the reason for this is the way in which some key sites, such as the Wag of Forse, Caithness, are actually complex, accumulated sites, with different structures built over each other in a tell-like fashion.[16] Indeed, the predominance of Iron Age structures on some wag sites makes a stronger case for discussing them under a late prehistoric rather than an early historic heading. The Wag of Forse illustrates these difficulties of categorisation. The earliest settlement detected on the site consists of small, cellular huts. These were replaced by what some see as a broch but others as a dun, a structure named by Curle as the primary dun. This broch or dun was, in turn, replaced by two rectangular longhouse dwellings that are characterised by having pillars on either side, presumably supporting their roof but possibly also acting as the basis for an aisled arrangement of their sides. These formed the basis of what Curle labelled the secondary wag. Beside one of them is a small cellular structure that he labelled as sub-secondary. Strictly speaking, it is the pillared or aisled longhouses alone that formed the wag.[17] Their post-broch stratigraphy suggests that some, especially those with longhouse-styled – as opposed to curved – walls, may have been post-Iron Age in origin,[18] that is, occupied at a time when those whom we know culturally as the Picts inhabited the area, though only effective dating will confirm this. Mention should also be made of the irregular fields that lie beside some wags.[19] In the case of those that lie to the south and east of the Wag of Forse, however, these are thought to be later in date than the Wag.

The meaning and distribution of the place-name element, *Pit-*, is a well-studied topic in Pictish studies. It is thought to be based on the P-Celtic term *Pett-*, meaning a share or piece of land but, in use, it is invariably combined with a Q-Celtic suffix.[20] Geographically, more than three hundred or so examples are found, largely across eastern Scotland. They are primarily concentrated in Fife, Angus with outliers in Perthshire, and along the east coast through Aberdeenshire to the lowlands around the Moray and Beauly Firths, from Moray itself round into south-east Sutherland.[21] It is a pattern that has notable gaps: the Buchan Plateau, between the Spey and Deveron and on the low ground to the east of Inverness, and, overall, there are more to the south of the Mounth than to the north.[22] Though many are located in the coastal lowlands, they do not occur on the coast itself .[23] Examples do occur far inland, along the Tay, Dee, Don and Spey, as well as in the Great Glen. Scattered outliers occur on Bute, in Glenelg and at the head of Lochcarron but there is none in areas such as the Hebrides or the Northern Isles. Analysis of the environmental setting of *Pit-* place names has established that, as well as their avoidance of coastal sites, *Pit-* elements had a preference for sheltered, south-facing slopes. When related to soils or to land capability, however, there emerged an interesting distinction between those south of the Mounth and those to the north. Those to the south displayed a preference for fertile brown forest soils, and avoided podzol soils, whereas those to the north were to be found more on podzols than on the brown forest soils.[24] The reason for such a difference may have something to do with differences in the chronology of settlement between the two areas or with the use of other place-name elements alongside *Pit-* elements in areas to the north of the Mounth. Whatever the reason, the concentration of *Pit-* elements on brown forest and podzolic soils suggests that the field economy of Pictish communities was as much to do with arable as with pasture.[25]

Land grants recorded in the *Book of Deer* and dating from the tenth century show the element *Pit-* used to denote a share or piece of land.[26] In this sense, the element *Pit-* was, as Whittington pointed out, a unit of land administration.[27] To this we can add Barrow's conclusion that the units of land denoted by *Pit-* elements may have formed constituent parts of a territorial unit known in parts of eastern Scotland as a thanage.[28] In character, multiple estates such as the thanage comprised a cluster of settlements scattered over an area which might be the size of a large parish or more. They would have been tied together by their dependence on a single lord, a lord whose power was focused on a single headquarters settlement, or caput, from which the thanage took its name. The functioning unity of the estate found expression in a number of ways. In most cases, for example, the various constituent settlements shared the same common grazings together. Where unfree communities were present, they were burdened with labour services on the demesne located at the estate's headquarters settlement. Meanwhile, where free communities were present, they either owed food renders to the lord or were burdened with obligations of hospitality for him.

The fact that *Pit*-based place names invariably combine *Pit*-, a P-Celtic element, with a Q-Celtic element suggested that such place names took shape or acquired this form over the ninth to tenth centuries AD. The trigger for this may have been the union of the Scotti or Dalriadic kingship and that of the Picts in AD 843. To account for the appearance of settlements combining *Pit*- with a Q-Celtic element, some have suggested that their union was underpinned by a subsequent movement of Scots eastwards into Pictish areas so that, for a time, one had Pictish and Celtic communities living side by side in the central and eastern Highlands, with Q-Celtic-speaking communities slowly becoming ascendant. Not all scholars have accepted that such migration took place.[29] Indeed, even if such a movement did take place, we would still need to explain why Q-Celtic communities should have adopted place names with a P-Celtic prefix if they were in the ascendancy, especially if there was a Q-Celtic equivalent in the form of *Bal*-. Any answers to this question needs to take account of how Whittington took the problem forward. He observed that outlying examples of *Pit*- place names in areas like Glenelg and south of the Forth suggest that, far from being an element that only came into use late on when the Pictish kingdom was confined to the east of Scotland north of the Forth, its use actually extended back to a time when the Picts were settled across a much larger area of Scotland. Their occurrence in western Scotland was probably reduced following their local replacement with the Q-Celtic element *Bal*-. Indeed, examples can be cited of touns that, even in the early modern period, switched between the prefixes *Bal*- and *Pit*-. Such a simple switch did not occur in the far north where the Norse appear to have introduced their own place names en bloc.[30] Whittington's case for an early switch from *Pit*- to *Bal*- in western Scotland is convincing but it does raise new issues. In particular, it poses the question of how did conditions differ between the western Highlands and eastern Highlands if, in the former, *Pit*- was replaced by its Q-Celtic equivalent or *Bal*- whereas, in the latter, *Pit*- was preserved but now hybridised by being attached to a Q-Celtic element? We could argue, as Nicolaisen suggested,[31] that any Gaelic speakers who had emerged in the eastern Highlands by the time the Scots and Picts became united into a single kingdom in AD 843, would have been comfortable with the equivalence between *Pit*- and *Bal*- and the appropriateness of using the former for an estate of status in that area. We must also keep in mind that between the early switch from *Pit*- to

Bal- in western Scotland and the emergence of Q-Celtic/P-Celtic hybrids in the eastern Highlands from the mid-ninth century onwards, the meaning of these terms may have shifted. For comparison, work on the element *baile* in Ireland suggests that its association with nucleated settlement or with some form of co-operative farming came much later than its initial meaning as share or piece of land.[32]

Given the link which may have existed between *Pit-* elements and the component estates that made up the thanage, any long-term continuity which we can attach to the former has implications for how we might read the long-term history of the thanage. Driscoll has summarised the case for seeing thanages as an archaic institution, one that stretches back far beyond the first documentary references to them in the twelfth and thirteenth centuries AD, but one whose character and precise purpose changed over time.[33] Clearly, if the estates represented by *Pit-* elements were old established, in existence by late prehistory, then it would help to make the case for the deep-rootedness of the thanage. A late prehistoric origin is a possibility but one that has still to be affirmed. A lot depends on how much of the formal structure of the thanage that we see in the twelfth to thirteenth centuries AD can be extended back. We can certainly expect the greater chiefs of the Late Iron Age to have had their greater status underpinned through the payment of gifts and renders but whether it was developed through formalised estates or components, as represented by the *Pit-* elements, as opposed to being simply a burden on kin groups, is impossible to say.

The South-west Highlands and Islands

It was a long-held belief that there was a movement of Scotti from the Irish kingdom of Dalriada or Dál Riata in northern Ireland into the South-west Highlands and Islands around AD 500 where they established a parallel kingdom of Dalriada. Recent work has challenged this view.[34] At most, it may have been no more than a dynastic shift, with a small elite moving into the region, one that played on the fact that there had long been regular cultural and political interaction between Northern Ireland and the South-west Highlands.[35] Such a revision has a bearing on how we read the emergence of the latter as a Q-Celtic or Irish-speaking area. It would suggest that, far from being the hallmark of new colonists, the presence of Irish-speaking communities may have been long-standing. Clearly, such a revision affects how we read attempts to show the outward spread of Q-Celtic names from the South-west Highlands, such as that by Nicolaisen. His mapping of the element *sliabh* or *Slew-*, meaning a hill or mountain, would still delineate the areas of Q-Celtic or Gaelic-speaking communities around AD 500[36] but they would not represent newly established communities. That said, his use of the more widely occurring element, *Kil-*, meaning a church, to depict the expanded area of Gaelic-speaking communities which emerged following the later union between the kingship of Dalriada and the Picts, would still be valid.[37]

The seventh-century *Senchus Fer nAlban* suggests that the Scotti of Dalriada were divided between three main kindreds that owed allegiance to the overking of Dalriada. Each kindred, reputedly based on the sons or brothers of Fergus Mor, held a different territory. The Cenel Loairn held Mull, Coll, Tiree, Ardnamurchan and Lorn. The Cenel nGabrain held Arran, Jura, Bute, Cowal and Kintyre, while the Cenel nOengusa held Islay. Like the notion of Dalriada being founded following a rapid conquest, this picture of a neat tripartite split of Dalriada into three territorially based kindreds has now been rejected as a fiction, a post hoc justification of how these areas came to be unified under the kingship of Dalriada.[38] The *Senchus Fer nAlban* also contains systematic data which, on the face of it, appear as a census of houses, or *treb*, broken down per tribal district and between each area or township in the case of Islay, Arran, Jura, Bute, Cowal and Kintyre but simply in terms of septs or nobles for Mull, Coll, Tiree, Ardnamurchan and Lorn. Bannerman has suggested that the reasons for the compilation of such a house or sept listing was linked to the payment of tribute to the king of Dalriada. House custom, or *bes-tige*, was a standard form of tribute paid by freemen in early Gaelic society.[39] The fact that the amounts involved were in multiples of a standardised amount makes it clear that this was a notional system. Notional levies based on the house occurred in different contexts across the Highlands, such as in a twelfth-century grant to Saddell Abbey of a penny out of each house from which smoke issued or, much later, the levy in the Central Highlands of a reek hen from houses out of which smoke issued.[40] Such payments were a levy on the house itself. In the context of the *Senchus* listing, though, the term *treb* may well have denoted a household, that is, those attached to a particular family or lineage segment, a reading that would sit comfortably with those assessments in the *Senchus* tied directly to septs or families. We should, however, be wary of supposing that, because tribute was based on the house or the household, then it was also, by extension, levied on the land, as if a nascent form of land assessment, a pennyland by another name, as Lamont supposed.[41] Payments of tribute in kin-based societies, the sort of payments that would have existed when the *Senchus* was drawn up, were first and foremost obligations on the person or kin group. They related to land only in so far as land provided the individual or kin group with the means to meet its obligations. Their transfer on to land represented a separate and, arguably, later stage of development.

Settlement in Argyll during the first half of the first millennium AD has been characterised by two forms touched upon in the previous chapter: duns and crannogs. Already featured in the landscape by the end of the Iron Age, both types were in use well into the first millennium AD. Numerically the most important were the circular ring forts known as duns. These varied greatly in size and embraced a range of architectural complexity (Figure 3.3).

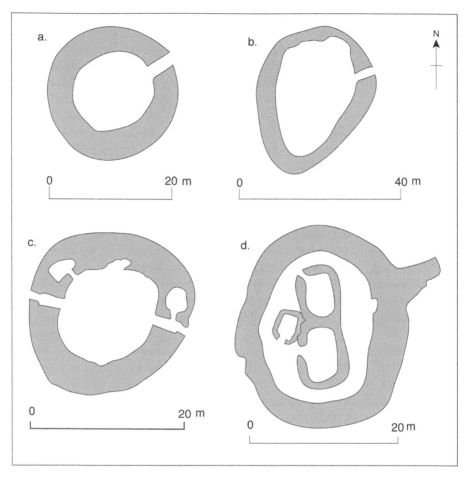

Figure 3.3 Duns in the South-west Highlands and Islands. Based on a.
Dun, Caisteal Suidhe, Cheannaidh, after RCAHMS, *Argyll*, vol. 2: *Lorn*
(Edinburgh, 1975), p. 81; b. Dun, Barr Mór, Loch Awe, ibid., p. 80; c. Dun
Kildalloig, Kintyre, ibid., vol. 1: *Kintyre*, 1972, pp. 88–9; d. Dùn Anlaimh,
Loch nan Cinneachan, Coll, ibid., vol. 3: *Mull, Tiree, Coll and Northern
Isles*, pp. 119–20.

The largest and most monumental were those built during the Iron
Age. Those constructed during the first millennium AD were smaller but
more numerous. In distribution, they occurred widely across the South-west
Highlands and Islands. For this reason, some saw them as associated with a
supposed Scotti settlement of Dalriada. Those that have been dated, though,
suggest otherwise. Many appear to have been constructed in the two or three
centuries before the emergence of Dalriada but, as Nieke and Duncan add,
the people we recognise as the Scotti are likely to have been already resident

in Argyll long before about 500 AD.[42] The majority of duns remained in occupation down into the mid- to late first millennium. In terms of location, the typical dun had access to some arable, with a few being clearly linked to adjacent or nearby field enclosures, though more analysis is needed before we can start to generalise about how their farm economy fitted into the wider landscape.[43] Crannogs not only existed alongside duns, with a particularly strong presence in the South-west Highlands, but they also shared their chronology. As made clear in the previous chapter, many were built and occupied during the Iron Age but some continued to be occupied, and even new ones built, well into the first millennium AD.

For all their widespread occurrence and prominence in the landscape, detailed field survey in Argyll has stressed the need to see duns and crannogs as being only part of the prevailing settlement pattern during the first millennium AD. Doubts exist over whether the larger duns, especially, could be built on a DIY basis by those who lived in them. Even crannogs have raised doubts over whether those occupying them could provide sufficient labour to build them, with one estimate putting the timber needed for a good-sized crannog at 650 trees. For some, this labour is more likely to have been provided by client groups of lesser status living in surrounding settlements. As Nieke put it when talking about the larger duns on Islay, they can be seen as 'the product of a system of labour mobilisation in order to bring the stone to the site, and actual construction of the homestead'.[44] Each dun and even crannogs would have had 'dependent clients . . . settled near the residence of their overlord for physical protection' and providing labour for them.[45] Some of these dependent settlements are likely to have been physically separated from the duns, positioned on their own sites, but there is also the possibility that some had a closer relationship, as with the so-called broch villages.[46] In other words, duns and crannogs need to be seen in a wider context in which settlements of different scale and social status existed.

Landscapes of the Norse

Though earlier raids were carried out by the Norse, their first settlements were not established until the late eighth century AD onwards when communities began to settle in the Northern Isles. The basic framework of their settlement can be studied through place names and surviving field evidence. Taking the place-name evidence first, their colonisation of the Northern Isles and the Western Isles involved the wholesale replacement of earlier place names, with little continuity of place identity from pre-Norse times. By comparison, the degree of place-name change brought about by them in the southern Hebrides and across parts of the Inner Isles was less. Some have argued that place names in these areas, too, were also changed comprehensively but a later resurgence of Gaelic place names meant fewer examples survived. Others, meanwhile, have concluded that their mix of Gaelic and Norse place names is because fewer Norse settlements were established in the southern Hebrides in the first place. The wholesale replacement of earlier place names in the Northern and Western Isles begs a number of questions. The most obvious is what it may tell us about the nature of Norse colonisation. For Wainwright, it signalled, in the Northern Isles at least, a settlement by 'mass migration'.[47] Along with the changes in landholding, building styles and subsistence patterns,[48] their comprehensive renaming of the landscape makes a strong case for the Norse colonisation being a point of discontinuity. For some, this break with what had gone before came about through a forceful usurpation by the Norse, one that involved the extermination of Pictish communities.[49] Others have raised the possibility that the Norse colonisation was comprehensive because they faced a relatively empty landscape, one that had already been thinned by a population decline.[50] Still others feel the Norse colonists followed a more accommodating strategy, allowing pre-Norse communities to live on alongside them and allowing some of their landholding institutions to survive unaltered.[51] There are still, however, serious problems to be confronted by those holding to a continuity or so-called 'Peace School' thesis, not least the question of why did Norse place names become overwhelming?[52]

Place names have long been used to map both the overall spread of Norse settlement across the region as well as to establish how it filled out locally. The most notable attempt to map its overall spread was that by Nicolaisen.[53] By mapping what he saw as key Norse habitative elements across the Northern and Western Isles and adjacent parts of the mainland, he hoped to isolate those that had an association with different phases of their colonisation. This was not a straightforward exercise. Work in Iceland demonstrated that, where they were attached to settlement, topographic elements, such as *dalr* meaning a valley, might enable us to isolate the earliest stages of Norse colonisation. When mapped in the Highlands and Islands, however, *dalr* occurs widely. Further, most examples are purely topographical rather than attached to farm names so can only map the ultimate extent of Norse influence, not a particular phase of their settlement.[54] If we restrict the analysis to the use of the *dalr*-element attached to settlement, there are too few examples to form the basis of a meaningful distribution. Other habitative elements thought to have been used in the earliest phases of colonisation (that, is *bær* or *-býr*) are also relatively few in number, some possibly disappearing following the division of estates between heirs and the adoption of wholly new names for the different shares.

Steering his approach around these problems, Nicolaisen proposed a sequencing of Norse colonisation based on three habitative elements: *staðir*, *setr* and *bolstaðir*. The element *staðir*, meaning farm or place, is often combined with a personal name, such as Grimsta in Shetland, which has the meaning of Grim's farm, or Tolsta, meaning Tholf's farm, on Lewis. The high percentage of *staðir* names combined with a personal name suggests that many

Norse farms were owned and occupied as separate farms. When mapped, they occur widely in the Northern Isles, with others in the Western Isles and on Skye. Their limited presence on the mainland led Nicolaisen to see them as indicative of the earliest phases of colonisation. The fact that many appear as small secondary farms set inland, however, suggested to others that they are more likely to belong to the middle to later phases of colonisation.[55] Initially, the elements *setr* and *seatr* were distinct, the former representing a permanent settlement and the latter a shieling or summer dwelling.[56] In practice, the two forms are difficult to differentiate. In all probability, its earliest meaning was as a shieling site, with its secondary use as a permanent settlement following from the all-year occupation of those shielings. A further point to be noted is that the use of *-setr* for a shieling gives way as one moves southwards through the Hebrides to the Gaelic word *airge*, even where the main settlements carried Norse names. In contrast to settlement names incorporating the element *-setr*, the element *bolstaðir* was associated with larger settlements located on favourable sites for farming.[57] When mapped in all its various forms, it is found in all the main areas settled by the Norse, across the Northern Isles and Hebrides, as well as Caithness and Sutherland.

While there is undoubtedly some value in mapping such elements, some have questioned how far they can be reliably used to study the overall spread of Norse settlement owing to the way naming practices differed across the region.[58] Such schemes are plausible when used to establish the evolution of local settlement patterns but are less reliable if used to develop pan-regional schemes. Marwick's work on Orkney was the first to use place names to construct a local sequencing of how Norse settlement unfolded.[59] At the heart of his argument was the simple assumption that colonisation began with a primary settlement, invariably sited in a coastal position and identified by the use of the elements of *bær, -býr, or bu of . . .*, or that of *skáili*. As a freestanding name, *bær, -býr* take the form of Bay or Bae. Where it occurs in compound form, it does so as *-by*. These were 'the original settlements of the Scandinavian *landnàmsmenn*' or land takers[60] though, as already noted, place names incorporating *bær, -býr* or *skáili* have not survived in any quantity owing to the fact that many were later subjected to splitting and their component parts renamed. Significantly, Marwick imposed a tight chronology on their formation, seeing them as in place by around AD 850. For Marwick, subsequent expansion not only saw the subdivision of these primary settlements through their inheritance by co-heirs, it also saw the formation of secondary settlements formed as outgrowths from the original settlement core. As well as being peripheral to the primary settlement, these secondary settlements were characterised by their use of place-name elements such as *staðir* or *-sta, bolstaðir*, -land and -garth. The imposition of skat, a tax, was crucial to how he dates these secondary settlements, for all lay within the skattald, the area burdened with skat. He assumed that skat had been imposed by about AD 900 and thereafter

remained unchanged, thus providing us with a *terminus ad quem* for the creation of these secondary settlements. His dating of when skat was imposed also
provides us with a date (after AD 900) for when settlements not burdened
with skat, beyond the outer limits of the skattald, first emerged, either along
its outer edge or free-standing on the hill ground beyond. These non-skatted
settlements bore elements like *quoy* and *setr* in their place names. Overall, the
prime feature of Marwick's scheme is the way it compressed the chronology
of Norse colonisation, emphasising its character as a fairly large-scale and
chronologically concentrated migration of people into the region.

In a more recent reading of how we might read the local evolution of Norse
settlement, Crawford took the same starting point as Marwick: the creation of
a primary settlement of *landnàmsmenn* in coastal areas. Similarly, she saw these
as bearing place-name elements like *-by*, *-sta*, *-ster*, *-skaill*, or a topographical
name. Over time, the expansion of this primary settlement led to the creation
of new farms out of the land or territory of primary settlement through inheritance. This expansion took different forms. The primary settlement itself may
have been split into different farms, or *houses*, with each *house* distinguished
from each other by a prefix such as North/South or Upper/Lower, and turning
the original settlement into linked but loose scatter farms around the original
-byr settlement. In some cases, such a cluster – the primary settlement and its
offshoots – carried the descriptive name of *Biggins*, with the grouped farms
or *houses* effectively forming a family estate or township, one that mapped
the genealogy of the family. On the edge of this primary settlement cluster,
but still within its skattald or taxable land, there also developed a range of
smaller secondary settlements identified by place names incorporating elements
such as *bolstaðir* or *-bister*, or *-land*. Beyond the limits of the skattald and the
head dyke that bounded it, Crawford – like Marwick – envisaged the eventual
emergence of other secondary settlements. The earliest may have been those settlements that developed just beyond or on the edge of the skattald, and which
used standardised affixes, such as *-qvi* or *-quoy* and *-garðr*, *gard* or *garth*. The
first was usually a form of pastoral holding, a fold, perhaps one that was closely
allied in character to *-setr*. The *-garðr*, *-gard* or *-garth* names represented a
small intake out of the non-skattald waste, what in other areas would be seen as
an outset. This contrasts with Marwick's scheme in which holdings with *-garðr*,
-gard or *-garth* names occurred within the skattald area. Chronologically later,
but probably still part of the early Viking phase of colonisation, were the *-setr*
or *-ster* sites, at least those attached to farm names. First emerging as temporary
summer occupation sites, some were later converted into sites of permanent
settlement as part of the infilling or intensification of settlement that followed
naturally from the arrival of extra numbers and local population growth.[61]

In a survey of early Norse place names published in the same year as
Crawford's, Thomson attempted a more revisionary approach. First, he questioned the assumption that colonisation began with primary settlements on

coastal sites and then spread outwards via a peripheral fringe of secondary settlements. His doubts rested partly on his belief that, when the Norse arrived in Orkney, they would not have found a *tabula rasa*, nor have created one, but one partly settled.[62] He reinforced his criticism of the primary–secondary/coastal–inland contrast by a simple locational analysis of place names, showing that whereas *by* and *-skaill* elements were associated with coastal settlements and setter and *staðir* elements were associated with inland sites, setter with marginal sites and *staðir* with inland arable sites, other key elements – such as boer, bister, land, garth and quoy – displayed no obvious locational preference. Second, he was also critical of Marwick's chronology, a chronology framed around his assumption that skat was imposed c.900, so that all settlements whose land was skatted – what amounted to the majority – had been formed before c.900. For Thomson, dating when skat was imposed is not so straightforward, given that there may have been later revaluations. His doubt over the imposition of skat as a fundamental reference leads to his claim that the use of key place-name elements was actually drawn out over a much longer period than Marwick allowed, to the extent that we cannot attach their use in place name formation to discrete phases.[63] In fact, his blunt conclusion is that using place-name elements to develop a chronologically framed interpretation of how early Norse settlement unfolded is fruitless, 'gone by the board'.[64] Third, he stressed the role played by the fissioning of farms, the carving out of new ones and the general intensification of settlement as key processes. These featured in what Marwick and Crawford proposed. What Thomson adds is the need to see phases of settlement fissioning and expansion as drawn out, but as happening at a quicker pace during times of rapid population growth, that is during the twelfth to thirteenth centuries.

The number of excavated Norse sites has increased over recent years, in the Hebrides as well as the Northern Isles. What is clear from these excavations is that, while the Norse were not averse to living beside earlier, pre-Norse structures, nevertheless, there was a clear break in building styles between the two.[65] Nothing in the pre-Norse period in the Northern and Western Isles compares to the longhouses that appeared with the Vikings. On the basis of his excavations at Jarlshof, Hamilton proposed a chronological division of Norse housing into a Viking phase, approximately AD 800–1100, and a Late Norse phase, c.1100–1500.[66] The typical house of the Viking phase was an elongated rectangular house, roughly 20 metres by 5 metres. Its walls were 1 to 1.5 metres thick, made of drystone and turf, and usually took a characteristic bow shape with slightly rounded ends. The interior was open, though we cannot rule out the possibility that some internal partitioning was present, or that the side walls of some were panelled. The roof was supported by a series of double posts that ran the length of the building. In style, this was the *skali* (fire house), or hall house, the term 'hall' capturing the openness of its interior. Excavated examples of a *skali* are provided by those at The Biggings on Papa Stour and

Figure 3.4 Original Viking longhouse at Jarlshof, Shetland.

at Jarlshof. The former is especially noteworthy because, as well as being a comprehensively excavated example,[67] it probably formed the site at which a thirteenth-century dispute that has a bearing on my later discussion was aired (p. 93). The example at Jarlshof formed the original or parent farmstead of the Viking settlement there. In design, it was a rectangular farmstead, 21 metres by 6 metres. Its walls were 1 to 1.5 metres thick, made out of drystone and turf filled. Only one side appears to have the characteristic bow shape of a Norse longhouse but both ends were rounded at the corners. Its interior was open but there were two functional spaces, a main living space with a central long hearth and, at one end, a bathhouse and one structure that may have been used as a byre (Figure 3.4).

Subsequently, a new farmstead was built at right angles to the original or parent farmstead.[68] Though the same size as the parent farmstead, and preserving its rectangular shape approximately 21 metres by 5 metres, its ground plan differed from the parent farmstead in having straight, rather than bowed, walls and more angular corners. It also had opposite entrances at either end, not across its middle as with the parent farmstead. Internally, the living space may have had benches installed along the wall on either side but the most striking difference from the original parent farmstead was the incorporation of a byre within one end of the building, making it a longhouse, with animals and humans under the same roof.[69] A further new farmstead, the third on the site, was built after AD 900. Slightly longer than the first two (22 metres), it again incorporated a paved byre.

During the late Norse period, around 1100–1500, some of the features that were found in the second and third farmsteads built at Jarlshof became more

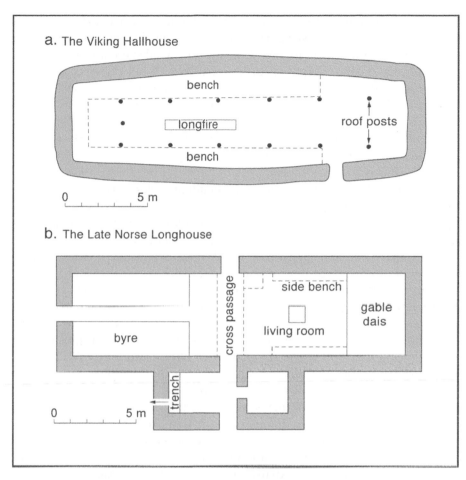

Figure 3.5 Bigelow's representation of how Norse dwellings changed between the Viking and Late Norse periods in the Northern Isles, Bigelow, 'Domestic architecture', pp. 26, 31.

widespread in the region. The Late Norse house was still built on a rectangular plan but its walls were now typically straight rather than bow shaped. More significantly, byres acquired a more prominent role, with some being added as an extension to existing buildings and others being incorporated into the ends of newly constructed farmsteads. Even at Jarlshof, where byres were already a feature of housing before the end of the Viking age, they acquired a more significant role. During the early part of the Late Norse period, or the eleventh century, the layout of the site at Jarlshof underwent change. One of the two secondary farmsteads which had been built during the Viking period was abandoned altogether and the other was converted into a byre. The parent farmstead, meanwhile, had a byre added to one of its gable ends and the other end was extended so as to increase the available living space, extensions that made it an extremely long structure. With its expanded living space, and both its adjoining and free-standing byres, it was now a substantial farm complex, with a greatly enhanced capacity for housing stock. While there was probably a phase when this was the only farmstead on the site, its solitary position did not last for long. By the end of the eleventh century, a pair of new, smaller farmsteads was added to the site. Initially, they were simply dwellings but, later, one acquired a byre extension. Continued modifications to the different farms were made down to the thirteenth century but, by now, the whole site was in decline as a farm complex.

Reviewing the overall transition of building styles between the Viking and Late Norse periods and, in particular, the addition of byres (Figure 3.5), Bigelow concluded that it reflected a significant expansion of dairy output, one probably driven by the uplifting of cheese and butter as tax and tithe.

Quite apart from the appearance and then the expansion of byres, his case rested on the presence of bone dumps at sites such as Jarlshof and Sandvick that suggested that there had been a selective culling of young stock. His interpretation was that, quite apart from the greater warmth of the byre increasing milk yields, Late Norse farmers were killing young calves so as to rear the surviving calves on the milk of more than one cow, with the surplus milk so released being used to expand the making of cheese and butter.[70] The byres that appeared at these Late Norse sites were not the first byres to appear in the Highland archaeological record. Nor were the bone dumps at sites like Jarlshof the first to suggest that selective culling of young stock had been practised. Comparable bone dumps occur at Late Iron Age sites. What was significant about how Bigelow interpreted the evidence was the way in which he combined the evidence for such bone dumps with that for an expansion of byre capacity to argue for an expansion of dairy output, to sustain lordly exactions as much as in-farm subsistence. His key assumption, that some calves were being slaughtered so as to release more milk for butter and cheese, was criticised on the grounds that cows would not suckle any calf other than their own. In fact, as we shall see in Chapter 5, it is given support by seventeenth- and eighteenth-century documentation for the region.

Until quite recently, what could be said about the Viking age in the Hebrides was based on only a handful of excavated sites. That has now changed. One of the first to be excavated was that at Drimore (South Uist). The site comprised a single house structure that was broadly rectangular in shape but with slightly bowed walls and rounded corners though, subsequently, one of its side walls was straightened and the corners at one end sharpened. In construction, its walls were built using drystone and possibly made use of a turf capping. Overall, the building was significantly smaller than the farmsteads at Jarlshof, being only 14 metres in length and 5 metres in width. Internally, its open plan and a long central hearth were characteristic of a *skali*.[71] It may have been modified and possibly kept in occupation until well into the late medieval period but the fact that it is thought to have been an isolated settlement means that it was atypical as a Hebridean Viking site.[72] The site excavated at Kilpheder on the machair of South Uist appears to have had a complex history in terms of occupation. The first Viking house recorded there was similar in size to that at Drimore but the Viking and Norse houses that were subsequently built on the site were smaller, a feature that the excavators explain not by invoking a shortage of timber for roofing but by relating it to the prevailing size of Norse families. Another aspect highlighted by the excavators is the fact that each new building was built at right angles to its predecessor but in a way that allowed the new house to be founded on part of the walls of the old or pre-existing house.[73] The consistent way in which this was done suggests that it may have served a symbolic purpose.

A still more complex site has been excavated at Bornais on South Uist. It consists of a small number of low settlement mounds on the machair that were occupied from the Middle Iron Age onwards. Each mound is a discrete form but, together, they form 'nodal points in an extensive settlement area which covers an area of 0.8 ha'.[74] Altogether, across three mounds and their apron of adjoining machair, eighteen separate buildings have been discovered. Provisionally, the two mounds that contain Viking structures were each seen as being possible sites for two single dwellings, and their associated outbuildings, that may have been modified and rebuilt over time. Two other areas, one within a mound and one lying between the two Viking mounds, provided geophysical traces of more complicated settlement foci, each trace hinting at a small cluster of buildings positioned around a courtyard. Brought together, these different sites suggested that the Viking and Norse settlement at Bornais may have amounted to a small, hamlet-like cluster. Closer dating of the different sites may tell us whether the different foci were all occupied from the outset as a single complex or whether, as with Crawford's model of settlement evolution, we are dealing with an initial primary settlement, that which acted as the focus for an elite family, that acquired a surrounding cluster of new farms or *houses* formed through the splitting of the primary holding.

As already noted, settlement mounds are a feature of both the Northern and Western Isles, with quite a number boasting a record of intermittent occupation from the Neolithic or Bronze Ages down to the early historic period, or even the eighteenth century in the case of The Udal.[75] Their occupation was not continuous but it was sufficient to produce a tell-like accumulation of debris. While there are signs on some sites that household waste and ash were used as field manure during the Bronze and early Iron Ages, less was recycled as such as the Iron Age progressed. 'Waste not, want not' might have been a maxim of communities across the Atlantic Province for most of prehistory, but they increasingly appear to have dumped their domestic refuse outside the door as they came to rely more on animal dung mixed with organic matter, such as turf

Figure 3.6 Soil mound, Sanday, Orkney.

and old thatch, for their field manure. The immediate dumping of domestic refuse meant that, on site after site, we find new or reconstructed buildings being set into former middens, to the extent that some became 'encased in their middens', as Parker Pearson and Sharples put it.[76] This last point leads us into the problem posed by the farm mounds of Orkney. They are different from the prehistoric settlement mounds found elsewhere in that most are still crowned by well-established farmsteads and their outbuildings (Figure 3.6).

Considered a 'striking feature' of the Orkney landscape, they are especially well represented on Sanday, where as many as fifteen have been identified,[77] as well as on South Ronaldsay, Papa Westray and Westray, though they also had their counterparts elsewhere in the North Atlantic Province.[78] In size, they were substantial. On Sanday, where they have been analysed in detail, they ranged from 50 to 205 metres across with a height of up to 4 to 5 metres (Figure 3.6), though the settlement or farm mound at Pool, a mound 100 metres across and 8 metres in height, has now also been shown to be anthropogenic.[79] Analysis of the pedogenic composition of the soils forming the mounds at Westbrough, Skelbrae and Langskaill, on the north side of Sanday, has confirmed that 'the mounds grew as middens'.[80] The midden material had 'major inputs derived from the byres (cut and dried turves, ash); [while] subsidiary inputs would have been derived from domestic refuse as well as from turves used for roofing or for windproofing flagstone roofs'.[81] In a breakdown of these results, the Westbrough mound showed differences between the basal and upper layers. Manure appears as an input throughout the profile but, between the lower and upper layers, there is a shift from peat and turf to ash as the major supplementary input.[82] All three farm mounds have been dated. Two appeared to have been formed after the seventh century AD and the third from the early part of the thirteenth century,[83] though the formation of the farm mound at Pool, on the south side of Sanday, appears to have started in the Iron Age.[84]

The build-up of deep midden deposits underneath settlements has its field corollary in the formation of plaggen, that is, soils that have been organically enriched and deepened by the addition of manures. These deepened topsoils are found in all parts of the Atlantic Province but the extent of their presence varied between the Western and Northern Isles. In the Western Isles, recent analysis of some field soils, such as on Baleshare, has shown a sufficiently deep anthropogenically generated build-up of organic matter in its soils for them to be classed as plaggen. In fact, all along the edge of the machair in the Uists, we can find instances of land whose soils have been organically enriched and instances of sandy soils stabilised by the heavy addition of midden, turf and especially peat, the last being available in quantity from the peatlands to the east. In fact, early modern sources for the Uists make it clear that bringing the machair and peat together was a prime form of land improvement, one that built on their natural mixing when sand was blown inland over more organic soils. The extent to which such plaggen-type soils have been found in prehistoric and early historic contexts suggests that adding turf, manure and peat was long-standing. On St Kilda, the switch to a field culture based on the heavy use of turf, peat and manure to replenish the organic content of soils was accompanied by the introduction of small, raised cultivation plots, an adjustment dated to just before, or during, the early first millennium AD.[85] The creation of deepened topsoils was especially marked in the Northern Isles during the Viking and Norse periods. One of the most anthropogenically transformed islands is that of Papa Stour in Shetland where one half of the island was effectively stripped of turf down to bare rock to provide manurial inputs for the arable on the remaining half.[86] Just as with midden soils, islands such as Sanday in Orkney experienced a particularly marked build-up of deep topsoils. In fact, there occurs across Sanday a type of anthropogenic soil, which has been labelled as the Bilbester series, comprising a deep topsoil whose depth exceeds 75 centimetres. Such soils merge into the deep middens that underpin the farm mounds. Their dating has produced rates of accumulation of 0.9 millimetres per year, with a starting point for accumulation at between AD 1000 and 1100.[87]

Lords and Landscape

When seen in the centuries immediately prior to the Clearances, the Highlands and Islands possessed an array of different land-assessment systems that were still being actively used to define and set touns. They were the basis on which the amount of land within a district or a toun was calculated, and the basis on which such land was leased out. Generally, those in use were adapted or derived from other measures, so that we find different systems of land assessment using different ways to calculate the quantities involved and employing different metrologies. Some appear based *ab origine* on cash burdens or dues; others appear to be linked to the amount of grain that could be sown or harvested; while still others were linked to the capacity of a plough team. What makes them of rich historical interest is the fact that the different systems had deep roots, taking us back into the early medieval period so that establishing their origins and intrinsic nature may shed further light on the early medieval landscape. Adding to this interest is the fact that the various systems of land assessment do not occupy discrete areas. In a number of areas, different systems were used alongside one another, enabling us to ask how they related to each other.

In looking closer at these issues, I want to draw out four aspects. First, I want to review the different systems that existed across the region and the extent to which they can be correlated. Second I want to offer some thoughts about how land-assessment schemes developed. Third, I want to comment on precisely how touns and their assessment were linked. Fourth, I want explore what assessment meant for touns.

Regional Patterns of Assessment

Early charters and later, rentals, show a system of land assessment in use across the main body of the Highlands based on a unit known as the *dabhach* or davoch. Though derived from a Gaelic term meaning a container or vat,[88] there seems, as Barrow observed, 'something inescapably Pictish about the use of the davoch of land'.[89] The earliest reference to it consists of an eleventh-century grant of two davochs in Upper Rosabard (Aberdeenshire) recorded in the *Book of Deer*.[90] When seen through sources for the sixteenth to eighteenth centuries, its core area of use was the eastern Highlands but there is evidence of its former use in areas along the western seaboard, in mainland areas such as Glenelg, Kintail, Applecross, Gairloch and Assynt, as well as on islands such as Tiree and Skye. Derived from a Gaelic word for a container or a vat, its use as a basis for land assessment has logically been seen as linked to the amount of land that could be sown with a designated vat of seed or, at the other end of the cropping cycle, the amount of harvested grain paid as render. Though the latter has tended to be favoured, we should not ignore the fact that a great deal of land in the Highlands was customarily set in terms of its 'bolls of sowing'. Reinforcing its interpretation as based on a measure of seed used or grain yielded is the fact that, in everyday practice, the davoch constituted a unit of cultivated land, one that could be recognised on the ground as having agreed bounds.[91] That it was an assessment based on arable is conveyed by the fact wherever the data allow us to see it in detail, the davoch appears subdivided into ploughs and oxgates.[92] Barrow made out a plausible case for seeing it initially, or in the earliest references, as consisting of only two ploughs.[93] By the seventeenth and eighteenth centuries, the data for the Eastern Highlands suggest that, by then, the typical davoch was subdivided into four ploughs. Gordon of Straloch made this clear in how he defines the davoch in Aberdeenshire. Each district, he wrote, was divided in 'daachs' or pagi, each of which contained as much land as could be cultivated with four ploughs in a year.[94] When we look at rentals for the eastern Highlands, such as those drawn up for the Gordon estate, they are replete with entries showing touns that comprised one, two, three or four ploughs alongside references to others assessed as a davoch, a half davoch or a quarter of a davoch, the two forms effectively being interchangeable.[95] Also worth noting is that, even where the davoch was subdivided into separate shares (that is, halves, two ploughs, and so on) and even where these subdivisions were clearly laid-out separate touns, each with its own place name, their affiliation to what we can call the parent or original davoch was still recognised, as with the 'quarter of a Davoch land of Strathconnan callet Knokaninroivick'.[96] As Barrow observed,[97] this sort of fractionalisation was inevitable given the scale of the davoch as a unit.

In a recent analysis based on data for Moray, Ross has suggested that the discrepancy between early references, implying a davoch of only two ploughs, and later sources suggesting that they comprised four ploughs could be because of a rerating in the shift from Old to New Extent. In this context, Old Extent dated from the second half of the thirteenth century and New Extent from c.1474.[98] In fact, there are hints of a rerating in what Gordon of Straloch had to say. The expansion of arable, he wrote at the end of the seventeenth century, brought about an increase in the amount of arable per davoch, a doubling in fact.[99] If so, more ploughs, and more ploughgates, would have been needed to maintain arable. The number of ploughs per davoch has a bearing on other aspects of Ross's argument. While acknowledging how their plough structure may have changed over time, even though the overall number of davochs present remained static, he questioned the basic assumption that the davoch was a straightforward unit of land assessment, one scaled through the amount of render uplifted from a give unit of land. Using data for Moray, he concluded that, when we reconstruct their layouts, especially those for which detailed perambulations are available, they embraced pasture and forest ground as well as arable.[100] This caused him to question whether, in its initial formulation, the davoch was derived from a grain- or seed-based render and, more importantly, to question its meaning as a unit of land or arable assessment, even going so far as to dismiss the idea as 'probably untenable'.[101] There is much to consider in this rethink but we must not be too hasty in dispensing with the davoch as a unit of arable assessment. In the first place, the importance attached to the careful recording of the davoch and its constituent ploughs in sources of the seventeenth and eighteenth centuries when it came to describing the working arrangement of touns can be explained only if they had been *customarily* used to proportion rights between touns and tenants, and to proportion the rents, renders and obligations burdened on them, which is precisely the function served by land assessments. Second, when we look at how the davoch was used in rentals and surveys, the assessment of arable appears at its core which is what one would expect given its structuring around ploughs and oxgates. No matter how we ultimately derive the davoch as a term, its subdivision into ploughs and oxgates surely means that it had something to do with the arable capacity of land. Third, the way some sources portray the davoch as embracing all available land, arable plus pasture and woodland, does not in itself rule out its primary function as a unit of land assessment directed at arable. All rights in arable carried with them proportionate rights in the surrounding pasture and waste. In this sense, land assessment had two layers of identity, one wrapped around arable specifically and the other also embracing the wider rights appendaged to it, so that, as with the bull of Ropnes on Westray, what might be prescribed as the nine pennyland in one source[102] is described as having entitlement to all 'fra the hiest of the hill to the lowest of the ebb' in another.[103] For whatever reason, early perambulations of Moray davochs appear to give emphasis to this wider frame of meaning.[104] We find a similar emphasis on outer limits of resource used in a reference to the four davochs of Assynt.[105]

Along the western seaboard and across the Hebrides, the davoch overlaps with the region's other large-scale unit of assessment: the ounceland or, to give its Gaelic form, the *tirunga*. Enough sources equate the davoch with the ounceland for us to be confident about their equivalence, as with the charter reference to 'the davoch called in Scotch the terung of Paible . . . the davoch called the terung of Balranald' on South Uist.[106] Interestingly, we find the circumstances in the West Highlands and Islands in which the standard subunit of the davoch, the ploughland, was also present. A 1642 rental for Colonsay rated all but one of its twelve touns as a 'half pleuche land', the exception, Ballanahard, being a whole 'pleuche land'.[107] Like the ploughlands of the eastern Highlands, Argyllshire ploughlands were further broken down into oxgate equivalents, described locally as horsegangs,[108] that could be equated with other, more familiar assessment units as with the tenants 'possessing half a merkland or a Horse gang' in a 1727 rental for the Barcaldine estate.[109] Some have suggested that the davoch could be equated to the *bailebiataigh* (or baile) and its subdivision into four quarterlands on Islay.[110] The quarterlands of Islay were undoubtedly a formal division of the *bailebiataigh*.[111] The fact that the Shawfield estate on Islay was said to comprise 134 quarterlands,[112] like references to 'Quarterlands' in seventeenth-century rentals for Tiree[113] suggest that it denoted a specific block of land and not just an aliquot portion of the baile.

Across the Western and North-western Highlands and Islands, the fine-scale forms of assessment comprised pennylands and merklands. How they related to the *tirunga* can best be understood by looking at the Hebridean evidence, starting with Tiree. From a late seventeenth-century memorandum on the island, we learn that Tiree comprised twenty *tirunga*, each of which comprised six merklands and twenty pennylands.[114] With 3.33 pennylands per merkland, these Tiree figures do not suggest that the island's scheme of pennylands and merklands had been designed with the other in mind. Its rating of pennylands per tirunga appears consistent with what we can find elsewhere for the southern Hebrides or for the western seaboard but its rating of merklands either per tirunga or per pennyland is not. If we take a wider view, we find that in nearby Morvern, as well as on Mull and Skye, a distinction was drawn between single and double pennylands. In the case of Skye, we even find them used side by side in a 1741 rental of Strath, with one toun rated at four 'pennies single' and another, Kyleakin, at five 'pennies Double'.[115] The same distinction was made by reference to 'great' pennylands as opposed to ordinary pennylands.[116] A seventeenth-century source document, ostensibly about Tiree but with an aside about Morvern, provides an explanation, declaring that a Morvern pennyland is 'a halfe merkland',[117] so that a double pennyland there was simply a merkland by another name.[118] The style can also be documented for North Uist, such as in a 1756 rental which refers to 'the five penny's Double Land' of Ballivannich.[119] We can reach a similar conclusion for Ardnamurchan and Sunart, by comparing their 1541 rental with a listing drawn up *c*.1723, with quite a number of the touns listed as 2½ merklands in the former being recorded as five pennylands in the latter, again suggesting a merk equalled a 'double' pennyland.[120]

In practice, mainland areas, such as Netherlorn and Ardnamurchan, along with islands such as the Uists, Skye and Harris, used pennylands for the day-to-day business of rating touns or setting them to tenants. Tiree, though, used a different system altogether. Summary rentals rated individual touns in merklands if not pennylands but the most detailed listings, those that specified what each tenant held, tended to employ a unit called the malie. Even Turnbull's survey of 1768 used malies not merklands or pennylands to assess touns. The 'Memorandum' mentioned above specified that each *tirunga* comprised 48 malies, giving 960 for the entire island. The way in which it also specified the rent liabilities attached to each malie in terms of grain or meal and other payments, including stock, suggests, as its name would imply, that it had something to do with the calculation of rent liabilities attached to holdings. Its occasional use elsewhere, such as in Netherlorn or on the Uists, also links it to the payment of rent, especially grain rents.[121] The Tiree data, though, takes us further by showing that what each *tirunga* was allowed in terms of bolls sowing (48 bolls of oats, 24 bere) or stock souming (72 soums) appears to be based on what they each contained as regards malies, not its pennyland or merkland rating, since the number of bolls or soums involved were simple multipliers or proportions of 48 (24, 48 and 72 = ½, 1, 1½). Before we assume that these were simply abstract calculations so as to proportion what each person could sow or stock and what, therefore, was owed in rent, we need to take on board another piece of evidence. In a late seventeenth-century comment on the island's landholding arrangement, it is clearly implied that malies were seen as physical entities, something that had a material reality on the ground, by saying that not all malies paid the standard rent set for them partly because 'some of the *malies* were not as good as others'.[122] In fact, when we look at the most detailed late seventeenth-century rental for the island, it refers to each tenant's allocation of malie or malies as their allocation of 'malie land'. In other words, malies were simply another form of land assessment.

More so than elsewhere, the Western Highlands and Islands had different systems of land assessment being used either side by side or at different times. The system of ouncelands and pennylands was probably introduced by the Norse, possibly by *c*.900. *Ab origine*, an ounceland was an area of land burdened with a levy or skat of 1 ounce of silver.[123] In the Hebrides, the ounce was made up of 20 pennies, hence the ounceland's subdivision into 20 pennylands. In the Northern Isles, meanwhile, it was burdened with an ounce of silver made up of 18 pennies, making it a unit of 18 pennylands.[124] While the ounceland and pennyland were seen as being essentially Norse assessments, McKerral followed Skene in seeing it as adapted to the pre-existing form of the Celtic townland or *bailbiataigh*. Guided by his reading of the *Senchus*, he saw pre-Norse settlement in the area of Dalriada as organised around townlands of 20 houses, each with an associated area of arable that formed what McKerral called 'the original dabhach or davoch'.[125] He saw the division of touns into halves, quarters and eighths as carried over into Norse assessments so that, when seen in terms of real touns, the 20 pennylands that comprised each ounceland appear as organised into 10, 5 and 2½ pennylands, though such subdivisions frequently embraced a cluster of townships rather than a single toun.[126]

In his detailed examination of assessments on Islay, Lamont argued that Islay stood out by not having a pennyland assessment, only an ounceland assessment. The lack of such a grass-roots system has allowed other forms of evidence to persist. Analysing the valuation of land embodied in the Old Extent, a valuation drawn up in the mid-thirteenth century, following the handover of the Hebrides to the Scottish Crown, he demonstrated that the house listing in the *Senchus* could be correlated to the later pound or merkland values that formed the basis of the Old and then, New Extent, with each house having as assessment of 10 shillings.[127] Easson reinforced this reading, arguing that the subdivision of the ounceland into twenty pennylands 'was not the result of Scandinavian influence'[128] but stemmed from the twenty-house system that lay at the root of earlier Dalriadic systems of assessment. In fact, there is a case for arguing that it formed the basis for a naval levy as early as the seventh century. Drawing this material together, Easson concluded that what the evidence for the South-west Highlands and Islands shows is that there was a 'considerable measure of continuity in the system of land assessment in the western seaboard and the Isles from the Dark Ages through to the medieval period'.[129]

As well as allowing us to look backwards, the data for the South-west Highlands and Islands also allow us to see how some assessments changed subsequently. Following the collapse of the Lordship of the Isles in 1493, what formed Old Extent was revalued into a New Extent,[130] one assessed in merklands. The standardised way in which New Extent was applied in areas such as Kintyre and Sunart, with four merklands being common in the former and 2½ merklands in the latter, suggests that New Extent was imposed in a quite systematic way. When we combine this with the way in which merklands were used to define the number of bolls sown by a township and the soums attached to it, we are left with the impression that New Extent may have served to encourage a uniform toun economy in those areas where assessments were stamped almost mechanically across parts of the countryside, as in Kintyre and Sunart. Even where merkland assessments were used simply to generate a rent per merkland, the effect would have been to shape the toun economy within set limits, especially where much of the rent was made up of payments in kind such as meal, barley and cheese.

Assessments in the Northern Isles, too, were based on ouncelands and pennylands but, there, each ounceland was made up of 18 pennylands. Significant differences exist, however, between Orkney and Shetland over how they were used. In Shetland, there are early references to both ouncelands and pennylands but the latter appear to have fallen out of use soon after.[131] By comparison, pennylands formed a persistent and actively used system of assessment in Orkney down into the early modern period, being used to define individual farms.[132] The ounceland in Orkney, meanwhile, had a purely fiscal connotation whereas in Shetland it was attached to the land beyond the head dyke, or the skattald. Thomson used this last point to explain how the ounceland came into being. In recent centuries, the skattald was an area of common grazings shared between a cluster of communities. When the Norse came to impose skat or tax, communities, bound together through their mutual interest in a block of common grazings, would have provided a basis for a large-scale fiscal levy such as the ounceland.[133] Others, though, including the present author, would see the original basis for units of taxation, such as the ounceland, as the skat land that lay with*in* the head dyke, not the grazing land – or what later became the skattald – that lay beyond.[134]

The Origins of Land Assessment?

It will already be clear from what has been said that the origin of land assessments has different dimensions to it. How assessments were derived as measures, and the system of metrology employed to formulate them, form only one dimension. We also need to consider the question of how they were first devised? Were they something wholly new, or were they adapted from other forms of assessment? Seen from a purely practical point of view, how exactly were they imposed over the landscape? Was it from the bottom up, matching assessment exactly to what existed on the ground or was it via a top-down approach, with levies and taxes being set at a territorial level and then being broken down in the most convenient way, first between districts and then between touns or groups of touns.

When we see them in the context of the sixteenth- to eighteenth-century sources, their prime use was to organise estate matters, apportioning out rent and other obligations and framing what touns could crop and stock. It is this essential, everyday use that ensured their survival. As we push back in time, though, we gain a sense of their innate variety, not just with regard to how they were defined but, equally, with regard to how they were used to apportion the rents, renders, dues, rights of hospitality, military levies, and so on. This multidimensional use is almost certainly the reason why we find different forms of assessment being used side by side, a sign perhaps that, at the outset, different forms of assessment were pointing in different directions. Transcending such changes and adjustments, however, was a more significant change. It is one already hinted at in the debate earlier regarding how the ounceland came to be computed at twenty pennylands in the South-west Highlands and Islands and at eighteen in the Northern Isles. Lamont, and later Easson, both argued that the twenty pennylands in the former were likely to have been rooted in the Dalriadic twenty-house system of assessment but neither elaborated on how such a 'house' system could become transformed into a 'land' system. The *Senchus* listing suggests the house system was the basis for burdening client freemen with the payment of levies, or house custom.[135] Even though a notional system, it suggests a world in which families or groups were the focus of assessment, with the house (=family, kin group) being the basic unit of apportionment between them, with each 'house' burdened with the same rating.

We need to grasp the difference between such a system and one based more directly on a *land* assessment. It is one that can be expressed through the ideas of the anthropologist, Sahlins, on the transition from kin-based societies to early states. Order in the former, he argued, was based on establishing territoriality through society but, in the latter, it was based on establishing order in society through territory.[136] Put another way, the former was concerned first and foremost with relationships between people, with issues of power, territory and landholding being consequences of their social identity and kinship. The latter, meanwhile, was concerned, essentially, with defining a more abstract jurisdiction or control over territory and its resources, and with using this as the basis for relationships between people, with lordship being exercised over people through its overriding control over law and land *within a defined territory*. Seen against this background, there is a case for arguing that assessments probably originated as renders, obligations of hospitality, and so on, that were burdened on the person, family or lineage, either directly or through the concept of houses, and which were paid to a chief or lord in acknowledgement of their greater rank or overlordship. Arguably, assessments may have existed in this form by the Late Iron Age. Their subsequent conversion into *land* assessments would have marked a moment of significant change, not least for landscape, creating as it did a different form of assessment. What had been burdened on the person or family now became burdened on the land, with the amount of land occupied serving as the basis for apportioning obligations, dues, renders and services and, for that reason, having to be quantified or defined on the ground. Such a shift formed part of a wider pattern of sociopolitical change bound up with the emergence of early states, one that saw the rise of a feudalised lordship and, ultimately, crown feudalism, with individuals now holding land from a superior in a more contractual way, owing services, obligations and dues in return for jurisdiction or control over a defined territory or block of land.[137]

Any discussion over when Highland assessments first originated, and whether their meaning shifted over time, must keep in mind the distinction between assessments burdened on the person and assessments burdened directly on land because they reflect different historical contexts. We can see this in relation to the debate over when the davoch first originated. Even among recent contributions, there is a wide divergence of opinion on the chronology of its first appearance. After arguing that the Norse ounceland in Orkney was developed around – and shaped by – the earlier or pre-Norse form of the davoch, Bäcklund concluded that the davoch must have been a well-established form of assessment before the Norse arrived.[138] Her basic assumption here has been criticised on the grounds that the davoch never existed in Orkney, and there are no reasons for even assuming that it did.[139] Yet her assumption that 'the Norse kept and later developed the native system of land administration in Orkney' not only follows the earlier conclusion of scholars such as Stiennes and Thomson, it also follows what Bangor-Jones concluded in a study of assessments in Caithness, an area where the davoch was well established[140] and, of course, echoes what McKerral and Easson argued for the South-west Highlands. In response, I would argue that, while native systems of assessment are likely to have been present prior to the Norse colonisation, they are more likely, as the *Senchus* strongly suggests for the South-west Highlands and Islands, to have been burdened on the person or family. In his response to Bäcklund's paper, Williams concluded that the davoch did not originate until after AD 900, with the emergence of Alba. Following Broun, he sees Alba as the reformulation of the Pictish kingdom under the Mac Alpin dynasty. He notes how the davoch accords with the areas covered by Alba and how its Gaelic ruling dynasty would explain why a Gaelic name might be used for what otherwise appears to us to be a Pictish assessment.[141] As an emergent early state, one in which territory mattered more than the cultural identity of those who occupied it,[142] then Alba after AD 900 would seem to be a more likely context in which a land assessment, as opposed to an assessment burdened on people or kin groups, might have emerged. Ross's still more recent study of the davoch in Moray provides us with yet another chronology. His analysis, using perambulations, established that, while they formed large, extended territories of pasture, woodland and arable, the majority of them appear to sit comfortably within the bounds of parishes, with only a few lying across parish boundaries. The impression given is that they had formed, in his words, 'the building block for the medieval parochial system in Moray', so that we can regard its creation as pre-parochial.[143] There is uncertainty over when parishes were created but they are likely to be at least earlier than twelfth century. The window of possible development formed by Williams's post-AD 900 date and Ross's pre-twelfth-century date would certainly fit well with when we might expect lordship to have been established through territoriality, with the davoch being part of the means by which 'extensive lordship' was, as Ross put it, imposed 'upon the landscape'.[144]

How Were Assessments Imposed?

Quite apart from questions over how land assessments originated, there are also basic questions to be answered about how they related to landholding and settlement when they were first introduced. If we work from the assumption that assessments were initially burdened on families or lineages and that they were transformed into assessments on land only when lordship became exercised over men through land and territorialised jurisdictions, then it follows that mature systems of arable, landholding and settlement must already have existed long before this point. This raises the issue of how the imposition of land assessments and what existed on the ground at this point were brought together. The easy answer would be to assume that the latter were arranged into recognisable touns, each of which was simply given an assessment based on a calculation of what it possessed. Such a view would, of course, link up with those who suppose that the farming toun possessed an innate unity that was deep-rooted, a unity built from the ground upwards through two millennia or more of co-operation and effort.

In fact, there are good reasons why this would be an unsatisfactory answer. Overall, the evidence favours a pattern of assessment that was imposed downwards over touns in an uncompromising way rather than one built upwards through an assessment of what holdings actually possessed. There are three reasons for taking this view. First, many parts of the Highlands and Islands possessed a *territorial* assessment, one that endured across the centuries, such as Tiree's twenty *tirunga*,[145] or Waternish's thirty-two merkland 'that is four daachs of land'.[146] It would be inconceivable for such rounded territorial assessments to have emerged upwards through the many vagaries of settlement expansion. Rather, is it far more likely that assessments were imposed at a gross level before being broken down and levied on the pattern of landholding as it existed at the point of land assessment, a disaggregation that was responsible for what Bangor-Jones likened to a hierarchy of assessments.[147]

Second, when we look at sixteenth- to seventeenth-century rentals, we find listings in which a striking number of touns within particular areas appeared to have similar, rather than random, assessments. The late sixteenth-century report on the Hebrides, the so-called Skene Manuscript, conveys for Islay this sense of a designed or planned system in a very powerful way. 'Ilk town in this isle', it said, 'is twa merk land and ane half'.[148] McKerral's analysis of Islay's 1507 rental actually shows fifty-five of the seventy-seven holdings or touns listed as being assessed at 2½ merklands.[149] The liabilities attached to each merkland also underscore their seemingly programmatic nature, with each toun paying 'yeirly of gersum [= an entry payment] at Beltane four ky with calf, four yowis with lamb, 4 geis, nine hennis, and 10s of silver' plus, as yearly rent for each merkland, 'three mairtis and ane half, 14 wedderis, 28 geis, 4 dozen and 8 pultrie, 5 bollis malt with ane peek to ilk boll, 6 bolls meill, 20 stone of cheis, and twa merk of silver'.[150] In all probability, some of the design principles behind this scheme were starting to drift at the point this manuscript was compiled in the late sixteenth century. Even so, the 1722 rental for the island still shows a significant percentage as still rated at 2½ merklands.[151] We can find a similar patterning of rent per unit of assessment elsewhere, such as in Ardnamurchan and Sunart,[152] and on Tiree,[153] one that could have been created only through the systematic breakdown of a wider territorial assessment rating.

Third, a striking feature of the assessment listings for areas such as Ardnamurchan, Kintyre, and on Islay is the way in which those recorded in rentals embrace more than one toun. In some cases, as many as four or five touns were embraced by a single rating of assessment.[154] In some cases, we find clusters bundled together under a single assessment even though each toun was named separately and held by a separate tenant. The impression given is that touns had been aggregated together to fit under a standardised levy of assessment, not vice versa. It would be wrong, however, to see all such entries as signalling an arrangement that takes us back to when land assessments were first imposed. Some entries covering multiple touns could have emerge later through a process of splitting (pp. 98–101) or through the creation of outsets or daughter settlements.

Seeing land assessments as imposed downwards in a best-fit sort of way, rather than built upwards through an exact calculation of what was actually occupied on the ground, raises further questions about how touns came into being. If assessments and the occupation of land were not, at the outset, aligned in a neat and tidy way, we need to understand how, and with what effect, the two levels of identity subsequently interacted? By nature, land assessments proved to be a deeply entrenched and heavily institutionalised feature of local landscapes. Most, if not all, appear to endure in an unchanged way from their initial imposition onwards, a survival aided by their use on an everyday basis for defining a toun's resources and its rent liabilities. By comparison, how land was actually occupied and worked was dynamic, with changes in who held what, in how landholding and settlement was ordered, and in the amount of land used for arable, meadow and pasture. How these two dimensions interacted, their differing capacities for absorbing change, create issues that I want to explore in the next chapter.

Notes

1. For example, Harding, *Hebridean Iron Age*, pp. 15–18.
2. A. Ritchie, *The Prehistory of Orkney* (Edinburgh, 1985), p. 1.
3. Ibid., p. 1.
4. C. Thomas, 'The interpretation of the Pictish symbols', *The Archaeological Journal*, 120 (1963), pp. 30–97.
5. A. A. Jackson, *The Symbol Stones of Scotland* (Edinburgh, 1984), pp. 60–92.
6. G. Whittington, 'Placenames and the settlement pattern of dark-age Scotland', *PSAS*, 106 (1974–75), p. 106; I.A.G. Shepherd, 'Pictish settlement problems in North-East Scotland', in J. C. Chapman and H. C. Mytum (eds), *Settlement in North Britain 1000 BC–AD 1000*, BAR, British Series no. 118 (1983), p. 328.
7. Whittington, 'Placenames', p. 106; Shepherd 'Pictish settlement problems', p. 328.
8. Ballin Smith, *Howe*, pp. 267, 272–3.
9. M. B. Cottam and A. Small, 'The distribution of settlement in Southern Pictland', *Medieval Archaeology*, xviii (1974), p. 49.
10. Shepherd, 'Pictish settlement', p. 331; see also, F. T. Wainwright, *The Problem of the Picts* (Edinburgh, 1955), p. 92.
11. J. W. Hedges, 'Surveying the Foundations', p. 31; Armit, 'Brochs and beyond', p. 68; J. R. Hunter, 'Pool, Sanday: a case study for the Late Iron Age and Viking Periods', in Armit (ed.), *Beyond the Brochs*, pp. 182–5.
12. Ritchie, 'Pictish and Viking-age farmsteads', pp. 174–227.
13. Ibid., p. 182.
14. Armit, 'Brochs and beyond', p. 66.
15. J. C. Barrett and J. M. Downes, North Pitcarmick (Kirkmichael Parish), *Discovery and Excavation in Scotland*, 1994, pp. 88–9; RCAHMS, *North-East Perth*, pp. 12–13; Atkinson, 'Settlement Form', pp. 324–5.
16. A. O. Curle, 'The Wag of Forse, Caithness. Report of further excavation made in 1947 and 1948', *PSAS*, 82 (1947–48), pp. 275–85.
17. J. Close-Brooks, *The Highlands* (Edinburgh, 1986), p. 150.
18. Morrison, *Dunbeath*, p. 59.
19. Ibid., p. 62.
20. W. F. H. Nicolaisen, *Scottish Place Names. Their Study and Significance* (London, 1976), p. 154.

21. Ibid., p. 153.
22. Whittington, 'Placenames', p. 100.
23. Ibid., p. 99.
24. Ibid., p. 100.
25. Ibid., pp. 100, 104.
26. K. H. Jackson, *Gaelic Notes in the Book of Deer* (Cambridge, 1972), p. 114.
27. Ibid., p. 105.
28. G. W. S. Barrow, *The Kingdom of the Scots* (London, 1973), pp. 7–68.
29. M. R. Nieke and H. B. Duncan, 'Dalriada: the establishment and maintenance of an Early Historic kingdom in northern Britain', S. T. Driscoll and M. R. Nieke (eds), *Power and Politics in Early Medieval Britain and Ireland* (Edinburgh, 1988), p. 20.
30. Whittington, 'Placenames', pp. 99–100.
31. Nicolaisen, *Scottish Place Names*, pp. 155–7.
32. L. Price, 'A note on the use of the word *baile* in placenames', *Celtica*, 6 (1963), pp. 119–26; G. Toner, 'Baile: settlement and landholding in medieval Ireland', *Èigse: a Journal of Irish Studies*, xxxiv (2004), pp. 25–6.
33. S. T. Driscoll, 'The archaeology of state formation', in W. S. Hanson and E. A. Slater (eds), *Scottish Archaeology: New Perspectives* (Aberdeen, 1991), pp. 81–111.
34. E. Campbell, *Saints and Sea-kings: The First Kingdom of the Scots* (Edinburgh, 1999), pp. 13–15; E. Campbell, 'Were the Scots Irish?', *Antiquity*, 75 (2001), pp. 285–92.
35. Nieke and Duncan, 'Dalriada', p. 9.
36. Nicolaisen, *Scottish Place Names*, pp. 121–48.
37. Ibid., p. 142.
38. Nieke and Duncan, 'Dalriada', p. 9.
39. J. W. Bannerman, *Studies in the History of Dalriada* (Edinburgh, 1974), pp. 131–8.
40. NAS, GCM GD44/51/745/1/3, 1683.
41. W. D. Lamont, '"House" and "Pennyland" in the Highlands and Isles', *SS*, 25 (1981), p. 65.
42. Nieke and Duncan, 'Dalriada', pp. 1–21.
43. M. R. Nieke, 'Fortifications in Argyll', pp. 139–40.
44. M. R. Nieke, 'Settlement patterns in the first millennium AD: a case study of the island of Islay, in J. C. Chapman and H. C. Mytum (eds), *Settlement in North Britain 1000 BC–AD 1000*, BAR, British Series no. 118, (Oxford, 1983), p. 304.
45. Nieke and Duncan, 'Dalriada', p. 17.
46. Ibid., p. 17.
47. F. T. Wainwright, 'The Scandinavian settlement', in F. T. Wainwright (ed.), *The Northern Isles* (Edinburgh, 1964), pp. 125–6.
48. Hunter, 'Pool, Sanday', esp. pp. 181–92; Armit, *Archaeology of Skye*, p. 188.

49. Wainwright, 'Scandinavian settlement', pp. 152–3. For a comprehensive review, one favouring a violent takeover, see B. D. Smith, 'The Picts and the martyrs or did Vikings kill the native population of Orkney and Shetland', *Society for Northern Studies*, 36 (2001), pp. 7–32.

50. Hunter, 'Pool, Sanday', pp. 192–3. For an alternative view, see Bangor-Jones, 'Ouncelands and pennylands', p. 20.

51. Ritchie, 'Pictish and Viking-age farmsteads', p. 192; B. E. Crawford, *Scandinavian Scotland* (Leicester, 1987), pp. 56–7; W. P. L. Thomson, *The New History of Orkney* (Edinburgh, 2001), pp. 43–7; Bangor-Jones, 'Pennylands and Ouncelands in Sutherland', in L. J. Macgregor and B. E. Crawford (eds), *Ouncelands and Pennylands*, p. 20; J. Bäcklund, 'War or Peace? the relations between the Picts and the Norse in Orkney', *Society for Northern Studies*, 36 (2001), pp. 33–47.

52. G. Fellows-Jensen, 'Viking settlement in the Northern and Western Isles – the place-name evidence as seen from Denmark and the Danelaw', in A. Fenton and H. Pálsson, *The Northern and Western Isles in the Viking World: Survival, Continuity and Change* (Edinburgh, 1984), p. 152.

53. W. F. H. Nicolaisen, 'Norse settlement in the Northern and Western Isles', *SS*, xlviii (1969), pp. 6–17.

54. Nicolaisen, *Scottish Place-Names*, pp. 95–6.

55. Fellows-Jensen, 'Viking settlement', pp. 158–9.

56. Crawford, *Scandinavian Scotland*, pp. 108–9.

57. Fellows-Jensen, 'Viking settlement', p. 160.

58. Crawford, *Scandinavian Scotland*, p. 108.

59. H. Marwick, *Orkney Farm-Names*.

60. Wainwright, 'The Scandinavian settlement', p. 139.

61. Crawford, *Scandinavian Scotland*, pp. 149–51.

62. Thomson, *Orkney*, pp. 51–2.

63. Ibid., p. 54.

64. Ibid., p. 53.

65. Crawford, *Scandinavian Scotland*, p. 140 *et seq.*; Hedge, *Bu, Gurness*, Part III, p. 42.

66. J. R. C. Hamilton, *Excavations at Jarlshof, Shetland* (Edinburgh, 1956), pp. 93–189. See also, G. F. Bigelow, 'Domestic architecture', pp. 23–38; J. Graham-Campbell and C. E. Batey, *Vikings in Scotland* (Edinburgh, 1998), pp. 155–205.

67. Crawford and Ballin Smith, *The Biggings*, pp. 62–238.

68. Graham-Campbell and Batey, *Vikings in Scotland*, pp. 155–60.

69. Ibid., p. 158.

70. Bigelow, 'Domestic architecture', pp. 23–38.

71. Armit, *Archaeology of Skye*, p. 191.

72. M. Brennand, M. Parker Pearson, and H. Smith, *The Viking Age Settlement At Kilpheder (Cille Pheadair), South Uist*, Excavations in 1997, Sheffield, n.d., p. 17.

73. Ibid., p. 17.

74. Sharples, 1996, no pagination.

75. Crawford and Switsur, 'Sandscaping and C14', pp. 124–36.
76. Parker Pearson and Sharples, 'South Uist in the Iron Age', p. 16
77. Davidson et al., 'Formation of Farm Mounds', p. 46.
78. R. Bertelsen and R. G. Lamb, 'Settlement mounds in the north Atlantic', in C. E. Batey, J. Jesch and C. D. Morris (eds), *The Viking Age in Caithness, Orkney, and the North Atlantic* (Edinburgh, 1993), pp. 544–54.
79. Hunter, 'Pool, Sanday', p. 182; Hunter, Bond and Smith, 'Some aspects of early Viking settlement in Orkney', in C. E. Batey, J. Jesch and C. D. Morris (eds), *The Viking Age in Caithness, Orkney, and the North Atlantic* (Edinburgh, 1993), p. 274.
80. Davidson et al., 'Formation of Farm Mounds', p. 58.
81. Ibid., p. 58.
82. I. A. Simpson, 'Stable Carbon Isotope Analysis of Anthropogenic Soils and Sediments in Orkney', in A. R. Hands and D. R. Walker (eds), *Paleoenvironmental Investigations: Research Design, Methods and Interpretation*, BAR, no. 258 (Oxford, 1985), p. 61.
83. Davidson, Harkness and Simpson, 'Formation of Farm Mounds', pp. 45–60.
84. Hunter, 'Pool, Sanday', p. 192.
85. Harden and Lelong, *Winds of Change*, p. 169.
86. D. A. Davidson and S. P. Carter, 'Micromorphological evidence of past agricultural practices in cultivated soils: the impact of a traditional agricultural system on soils in Papa Stour, Shetland', *Journal of Archaeological Science*, 25 (1998), pp. 827–38.
87. Simpson, 'Stable Carbon Isotope Analysis', p. 55.
88. Jackson, *Gaelic Notes*, p. 116.
89. Barrow, *The Kingdom of the Scots*, p. 273.
90. Ibid., pp. 267–8; Jackson, *Gaelic Notes*, p. 114; Whittington, 'Placenames', p. 105.
91. A. McKerral, 'Ancient denominations of agricultural land in Scotland', *PSAS*, lxxviii (1943–44), p. 52; Barrow, *Kingdom of the Scots*, p. 270.
92. Ibid., pp. 270–1; NAS, GCM, GD44/51/732/54, Rental of Badenoch 1595.
93. Barrow, *Kingdom of the Scots*, pp. 270–1; A. Ross, 'The dabhach in Moray: a new look at an old tub', in A. Woolf (ed.), *Landscape and Environment in Dark Age Scotland*, St John's House Papers no. 11 (St Andrews, 2006), pp. 63–4.
94. McKerral 'Ancient denominations', p. 52.
95. NAS, GCM, GD44/51/745/2/1.
96. NAS, FE, E655/1/1; see also ibid. E746/70; GD305/1/18/2; J. Anderson, *Calendar of the Laing Charters A.D. 854–1837* (Edinburgh, 1899), p. 562.
97. Barrow, *Kingdom of the Scots*, p. 271.
98. Ross, 'The dabhach', pp. 63–4.
99. J. Robertson, *Illustrations of the Topography and Antiquities of the Shires of Aberdeen and Banff*, Spalding Club (Aberdeen, 1862), iv, p. 173.
100. Ross, 'The dabhach', pp. 64–5.

101. Ibid., p. 65.
102. A. Peterkin, Rentals of the Ancient Earldom and Bishoprick of Orkney (Edinburgh, 1820), p. 84.
103. A. W. Johnson and A. Johnston (eds), *Orkney and Shetland Records*, vol. II, *1623–1628* (London, 1942), i: p. 94.
104. We find a similar definition as regards the four davochs of Assynt, C. Innes (ed.), *Origines Parochiales Scotiae*, Bannatyne Club (Edinburgh, 1851), ii: p. 692.
105. Ibid., ii: p. 692
106. Ibid., ii: part i, p. 374; A. R. Easson, 'Ouncelands and Pennylands in the West Highlands', in L. J. Macgregor and B. E. Crawford (eds), *Ouncelands and Pennylands (St Andrews, 1987)*, p. 2.
107. IC, AP, NE11, Colonsay, 1642.
108. NAS, GD112/9/35.
109. NLS, Adv. 29.1.1, vol. viii, pp. 65, 163.
110. W. D. Lamont, 'Old land demoninations and Old Extent in Islay: part i', *SS*, 1 (1957), p. 189.
111. D. Gregory and W. F. Skene (eds), *Collectanea de Rebus Albanicis*, Iona Club (Edinburgh, 1847), p. 178; McKerral, 'Ancient denominations', pp. 43–4.
112. NAS, E729/9/1.
113. IC, AP, Rentall . . . 1652; Ibid., Memorandum . . . Terii 1662.
114. IC, AP, Box 2531, Memorandum . . . Tirie.
115. DC, MDP, 2/492/1; Ibid., 2/492/2.
116. Innes (ed.), *Origines Parochiales*, vol. ii, part i, pp. 191, 314.
117. IC, AP, Memorandum anent Tirie 1662.
118. IC, AP, Bundle 1009, 1758; Innes (ed.), *Origines Parochiales*, 1851, vol. ii, part i, pp. 191, 314.
119. CDC, LMP, GD221/1//351/6.
120. P. McNeill (ed.), *The Exchequer Rolls of Scotland, XVII*, A.D. *1537–1542* (Edinburgh, 1897), pp. 622–5
121. For example, NAS, GD112/9/22.
122. IC, AP, Bundle 494, Letter, 1 June 1681.
123. McKerral, 'Ancient denominations', p. 54.
124. Thomson, *Orkney*, pp. 216–17.
125. McKerral, 'Ancient denominations', p. 55.
126. Ibid., p. 55.
127. W. D. Lamont, 'Old land denominations and Old Extent in Islay. Part ii', *SS*, 2(1958), pp. 93, 96.
128. Easson, 'Ouncelands and Pennylands', p. 9.
129. Ibid., p. 9.
130. Lamont, 'Old land denominations . . . part ii', p. 92.
131. Thomson, *Orkney*, p. 26.
132. Ibid., pp. 24, 26.
133. Ibid., p. 34.

134. Crawford, *Scandinavian Scotland*, p. 108.

135. Bannerman, *Dalriada*, pp. 132–46.

136. M. Sahlins, *Tribesmen* (Englewood Cliffs, 1968), p. 5.

137. For further discussion, see R. A. Dodgshon, *The European Past* (London, 1987), Chapters 5 and 6.

138. Bäcklund, 'War or Peace?', pp. 33–47.

139. G. Williams, 'The *dabhach* reconsidered: Pre-Norse or Post-Norse', *Society for Northern Studies*, 37 (2003), pp. 17, 19.

140. Bangor-Jones, 'Ouncelands and pennylands', p. 18.

141. Williams, 'The *dabhach* reconsidered', p. 26.

142. Ibid., p. 26; D. Broun, The origin of Scottish identity in its European context', in B. E. Crawford (ed.), Scotland in Dark Age Europe, 1994, St Andrews, pp. 21–31.

143. Ross, 'The dabhach', p. 70.

144. Ibid., p. 71.

145. IC, AP, Memorandum . . . Tirie, 1662.

146. Mitchell (ed.), *Macfarlane's Geographical Collections*, 1907, II, p. 532.

147. Bangor-Jones, 'Ouncelands and pennylands', p. 18.

148. G. G. Smith, *Book of Islay* (Edinburgh, 1895), p. 477.

149. A. McKerral, 'The tacksman and his holding in the South-West Highlands', *SHR*, xxvi (1947), pp. 13–14.

150. Smith, *Islay*, p. 477.

151. Ibid., Appendix A.

152. McNeill (ed.), *Exchequer Rolls*, pp. 622–5.

153. Ibid., pp. 647–8.

154. For example, ibid., pp. 634–50.

THE LATE MEDIEVAL AND EARLY MODERN LANDSCAPE: STASIS OR CHANGE?

As the medieval period unfolded, documentary sources became available that shed a more direct light on Highland farming. We are in a better position to interrogate how it made an impact on landscape through its prime institutions and the processes shaping their development. I want to organise my discussion of these aspects into four sections. First, by way of background, I want to comment on what were the drivers of change during the medieval and early modern periods. Gaps in the historical record prevent us from taking anything other than a cursory approach to these drivers but any review would be incomplete if it did not mention them in outline. Second, I shall examine the institutional character of the farming landscape and how key elements of this institutional character – land assessments and the farming toun – interacted. Third, I shall look at how the institutional underpinnings of the toun shaped its character as a unit of landholding. Lastly, I shall consider how these underpinnings affected the way in which farming developed.

The Drivers of Change

Demographic Drivers

Landscapes evolve through the interaction of a number of basic variables, including shifts in population, long-term environmental change, changes in the sociopolitical character of society and market forces. Among these, how demographic inputs shaped the Highland countryside are relatively straightforward to define but difficult to quantify in the context of the medieval and immediate post-medieval period. Indeed, of all the gaps that beset our view of Highland history during this period, the absence of empirically based demographic data is the most important. Even when parish registers become available in the early seventeenth century, only one Highland parish, that of Kilmorack, provides a long-run sequence,[1] though what survives generally still has value.[2] Faced with this lack of data, we can only surmise that the Highlands shared in some of the broader, pan-European trends in population, trends rooted in a mix of disease outbreaks, climate shifts and changing economic opportunities.[3] On this basis, Highland populations may have declined during the mid- to first millennium AD,[4] with a gradual recovery in numbers in progress by about AD 1000. This recovery would have peaked during the medieval climatic optimum of the eleventh to thirteenth centuries AD, when higher summer temperatures increased farm output on existing land and encouraged fresh colonisation. By 1300, though, the cooler, wetter conditions which set in with the start of the Little Ice Age increased the frequency of crop failure, lowered output and reduced the viability of marginal sites.[5] Over and above the impact that the worsening of climate after 1300 may have had on population, we also need to allow for how far the region shared in the pan-European collapse of population brought on by the bubonic plague outbreaks of the 1340s.[6]

As commentaries become more freely available and parish registers start to cast light on the problem during the seventeenth century, we can begin to be a little more specific about demographic trends. Fresh outbreaks of bubonic plague during the mid-seventeenth century had some effect locally, such as in Kintyre,[7] but they are unlikely to have been on the scale of those in the mid-fourteenth century. More important at this point, was the impact of one of the more severe cold phases of the Little Ice Age, known as the Maunder Minimum, 1645–1702. Its physical impacts are reviewed in the next section. What we need to note here is its impact on local populations. In a recent study of the particularly poor harvests of the 1690s, during King William's so-called lean years, Cullen has provided us with a secure foundation for understanding how the worst years of the Maunder Minimum may have affected population through the sequence of poor harvests, famine, disease and mortality, with an estimated overall decrease of as much as 15 per cent, although that figure included those who emigrated to Ulster. She concluded that the subsistence crises of the 1690s affected the north of Scotland more than the east or south and that, locally, we can expect mortality to have been 'pushed higher' in the worst-affected areas.[8] Nationally, this subsistence shock was followed by a short phase, lasting a few decades, when no national harvest failure was recorded,[9] but lack of sources prevent us from saying precisely how Highland population responded.[10] Only by the 1740s can we start to chart a more certain path for Highland population. The region was certainly affected by the poor harvests of 1739–41 and the subsistence crisis that followed but, as elsewhere, population growth recovered quite strongly during the 1750s to 1760s. Population in the Highlands shared in this post-1750s growth but the region suffered a short but acute subsistence crisis, with spikes in local mortality rates in the Highlands and parts of the Hebrides, following the poor harvests of 1782–3,[11] a crisis then made worse, but not initiated, by the Laki eruption in Iceland in 1783.[12] By this point, rates of population change in the region were being underpinned by a strong geographical component. The northward spread of sheep and the clearance of farm communities meant that some inland areas were experiencing falls in population. By contrast, the resettlement of some communities in coastal areas meant that these experienced above average growth. In fact, by the 1790s, when growth elsewhere in Scotland appears to falter, if only temporarily, northern and western coastal areas in the Highlands began a phase of particularly rapid growth for other reasons.[13] The rise of kelp production, the greater adoption of the potato and greater investment in facilities for fishing enabled many more people to be crowded on to the coast. Just how much growth was experienced is demonstrated for us by Tiree, its numbers almost trebling, from 1,509 to 4,391, between Dr Webster's census in 1755 and the 1841 census.[14]

Environmental Drivers

As already intimated, long-term swings in climate were also a prime driver of landscape change over the medieval and early modern period. Just as the Medieval Climatic Optimum of the eleventh to thirteenth centuries would have seen an expansion of farm output and settlement,[15] the onset of climatic deterioration by 1300 led to a decline of output and the contraction of settlement. In fact, increased climatic variability over the course of the thirteenth century may have caused problems for Highland farming even before 1300, with reports of more extreme weather events, such as floods, very cold winters and hot, dry summers.[16] This marked the start of the Little Ice Age, a phase of poorer climate which stretched from the beginning of the fourteenth century down to the end of the nineteenth. It was not a period of sustained climatic degradation. Periods of lower temperatures and increased storminess were punctuated by periods of climatic amelioration.[17] There were two particular phases of colder, stormier conditions. The first was during the early fifteenth century. The second was during the so-called Maunder Minimum 1645–1702. Lamb argued that, during the latter, climate over the north and north-west of Scotland changed in two ways. A migration of Arctic sea ice southwards meant that any air movements moving in off the north Atlantic – the prevailing trend of air movement for northern Scotland – would have caused a shift to colder conditions in the Highlands. In addition, he saw the seventeenth to nineteenth centuries as a time when Scotland was afflicted by an increased frequency of severe storms, with a number of particularly damaging storm events occurring during the second half of the seventeenth century.[18] If we examine rentals for areas that might have suffered from these lower temperatures during the Maunder Minimum, such as Rannoch, we can find instances of whole touns reported in rentals as lying waste. What stands out about such touns, however, is the fact that they only lay waste for a couple of years or so before being reoccupied.[19] Further, on closer inspection, we find that their tipping point was not simply when their arable ceased to produce but the point when they failed to produce enough output to cover both their subsistence and their rent in kind, with tenants accumulating 'rests of rent'. In some cases, especially with farms on higher ground, we are dealing with a downgrading rather than an abandonment, with a permanent all-year settlement giving way to a seasonal shieling site. Thus, Sallachul and Clushfern, two small touns sited high on the hill ground above Loch Leven, are recorded on Pont's map, drawn up in the 1580s/90s, as occupied settlements or touns but, when we see them in a mid-eighteenth-century survey, they were used only as temporary summer shielings for low-ground settlements beside Loch Leven.[20]

Any fall in mean annual temperatures was only part of the problem. Lamb noted that 'with the Arctic sea ice in advanced positions near Iceland and the polar water extending farther south, the thermal gradient between latitudes 50°N and 60–65°N in the eastern Atlantic sector was increased compared with most later situations in the pattern of air circulation'.[21] So, not only were the main storm tracks now across Britain, their severity increased, a conclusion given stronger support by the recent work of Dawson et al.[22] Areas along the western seaboard and across the Hebrides would have been fully exposed to this increased storm risk. The increased storminess may account for the evidence that is available for the temporary abandonment of touns along the western seaboard and in parts of the Hebrides. For example, rentals for the 1660s show a significant number of touns on Mull[23] and Tiree (Figure 4.1) as abandoned, or 'lying waist', either wholly or in part.

In each case, the abandonment of touns or holdings was temporary, lasting no more than a few years before reoccupation took place. The interpretation of these data is not straightforward as both islands suffered considerable political turmoil, and the presence of waste townships could be a symptom of such upheaval. Given that we are dealing with the early years of the Maunder Minimum, however, we cannot rule out the possibility that these short, temporary abandonments may be rooted in adverse weather conditions. The greater frequency of storms and the wetter summers would have affected the quality and abundance of harvests in these western areas. In comparison, when we come across 'waist' townships recorded in rentals of the 1690s for Netherlorn, an area that comprised small islands such as Seil and Luing, as well as the adjacent mainland, contemporary comments in accounts and letters make it clear that bad weather and poor harvests were the cause.[24]

Further south, on the mainland, a significant number of touns in both North and South Kintyre were reported as lying 'waist' in rentals for the 1590s, early 1600s and the early 1650s. To take North Kintyre as an example, twenty-one of its touns were wholly waste in 1596 and thirty-nine in 1605.[25] The scale of disruption was comparable during the crises of the 1650s, with twenty-nine

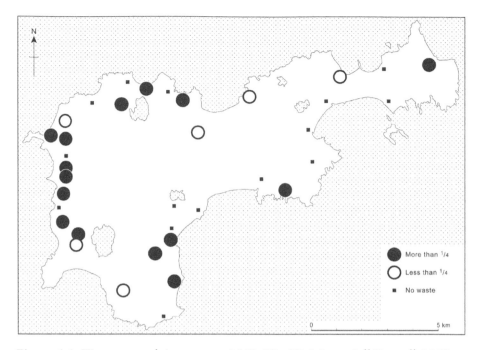

Figure 4.1 Tiree, touns lying waste, 1662, IC, AP, Memoriall Rentall 1662.
After Dodgshon, 'Little Ice Age', p. 331.

holdings in the parishes of Kilcolmkill and Kilbaan listed as wholly waste in
a 1651 rental and thirteen as partially waste.[26] As with Tiree and Mull, the
context in which we might interpret the reasons for these waste touns is not
straightforward. The 1590s and early 1600s was a time of political disruption
for Kintyre when we might expect a turnover in occupancy. Yet, as on Tiree
and Mull, even when individuals were absentee, they would still normally be
recorded in rentals as the tenants. For land to be deemed as 'waist', it had to
be both out of cultivation and out of set. In the circumstances, the fact that
the years 1592–99 were a time when poor weather led to some of the worst
harvests and famines of early modern Scotland may have been a contributory
factor. To complicate matters, though, Kintyre was affected by a serious local
outbreak of disease five or six years before these rentals recorded 'waist' touns
there[27] but local tradition also makes it clear that, a few years after the outbreak of disease, there then occurred a short sequence of failed harvests owing
to bad weather which caused severe famine across the peninsula.[28]

The increased frequency of severe storms affected settlement in the
Hebrides in other ways. Many western coastal areas were fringed by machair,
a pasture developed over calcareous sand. Even without human interference,
it was prone to erosion. Where communities ploughed machair soils, or used
plants such as bent for roping, the risks increased. When we add the greater

frequency of severe storms during the Little Ice Age, we can understand why instances of sandstorms overwhelming settlement and arable in the Hebrides were frequent. The islands that suffered most were the Uists, Harris, Pabbay, Bernera, Tiree and Coll, all of which had extensive machair. Dr John Walker's report on the Hebrides (based on visits in 1764 and 1771) recorded 300 acres (152 hectares) of former arable and pasture on the south-western side of the island of Pabbay as now covered by sand, with the sea flowing 'for a great Space, where many People still alive have reaped Crops'.[29] A 1772 estate survey makes it clear that this 300 acres (152 hectares) involved the loss of whole touns.[30] Much of this reported damage is probably attributable to the severe storms of the 1750s. Some, though, may have been caused by earlier damaging storms for the MacLeod estate reduced Pabbay's pennyland assessment in the late seventeenth century.[31] The 1772 survey also survey noted the 'great devastation' that had been caused to west-coast touns of the mainland, such as on the two Scaristas. In fact, its report of the damage here was as much prospect as retrospect, their arable being written down further between 1772 and an 1804–5 survey. In 1772, Little and Meikle Scarista had 371 and 428 acres (188 and 217 hectares) of arable respectively. By 1804–5, they had only 108 and 145 acres (55 and 74 hectares) of ploughed arable and 18 and 150 acres (9 and 76 hectares) respectively of spaded arable mixed with pasture.[32]

To the south, the Uists were equally afflicted by storm damage because many of its coastal areas were 'little raised above the ordinary Level of the Sea' so that they 'suffered greatly by extraordinary Tides, which are frequently occasioned by the great Violence of the South West Winds'.[33] The Uists suffered in the storms of the 1690s but the storm that has produced most documentation was that of 1756. Walker reported that, during this storm, the peninsula of Eachcamais, on which the touns of Baleshare and Illeray are located, was breached, destroying arable and pasture in the process.[34] The scale of damage caused can be illustrated by Illeray. Today, the site of the pre-crofting toun is represented by a scatter of house platforms surrounded by tidal channels and inlets (Figure 4.2).

Whereas the 1756 storm covered the houses of Baleshare 'upto their roofs',[35] it stripped the soil off Illeray's arable. This arable probably stretched to the north of its former house sites, out across where there are now simply sand flats, as well as across the fretted area of tidal marsh immediately around the house sites. The response of the inhabitants at Illeray was simply to move their settlement and arable. Tracked through rentals, the toun appears as a significant one in 1718, with five tenants listed.[36] Barely eight years after the storm, it was still being let but this time the rental only lets us glimpse the tacksmen, or lead tenants.[37] When the island was surveyed in detail by John Blackadder in 1799, it appears again as a thriving community, albeit now displaced eastwards.[38]

Given its low-lying character and the extent of its machair, it is not

Figure 4.2 Former settlement at Illeray, North Uist.

surprising that Tiree shared in this storm damage. The depth of its rental data enables us to detect probable signs of storm damage back in the sixteenth century when two touns, Baa and The Reef, were initially reported as lying waste then entirely absent from subsequent rentals.[39] Given that The Reef suffered repeatedly from sand blows, we need not search far for an explanation. Later, eighteenth-century documentation provides further evidence of the instability of the island's machair in the face of severe storms, with a note to the Duke of Argyll's chamberlain in 1750 noting that farms 'had suffered considerably' from sand drift.[40] Whatever the role played by the storminess of the times, the duke was convinced 'the Irregular Cutting of Turff [for housing] has been detrimental',[41] as was the cropping of 'light sandy soil to an injurious and dangerous degree'.[42] The estate's concern is highlighted by the way James Turnbull's survey in 1768 measured the amount of 'blown sand' in each toun. Predictably, his survey shows the problem was more acute in the west of the island than in the east, with 1,318 acres (670 hectares) affected to the west of The Reef and 347 acres (176 hectares) to the east (Figure 4.3).[43]

Not surprisingly, the increased risk of storm events during the Little Ice Age brought with it an increased risk of flooding. Work on the long-term history of flood events has suggested that the seventeenth to eighteenth centuries experienced a high frequency of such events.[44] In the Highlands, the damage caused was accentuated by the fact that, overall, these centuries saw an expansion of arable. In the central and northern Highlands, this expansion pressed against low-lying haugh land that, in normal circumstances, would have been kept under meadow. Estates tried to minimise the risk by putting in place regulations that banned tenants from cultivating land within a set distance of rivers

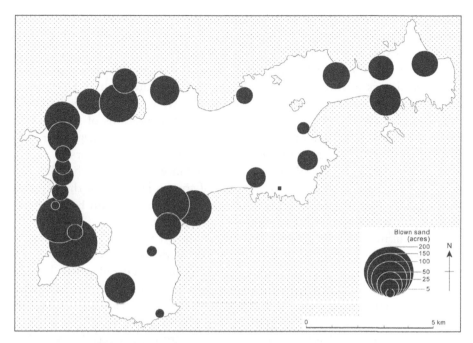

Figure 4.3 Land lost through sand blows, Tiree, 1768 based on IC, AP, vol. 65, Contents . . . Tiry, 1768. After Dodgshon 'Little Ice Age', p. 332.

or burns[45] but the frequency with which damage to arable was reported suggests that such regulations were not always a match for the problem. Most of the major straths in the central and northern Highlands suffered. A 1637 entry in the Wardlaw diary records that 'such an inundation happened here after harvest that all rivers and brookes run over their banks . . . The river of Connin did great hurt uppon bordering lands; the water of Beauly much more up about Maines and Agasis.'[46] Farm surveys which become available during the mid-eighteenth century are full of comment about the local risks of flooding. On the Spey, virtually every toun on the Cluny estate that fronted the river had problems with its spates. A survey of 1771 has an entry for Gaskinloan that speaks for all the Speyside townships. The Spey, it said, 'does unspeakable damage both in seed time and Harvest when swelled with heavy rains' as in the autumn of 1771.[47] To the south, a late eighteenth-century report of a severe storm in Glenlyon said it left only house tops visible, caused the loss of many stock and, further downstream, left the whole plain of Appin of Dull under water.[48]

The flooding of the larger rivers was only part of the problem. The slopes which bordered many Highlands touns were what contemporaries described as 'spouty', meaning that they had numerous small burns rising on them and creating a slopewash that, in times of spate, flooded out over adjacent arable or meadow.[49] Some areas were especially prone to this. A survey of the Struan estate in Rannoch described almost every toun as liable to flooding from streams crossing their slopes.[50] The scale of the problem is underlined by the regularity with which local courts bound tenants to channelise rivers or to join with neighbours in times of spate. On the Struan estate itself, a 1678 act of court reminded all tenants 'burnes and wateris' were to 'be mended and put in their owne right channel',[51] while, to the west, conditions of set issued in 1769 for farms in Lochaber required tenants to show good neighbourliness 'in time of speats . . . to manage and secure the course of the waters so as they may do less damage', describing it as the 'assistance of neighbourhood'.[52] Yet farmers gained a lesson from the flooding of spouty land which they applied to advantage with some, on estates such as those of Struan and Breadalbane, deliberately watering pasture to improve spring output.[53]

Different in character was a type of flood known as a *scriddan*. It occurred when sustained heavy rain, perhaps combined with melting snows, caused a build-up of water in a raised bog or moss. Flooding occurred when the edge of the peat was breached, producing an effect like a bursting dam. The processes involved are captured well in a description of a *scriddan* in Kintail in 1745. 'The farms', said its *Old Statistical Account* report, 'which are bases to high mountains, as in Kintail, suffer great losses from what is called Scriddan, or mountain torrent.' After heavy rains, the summit of the hills become 'so impregnated . . . till an aperture is made by chance somewhere on their sides . . . Gravel and mossy stones roll together, and desolate the fields beneath.' It gave the specific example of Auchuirn, in Glenelgchaig, which in 1745 was 'rendered uninhabitable, and is since converted to a grazing farm, by an awful Scriddan'.[54] Clearly, it was a catastrophic flooding event that was sufficiently familiar to have its own descriptive term. The toun of Auchuirn is listed in a seventeenth-century rental for Glenelg as Achichuirne.[55] To judge from its entry, Auchuirn was located somewhere between Scallasaig and Beolary. By 1773, its land had been combined with Beolary.[56] Elsewhere, we can also document a similar *scriddan* event at Lockeck in Strachur and Stralachlan parish, Argyll.[57]

Sociopolitical Drivers

A range of disparate factors can be bundled under the heading of sociopolitical drivers. Their prime influence was in shaping the institutional context of change. They did so in two ways. First, different sociopolitical systems employed different systems of landholding. We can only speculate about the range of landholding systems present in the region during much of the medieval period: native forms of kin-based tenure around multi-generational lineages; Norse forms of kin-based tenure deriving from Udal law; clientage; the holding of land by unfree groups under native forms of lordship; its holding under chiefly or clan-based systems of lordship; and patterns of landholding that had developed under feudalised forms of landholding. The greatest uncertainty surrounds the first of these tenures, or native forms of landholding based on some form of kin-based tenure. Given the close interaction between Ireland and the South-west Highlands and Islands, we might reasonably assume that a form of tenure may have been present during the early medieval period that was akin to the *finé* of Ireland: that is, an estate held jointly by a kin group whose members were linked together across a number of generations.[58] Individual members of such multigenerational kin groups would have occupied individual holdings but their right of possession was defined within the framework of the wider kin group or *finé*. If such a form of tenure did exist, then it might explain why a much diluted form of landholding based on kinship emerged later in the form of the clan system and why there was a deep-rooted assumption underpinning the clan system that chiefs only held land on behalf of clansmen and that the latter had an hereditary right, the *dutchas*, to their holding.[59]

We are on firmer ground when dealing with the kin-based landholding of the Northern Isles. Based around Udal tenures, these would have been present by the ninth century AD. Though they are most closely associated with the Northern Isles,[60] we would also expect them to have had a presence in the Hebrides from the ninth century down to the handover of the Western Isles to the Scottish Crown in 1266. They were kin based in that access to land was through inheritance but inheritance does not appear to have led to co-ownership by a shallow descent group as with the *finé*. The way separate *houses* emerged from the initial division of primary estates among co-heirs hints at this. The survival of Udal tenures was not a smooth affair. There was a phase between the late ninth and late tenth centuries when the earl held the Udal rights over land in Orkney and again, in the mid-twelfth century, when the earl contrived a situation in which Udallers had to pay to recover their rights from him.[61] Whatever the full story behind the repurchasing of Udal rights from the earl in 1137, it is clear is that there was a gradual erosion of Udal tenures in the centuries that followed.[62] Nevertheless, many survived despite further attempts during the seventeenth century to force those who still held by Udal tenure to accept feu tenure in lieu.[63]

As detailed farm-level data become available over the medieval period, some of the uncertainties over the nature of landholding disappear. Overall, two major changes affected landholding. The first stemmed from the spread of feudalised concepts of landholding, with all land being seen as held from a superior. Its most developed form was that of crown feudalism, with all land held ultimately from the crown. Land was granted from the crown to its major crown vassals or territorial lords who, in turn, leased out what they were granted to their vassals, tenants and their undertenants. At each level, land would be given out in return for services, rents, dues and obligations, with the major territorial vassals providing military services and those who actually worked the soil owing labour services, dues and other burdens. In the process, the possession of land became contractual, with a given quantity of land given in return for specified services, payments in kind and other dues. Assessments which had earlier been used to measure the payment of renders, dues and obligations owed by individuals or kin groups were now converted into measures of land and used as the basis for defining the amount of land held in return for rent and services. The spread of crown feudalism under the Anglo-Normans is a reasonably visible process. The areas around the eastern and southern edges of the Highlands were the first to be affected by it, either through the implantation of Anglo-Norman families, such as the Stewarts, Morays and Chisholms, or by means of a *fief de reprise* whereby existing landowners acknowledged the crown as their superior, processes that were underway by the thirteenth and fourteenth centuries.[64] There is also the possibility, however, that forms of lordship over men through land may already have existed in parts of the Highlands at this point so that the implantation of Anglo-Norman families simply amounted to a reorientation of the system.

The second change that affected landholding during the medieval period was the emergence of the clan system. Control over land was at the very heart of the clan system but it was a control exercised via ties of kin, real or assumed, or through alliances. The saying that a clan without land was a broken clan, just as a man without a clan was a broken man, makes this clear. What distinguished their control over land was the fact that it was let out first and foremost to kinsmen of the chief, real or assumed, and to members of affiliated clans. There was a hierarchical order to how they set out their estates, with local territories or clusters of touns being granted to senior members of the clan, or heads of lineages, who, in turn, sublet land to their own close kinsmen or to those of affiliated clans and so on, down to the level of those who worked the soil, whether tenants, subtenants, crofters or cottars. This packing of estates with kinsmen meant that the clan system tended to control land through its control over people, with the territory of the clan being the territory that was occupied by its members. This was a sociopolitical as well as an economic system, with those let whole districts or whole touns as tacksmen having a key military and political role within the clan, not least by subletting to those who, directly or through the sons of those leasing from them, could provide military support when required.[65] When they come into view through early rentals, the rents in kind levied by clan-based estates in the west and across the Hebrides were not materially different from those uplifted by estates to the east and south but, in some cases, they were uplifted in a different way. The *c.*1574 report for the Hebrides affords us a glimpse of how they differed, with chiefs burdening touns with heavy but irregular demands for hospitality or *cuddiches* and food renders for their household, retinues and fighting men.[66] Yet, while we can discern ways in which these clan-based estates in the far west were different from the more feudalised estates of the east and south, we must not overstate the differences. There is a sense in which the medieval landholding systems of the Highlands are best seen as neither fish nor fowl. The clan system was not something that had long existed. It emerged over the thirteenth and fourteenth centuries, probably in response to the emergence of the large and powerful estates in the east and south in the hands of crown vassals. In its response, it may even have adopted some of the organising principles of the latter, with chiefs asserting a more feudalised lordship over their kinsmen, replacing earlier forms of kin-based tenure with one now polarised around their control while preserving a presumption of kinship, real or otherwise, as the sinews that bound the system together.

Economic Drivers

Even during medieval times, the Highland toun economy was not a subsistence economy, pure and simple. Extractions in kind by lords or chiefs meant that, over and above subsistence needs, there existed a level of demand that could both shape and pressurise the toun economy. The high proportion of the bere crop that was handed over as rent in Hebridean areas, and initially consumed in displays of feasting and hospitality, is one indication of how such demands shaped the toun economy. Bigelow's use of increased lordly exactions to account for the expanded dairy output, evident from the enlarged byres of Norse sites in the Northern Isles, provides another example. Sixteenth-century crown rentals suggest that lordly exactions may have affected diary production elsewhere in the Highlands. In some areas, levels of butter and cheese extracted as rent were modest. In other areas, though, such as Kintyre, dairy production seems concentrated, with substantial amounts being extracted from a few specialised touns.[67] In still other areas, such as Ardnamurchan and Sunart, all touns were burdened with the payment of quite significant amounts[68] though not as much as the specialist touns to be found in Kintyre. Clearly, different estates had different strategies. Environmental opportunism played its part as much as social needs.

Looking forward from this point, we can identify three factors reshaping the economic landscape of touns. The first was the bundle of regulations targeting West Highland chiefs that were put in place by the Statutes of Iona (1609). Among them were regulations limiting their display behaviour (that is, feasting) and the size of their household, as well as requiring them to uplift their rents on a fixed and regular basis.[69] The statutes served as a push factor, with those chiefs affected having to find other ways of capitalising on the large amounts of rents in kind (that is, meal, bere, stock, and so on) gathered in. Marketing produce was the obvious solution. The second change was the rising demand from markets to the east and south. Those estates positioned along the eastern and southern edges of the Highlands, especially, were already engaged with market opportunities before the Statutes of Iona were enacted.

As these market opportunities expanded over the seventeenth century, more estates responded to their pull.[70] Grain formed an important part of the expanding flows, large amounts being shipped south by sea, including from the far north.[71] Even in the west, where sea transport was also an option, sizeable quantities of grain, gathered in as rent from areas such as Netherlorn or from touns along the western side of Harris and offshore islands such as Pabbay, also found their way to market.[72]

For a time, there was something of a paradox about marketing in the region. Estates lying close to markets saw the self-interest of continuing to uplift their grain rents, or at least have tenants deliver them to the market or nearest port. In contrast, estates in some of the more remote areas to the west and north-west, areas from which marketing was a challenge or involved low thresholds, tended to take quicker strides towards the monetisation of their rents in kind. This conversion provides a third driver of change but we need to see it as something that was flexible. Faced with tenants who were not in a position to market produce easily, estates began to negotiate over what could be paid in lieu. In effect, a form of barter was introduced to the estate economy. In one sense, this might seem to be a case of *plus ça change*, with rents in kind simply being replaced by rents in kind that were now convertible. In another sense, though, it marked a profound change. Once rents were monetised, it enabled tenants and landlords to shift the burden of rent, creating a new flexibility in the toun economy. At a time when market demand for cattle was growing, it was logical that cattle became a favoured item for tenants to set against their cash rents. It was a commodity that could be bartered and droved out of the most distant and difficult areas. As an item set against cash rents, we find estates themselves were behind a fair proportion of the droving that gathered momentum during the seventeenth century. But there is another point that must not be overlooked. From a tenant's point of view, the use of cattle as 'money-stuff' shifted the rent burden not just from grain on to stock but also from arable to grass. For communities whose numbers were growing but whose arable was scarce, this meant arable could be loaded with more of the demands of subsistence and the more extensive reserves of grass with the demands of rent. In time, other sources of flexibility emerged. Once chiefly displays were restricted, it suited some estates to convert bere payments into cash. On islands such as Tiree and Jura, tenants responded by converting the bere crop, which they now had in hand, into malt for brewing and distilling. Looking further forward, the take-up over the eighteenth century of the potato as the basic subsistence crop would not have been so straightforward if the toun economy had still been straitjacketed by fixed payments of meal and bere.

The Medieval and Early Modern Farming Toun

Possibly the most fundamental question that we can ask of the medieval coun-tryside is what shaped or defined the farming toun? Was it simply the resource set which particular communities happened to have brought into use through the labour of centuries or was it constituted in some other, more formalised way? I want to argue that the answer lies in how land assessments were imposed over what was already a settled landscape. It was not a case of their imposition helping to consolidate or to add further cohesion to a pattern of touns that already existed at this point. Instead, we need to see their top-down imposition as playing a formative role in forging a unity out of local systems of landholding that, beforehand, may have been more disaggregated. The key to understanding the impact of their imposition lies in the fact that blocks of land assessment did not make for an exact fit with what was on the ground, in terms of holdings and settlement, nor did they have the same flexibility compared to what was on the ground. We need to see assessments and what was on the ground as different layers of order, each with its own characteris-tics and history, rather than inseparable dimensions that emerged together as matched features. Though we can find hints of some readjustment during the early phases of their use,[73] assessments proved to be a conservative feature in the landscape, persisting unchanged for centuries.[74] In these circumstances, the ongoing integrity of the original assessment would have stood for the integrity of the estate. The dynamics of what was happening on the ground, in terms of landholding and its arable, had to fit in with this conservatism.

We can see this by looking at how assessments helped to structure touns. The way in which their imposition involved aggregated territorial assessments being broken down over the settled landscape via a top-down process resulted in the settled landscape being structured around fairly large, unitary blocks of assessment, such as the davoch, *tirunga* or Five Pennylands. At the point when their imposition took place, it is unlikely that the scale of the typical family holding could have matched that of the davoch, *tirunga* or standard ratings such as a Five Pennyland. Indeed, holdings of this scale would be a match for only the largest of eighteenth-century pre-Clearance touns. More to the point, what field data we have for late prehistoric and early historic landholding suggest that holding size then was far smaller than what would be encompassed by a davoch, *tirunga* or Five Pennyland. In other words, there are grounds for supposing that, when such assessments were first imposed, they are likely to have covered, umbrella-like, a number of holdings, each with its own settlement and patch of arable, some packed together but others detached. We can support this point in a number of ways.

Descriptions of how traditional pre-Clearance touns were arranged on the ground tended to portray their dwellings, outbuildings and tied enclosures or kailyards as arranged into a single, amorphous cluster. Later cartographic

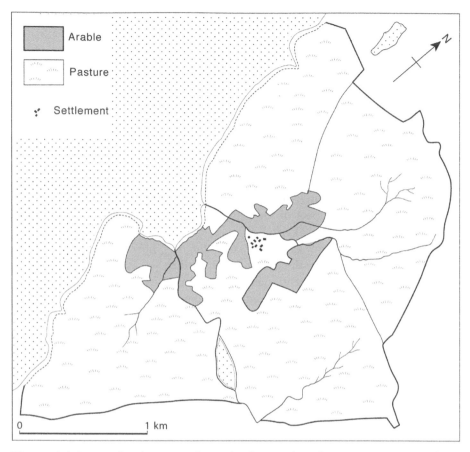

Figure 4.4 Example of toun with single clustered settlement, Ramasaig, Skye, based on plan by J. Chapman, surveyed 1810, redrawn by A. Cameron, 1848, archived at Dunvegan Castle, Skye.

surveys certainly yield examples of such an arrangement, such as at Ramasaig, Skye (Figure 4.4).

We also find touns, however, where settlement is arranged around a number of different foci, such as with Ardvoile and Ballemore, on the north side of Loch Tay (Figure 4.5).

So also was Bragar, a toun on the north-west coast of Lewis that comprised a north and south toun. The two discrete settlement sites on North Bragar were about 600 metres apart, while walking from the eastern edge of South Bragar's settlement to that on the west bank of Loch Ordais involved a journey of over a kilometre. Traces of these different settlement nuclei at Bragar can still be found.[75] By any standards, the touns along the north-west coast of Lewis were large. Their size may have had an influence on their internal organisation. In a

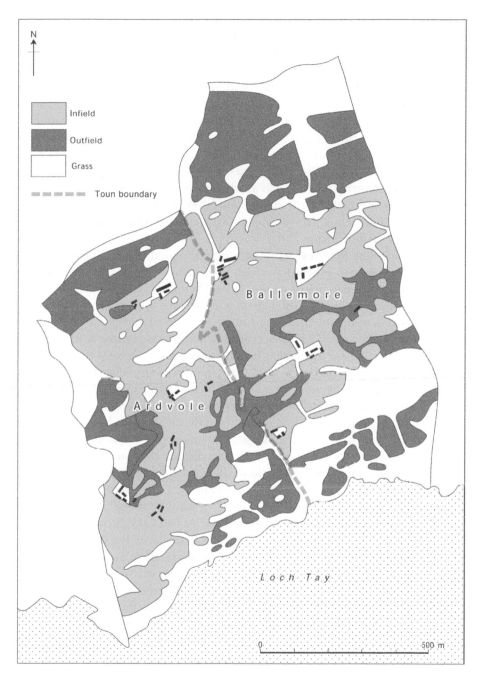

Figure 4.5 Ardvoile and Ballemore, Lochtayside, based on Farquharson's maps of Lochtayside, 1769, NAS 973/1. Mapped from a rectified version of Farquharson's map kindly provided by the RCAHMS, Edinburgh.

short essay on Lewis, Geddes noted how one of the rentals drawn up for Lewis in 1718 appeared to attach portions of the rent owed to groups of tenants. He reasoned that this might suggest that touns such as Bragar and its neighbours may have been organised internally around small, localised groups of tenants, each working its own block of arable within the toun and charged with a specific portion of the rent, usually a pennyland's worth.[76] Shaw set these 'hamlet farms, as she called them, in a wider context, documenting other examples from Harris, Netherlorn, Islay and the Northern Isles, but adding that the size of groups involved were more varied in terms of the amount of land held than Geddes allowed.[77] She supported her case by also noting that quite a number of touns in the Northern Isles, such as Hensbister and Orphir, both in Holme parish, Orkney, were physically subdivided into sub-touns on the ground, each with its own place name, providing us with a more formal arrangement of 'hamlet farms' within a toun.[78]

We can account for these seemingly disaggregated touns in a number of ways. Some may have evolved through expansion from a single core settlement, with different patches of workable land being progressively brought into use. Gailey thought this was the reason why Kintyre touns had multiple foci, though he had in mind only the expansion that occurred during the seventeenth and eighteenth centuries.[79] His observation was directed at those Kintyre touns with multiple settlement foci. What he did not comment on were those Kintyre touns that appear in sixteenth- and seventeenth-century rentals as grouped together under the framework of a single block of assessment, despite having separate names and separate tenants. Some may have originated as later outsets but, given their number, it is difficult to avoid the conclusion that some existed at the point when their covering assessment was imposed. Some of the issues posed by such clusters are captured by the dispute that surfaced in a 1299 letter relating to Papa Stour in Shetland. It centred on an attempt by the earl, or his agent, to extend the assessment of Da Biggins, the parent toun for Papa Stour, so as to embrace its daughter touns, or outsets, though the letter itself related to what a certain Ragnhild Simunsdatter had said during the dispute. This extension of what was covered by the assessment of Da Biggins (and, as a consequence, its skat) was clearly seen by the earl's agent as within the rules, though that was not how Ragnhild Simunsdatter saw it.[80]

Accepting that, at their point of imposition, assessments may have encompassed what had been separate holdings has implications for how the development of such holdings unfolded. By virtue of the fact they were now burdened with a common assessment for renders, dues, rents in kind, cash payments, and so on, we cannot be surprised if forms of co-operation emerged between such holdings over grazing, harvesting and even ploughing. The degree to which co-operation took place could have varied according to the extent to which arable was compact or fragmented and the degree to which occupiers were separated by social differences. When we place the separately named touns clustered under a single assessment in Kintyre rentals alongside those on Lewis, or in Shetland mentioned above, that were internally differentiated either around work-team groups or into sub-touns, and then add fully integrated touns such as Ramasaig, we are faced with the possibility that, over time, multiple holdings covered by a single assessment may have moved at different rates towards being fully integrated touns.

That holdings covered by a single assessment could shift from one end of this spectrum to the other is demonstrated by Thomson's work on the house system in Shetland. When looked at through seventeenth- and eighteenth-century sources, Shetland touns appear to be organised around what he described as a 'confederation' of houses.[81] These comprised separate landholding or farm units within the assessment framework of a larger parent toun. Houbie on Fetlar, for instance, was organised around five separate houses, while Norwick had four.[82] Some were free-standing but others, such as those in Norwick, had become intermixed by way of runrig with other houses.[83] Everywhere, though, each house had a separate identity highlighted by the fact that it bore a separate name: those at Houbie, for instance, were called Digron, Turvhoul, Kirkatoun and Sandeal.[84] Early sources describe landholders said to have their land 'under the *house* of . . . X or Y'.[85] What makes the house system of interest to the current discussion is that, initially, they were probably free-standing or consolidated holdings. Drawing on the evidence for Papa Stour, Crawford and Ballin Smith suggested that its house structure had probably been formed by new holdings being carved out of the primary or parent settlement to provide an inheritance for co-heirs, with the new holdings so formed, as elsewhere in the Northern Isles, incorporating the element *hus* in their names: those on Papa Stour being called North-house, Up-house and Souther-house.[86]

Over time, the organisation of Shetland houses was shaped by two processes. First, they became further subdivided between co-heirs and between tenants, some becoming runrig touns in the process. Second, as pressure on arable and grazing developed, houses expanded up against each other so that not only did they become runrig touns within themselves but parts became intermixed with the arable or grazings of other adjacent houses. Thomson has documented how houses whose land or runrig overlapped in this way took steps to rationalise their layouts into more integrated or combined runrig systems via a fresh division. By the eighteenth century, some of these redivisions were being carried out so as to form still larger 'whole-toun' runrig divisions, ones that combined the land of all houses within the toun (pp. 130–2). Like Crawford and Ballin Smith, Thomson saw houses as the fundamental building blocks of the Shetland toun. He went further, however, by arguing that something akin to the house structure of the early Shetland toun was to be found over a far wider area than that associated with Udal tenure. Udal tenure mattered, at least in Shetland, only in that it fostered its late survival, enabling us to glimpse how the integrated toun was fashioned from the coming together

No Stone Unturned

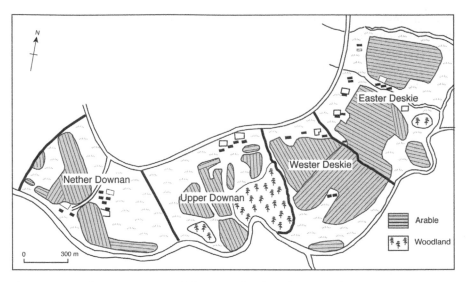

Figure 4.6 Split touns, part of the Daugh of Deskie, Strathavon, 1761, RHP1774.

of what had started out as separate farms or houses.[87]

The history of medieval and early modern landholding was not about the working out of consistent processes. Just as farms and touns could advance or retreat according to demand, and just as there were forces helping to aggregate separate holdings into larger, more integrated units, so also do we find forces at work that brought about the break-up or splitting of touns. Understanding how the medieval and early modern landscape unfolded requires us to appreciate how both these seemingly contrary processes had their place. The evidence for toun splitting is widespread.[88] In the first place, it is provided by the many touns throughout the region that were named as if they were part of a linked or paired cluster, with the cluster having what we can call a common surname but with each sub-toun in the cluster distinguished by prefixes such as Easter/ Wester, Nether/Over or affixes such as *Mòr/Beag*, with some having a third component, tagged by prefixes such as Mid (Figure 4.6). They were clearly touns that shared a common ancestry.

This common ancestry is particularly underlined by the way some of these touns were referred to after the style of 'The Two Botuaries' or 'The Ranoquhennis' in Sunart[89] or the 'two Oyngadellis', 'the two Sandis' and 'the two Nastis' in a 1638 rental for Gairloch,[90] as if sibling settlements. In some cases, touns which have the appearance of being split, or distinguished by prefixes such as East and West, may have been multiple touns covered by a single assessment rating at the point when assessments were first imposed. Others were created by a division or split after assessments were levied.

Such splitting was rooted in a number of processes. The most potent was the splitting of touns for family settlements, or inheritance by co-heirs. In the Northern Isles, where hereditary tenures based on Udal tenure became widespread with the Norse colonisation, Storer Clouston saw partible inheritance as a key reason why many touns became organised into Easter/Wester and Nether/Upper portions.[91] The division of the so-called 'family township' of Grimbister into Over, Mid and Nether Bigging, for instance, was the by-product of a division between co-heirs.[92] Elsewhere in Orkney, during a 1601 division of a holding at Corrigall, it was agreed that James and Robert Corigall were to 'have thair entres and house fredomes on the west syd of their houssis with barne and corneyaird . . . and John Corigall to have his entres and hous fredomes on the eist syd of his hous'.[93] For Thomson, the intensification of settlement through the splitting of touns via inheritance was the most powerful source of landholding change on Orkney during the medieval period, one manifest through 'a whole host of division-names such as Everbist, Midgarth, Nears, Nistigar, Uttesgarth, North Setter, Symbister and Isbister (respectively upper, middle, lower, lowest, outermost, north, south and east)'.[94] It was not, however, a process confined to the Northern Isles. Many instances of split touns, either because of inheritance or sale, can also be documented for the main body of the Highlands. Drummy, lying between lower Glenalmond and Strathearn, was divided into Easter, Wester and Middle Drummy by a decreet of division in 1573.[95] Moniack, in The Aird of Inverness, was subject to a double splitting process. At some point, the touns of Easter and Wester Moniack had already been split. In the early seventeenth century, the two portioners who shared what had become Easter Moniack decided to split their respective halves in 1608, creating two separate units 'which are hereby called the Easter and Wester half of the Toun & Lands of Easter Moniack'.[96] At about the same time, Sir Robert Chychtoun wadset the sunny runrig half of the toun of Baldornoch 'otherwise Admeston', near Dunkeld, to a James Stoup. This was followed over twenty years later, in 1629, by a division of the two runrig halves into two separate touns.[97] Not all examples of splitting between different heritors are so specific. Following a settlement in 1519 between the two sisters of Donald Gallda of Lochalsh, the lands of Lochbroom became equally divided between their respective families: the Macdonalds of Glengarry and the Dingwalls of Kildun.[98] In 1543, Thomas Dingwall sold his half of Lochbroom to John Mackenzie of Kintail, a half that amounted 'to a half share of each township'.[99] With a 'half share of each township [or toun]' in the hands of different heritors, there would have been a strong incentive for their tenants to separate their respective halves into distinct touns.

Family settlements or sales were not the only reasons for the splitting of touns. Some were a response to the problems of scalar growth, either having become too large in terms of tenant numbers, or too sprawled in terms of arable. We can glean from rentals circumstantial evidence for these sorts of splits. Where we have access to long-run rental sequences, such as for Tiree,

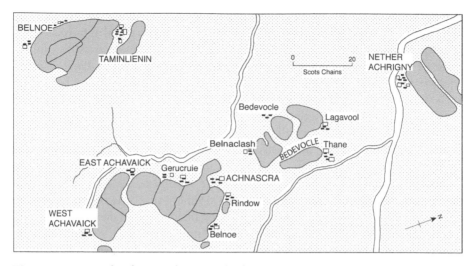

Figure 4.7 Daugh of Taminlienin, Glenlivet, 1761, NAS, RHP2487/6.

Ardnamurchan and Kintyre, we can find a few instances of split touns emerging *de novo* over the sixteenth and seventeenth centuries, without any evident change of ownership. The process is more graphically captured for us by what Gordon of Straloch had to say about touns in the eastern Grampians in the late sixteenth century. This was an area where the davoch reigned supreme and where touns divided into ploughs, or into Easter/Wester portions, were abundant. Offering a rare insight into how early toun layouts were open to revision, he declared that concern over the increasing time spent in moving back and forth across an expanding network of fields, and the problems created by having to negotiate with an increasing number of fellow tenants, led proprietors to split touns into smaller units.[100] Nor was it simply a shift in the layout of fields. There was a 'migration from the villages into the fields . . . And here their abodes were fixed.'[101] The impact of this reorganisation is not easily recovered through rentals. The 'Manes of Boigegeyght' in Bellye parish provides a case in point. Rated as a single plough, it was subdivided into thirds, with the 'Ouer Third' set to three tenants and a cautioner, the 'Mid Third' to four tenants and a cautioner, and the 'Nather Third' to two tenants.[102] These three shares could have been worked as a single runrig toun, since the 'Manes' was a plough, but, equally, they could have been arranged – as the style of their naming hints – into separate touns. Just how fractionated holdings had become through splitting is revealed by mid-eighteenth-century surveys for the eastern Grampians. The davoch of Taminlienin, for example, like other davochs, was divided into four ploughs, with each plough being a tenurially discrete unit and set at a distance from the others (Figure 4.7). When we focus on these individual ploughs, we find that, far from each being simple, integrated touns, they were themselves complex groupings of small sub-touns. One of the most complex, the plough of Achavaick and Achnascra, for instance, consisted of six separate touns: West Achiavaick, East Achavaick, Gerucruie, Achnascra, Belnoe and Rindow.

Examples such as the davoch of Taminlienin show that the splitting of touns has another more everyday context. By their very nature, multiple-tenant touns were routinely faced with a choice over how to divide their shares. Splitting the toun into smaller sub-touns or into several holdings was always part of that choice, with the sprawl of a toun's arable, its social scale, the compatibility of those involved and the attraction of keeping ploughs separate all shaping the choice. That it was a real choice is shown by those instances where documents actually refer to the need to cast shares into holdings on the ground. Coupar Angus abbey provides us with a number of explicitly documented instances of touns on the south-eastern edges of the Highlands being subdivided into small units during the fifteenth and sixteenth centuries. A tack issued in 1473, for instance, required the eleven tenants to divide the toun 'into four touns, or at least three',[103] as if the precise number was optional. In other cases, we can conclude that the option to divide shares into separate touns had been exercised from the way in which what appeared previously to be a consolidated toun appears in rentals as a split toun, like Lianach (Perthshire), which suddenly appears in a late seventeenth-century rental as Easter and Wester Lianach,[104] though there is always the possibility that such rental entries were simply formalising what had long been an informal arrangement. In terms of the everyday routines, splitting was not an exceptional event. It was a possibility whenever multiple tenants confronted the need to divide out shares. For the same reason, just as some tenants may have opted to split the toun, others may have taken a previously split toun and laid it out as an integrated one. In the case of the split Gairloch touns mentioned earlier, for instance, a rental dated to between 1660 and 1670 still refers to 'Engdadill more' and 'Engdagill glas' and to 'Sand, Little and Meikle' but the two 'Nastis' are now listed as a single toun, called Naist, held by three tenants and a cottar.[105]

The Problem of Runrig

Once they had become the basis for granting land, assessments were not simply a means of assigning crown rents and burdens to landowners. Because external liabilities had to be broken down and burdened on those who actually worked the soil, assessments became internalised into the organisation of estates as the basis for setting land to tenants. The large, standardised scale of most blocks of assessment meant that few were both set entirely to, and worked entirely by, one tenant, at least not once population growth increased the demand for land. In practice, many soon became shared between tenants, subtenants and cottars. Further, the fact that they constituted a finite or assessed amount of land meant that this multiple occupancy involved the interest of each tenant being set against those of others present, with each seen as having an aliquot share. This was the case whether tenants held a number of units of assessment out of a larger block of assessment (for example, a pennyland out of a Five Pennyland block or a ploughland out of a davoch) or explicitly as a proportional share of the toun (that is, a quarter, third, half, and so on). The combination of these two factors, the large scale of assessments and their occupation by multiple tenants through aliquot shares, are why assessments provided the seed out of which the farming toun grew.

We can look more closely at the internal character of touns by using estate rentals that become available from the fifteenth century. They are not an ideal source but they are an available source. Their prime weakness has to do with the fact that they are not always complete as regards occupiers, owing to the prevalence of subletting. The practice was present everywhere but it was especially significant in those parts of the Hebrides and along parts of the western seaboard where touns were set to a tacksman who took responsibility for setting the toun or touns out to tenants and collecting the rent on behalf of the landowner.[106] Only where listings of all occupiers are also available are we able to see just how many may have been present. This potential understatement by rentals clearly has a bearing on how we read any systematic data gleaned from them. Overall, if we take a sample of sixteenth-century rentals, such as those available for the South-west Highlands and Islands, and if we take their data at face value, it would actually suggest that touns were overwhelmingly held by single tenants, the only exceptions being those on Tiree and, to a lesser extent, those in Aros on Mull.[107] Given the presence of tacksmen, however, we must allow a significant proportion of those touns recorded as let to single tenants as sublet, with a hidden community of tenants and cottars who worked the soil. Even those touns kept 'in hand' by tacksmen would have had parts that were sublet. A 1652 rental for Tiree reveals the sorts of practices involved, with touns such as the *tirunga* of Hynish set to one tenant but 'a quarter set on subtenants' while the half *tirunga* of Mannall was likewise set to one tenant and the 'oyr quarter was on small tenants',[108] the sub- or small tenants providing labour for the tacksmen. In other words, what appears in rentals as held by a single tenant may, in reality, have been a hierarchy of tacksmen, principal tenants, subtenants and cottars.

If we take those visibly in the hands of multiple tenants, we can find a mix of touns, the average having between two and six tenants, plus subtenants and cottars, and the larger ones having as many as ten to twenty tenants plus subtenants and cottars. On balance, larger touns were more numerous in the Hebrides, especially on Tiree and in the Western Isles, than on the mainland, and certainly more than in the Central Highlands.[109] Establishing the importance of multiple tenancy is a necessary first step in understanding the role played by runrig in the traditional Highland landscape. As noted above, the presence of two or more tenants or subtenants in a toun meant that each person's portion was seen as a *proportionate* or aliquot share of the whole, with each share matched not just as regards its extent but also as regards its value. This notion of each unit of land assessment having a double-edged equality was implicit in the nature of land assessments. It is a point made not just by the common use of aliquot shares to set touns but also by the way written tacks refer to shares being 'just and equal' in extent and value.[110] In the Northern Isles, there was even a special term, *yarromana*, that captured the expectation that each portion should be equal 'in regard to its extent and produce'.[111] It was an interpretation that divisions invariably upheld, as when a share of two merks was 'designit, meithit, and merchit' in the toun of Buay (Orkney) 'als mekill in quantitie and qualitie' in 1612–13.[112] Of course, this does not mean that touns were communities of absolute equality with each landholder holding the same amount of land. Yes, we can find instances in which each landholder did hold an equal amount, be it a half, third, or quarter each, and so on[113] but we also find many examples in which holdings represented unequal shares of the toun. On Tiree, for instance, where some rentals show touns set in terms of a unitary measure known as the mail, the 1662 rental lists the five tenants at Killenarg as holding 8, 6, 3, 2, and 1 mail, while the thirteen tenants listed as holding Caillig held 12, 12, 8, 5, 1½ , 1, 1, 1, 3, 1½, 2, 2½, and ½ each.[114] In these circumstances, what mattered is that a person holding, say, a half would always be seen as holding twice as much as someone holding a quarter, twice the extent and twice the value.

The laying out of holdings in the form of intermixed strips, or open fields, stems from this shareholding basis to multiple-tenant touns. The use of terms such as runrig, *yarromana* and rendal has encouraged the view that there must be something distinct about the layouts created. The probable explanation lies not just in the way the tenure of multiple-tenant touns was structured around shareholding but also in the fact that the majority of touns were held by tenants on short, temporary tenures. Many along the western seaboard and in the Hebrides actually had no written tack but simply held land on a year-by-year basis by verbal agreement, or were tenants at will. Further, there were regular changes over who had shares, and how much. In other words, the question of how notional or abstract shares were to be cast into real holdings on the ground was continually revisited. It was this continual renewal of runrig, as shares divided out on the ground, that explains why terms such as runrig had such currency.

Most communities used formalised procedures for allocating shares. One such system stands out because of its symbolic character, that of sun division. This was a method for implementing a division, either into separate holdings or into runrig, that was based on the movement of the sun. It was found along the eastern and southern edges of the Highlands as well as in the adjacent Lowlands.[115] It is manifest in early charters and rentals through references to sunny and shadow shares, as with the 1592 reference to 'the west half of the sunny half of the lands of Wester Ennoch, extending to a fourth part of the lands of Wester Ennoch' (Balmacreuquhy barony),[116] or the grant for 'six bovates of the sunny half of the lands of Crabstoune', Strathdon.[117] The side-by-side use of descriptions such as 'sunny' with 'wester' is not unconnected. In practice, prefixes such as sunny/shadow, easter/wester, over/nether, great/little or their Gaelic equivalent *mòr/beag* were used in an equivalent way. If we want to understand why prefixes like easter/wester and nether/over were so widely used in Scotland, we need to understand how it fitted in to this symbolic scheme of land division.[118] This is confirmed by early law codes that make it clear that the sunny/shadow, easter/wester, over/nether and great/little or *mòr/beag* shares of a toun were one and the same distinction and involved the same process. When used to divide land, whether into separate touns or into a runrig layout, those with sunny/easter/over/great shares were given the strips or land that lay to the east or south in any allocation while those with the shadow/wester/nether or little portions were given the strips or land to the west or north.[119] The extent to which these were equivalent pairings in the mind of communities is underlined by the examples that can be found of switches or shifts between them. Often, the choice occurs in a single reference, such as with the 1698 reference to 'Easter Ortoun or Little Ortoun' in Elgin.[120] Sometimes, it can be pieced together from rental sequences. The two Scaristas on Harris are listed in rentals as Scarista More and Scarista Beg but also in their English forms of Scarista Meikle and Little or North and South Scarista. One, that for 1724, makes their equivalence clear, referring to 'Scarista North or Meikle Scarista'.[121] Likewise, in Assynt, what is referred to as Upper and Neather Inverkirkaig in 1681 was recast as East and West Inverkirkaig by the early 1700s.[122] It was not even necessary for tenants to share the same form of opposition, as when one tenant was set the 'western toun' of Achinarrow (Strathavon) in 1712 rental, while the tenant of the other half was set the 'upper toun'.[123] Consistency was restored with the toun's survey in 1761, with the two separate halves now being 'Upper Auchnarrow' and 'Nether Auchnarrow'.[124]

Assessments: an Institutional Framework for Farming?

Though some have offered a contrary view,[125] the interpretation upheld here is that the land assessment of touns referred first and foremost to their arable. When we find references to the 'meithing and marching' of land assessments, it referred to the 'meithing and marching', the 'lining out', of their arable. Access to other resources, such as pasture, were appendaged rights, exploited through and, therefore, proportioned to, the possession of assessed arable. Of course, given the nature of many Highland ecologies, it would have been difficult to lay assessments as a continuous block of arable. In fact, quite apart from grassy balks between rigs, Highland arable was usually very irregular, interspersed with patches of ground too wet or stony to be cultivated. This is why in some areas we find assessments interpreted not as denoting a specific acreage but as so many bolls sowing. Inevitably, in such ecologies, any head dyke thrown around such arable necessarily embraced patches of non-arable. The inclusion of grass within the head dyke, however, may have been prompted by more than the practicalities of enclosing arable and arable alone. Touns need to protect meadow and wintering ground no less than arable by excluding stock. In a comparison of touns in south-west Norway with those of the Northern Isles, Stylegar drew on Rønneseth's work to make the point that, when early touns there were 'matriculated' (= assessed), a process that saw the introduction of open fields, head dykes were arranged so as to embrace not just assessed arable but also the meadow needed to sustain arable via nutrient transfers.[126] Highland touns, as Stylegar makes clear, would have had a similar need for winter feed.[127]

In other words, the precise meaning of assessments as the core arable of a toun, its infield, may not have translated easily into 'the land within its head-dyke'. Assessed land may have been meithed and marched as the land that was 'plowable and bedewable' as one Caithness source put it,[128] but, once a head dyke was thrown around it and around any pasture that lay among it, it lost something of its precision. Later sources acknowledged this potential lack of clarity. As the Napier Commission recorded on Shetland, the 'merks of Shetland are generally considered – though the point is not free from doubt – to have been originally the arable only'.[129] The lack of clarity worked itself out in other ways. If anyone 'buy a piece of Land', wrote Brand about Caithness, *c.*1700, 'only what is Arable is Accounted for, as what serveth for Pasture, they were not to take notice of'.[130] What he meant is not that pasture did not have value but that one's use rights in it were calculated through possession of arable.

It is this framing by assessments, and not just the fact that it was manured with winter dung and cropped continually, that makes the late medieval or early modern infields different from what I defined earlier in a late prehistoric context as proto-infield. More than any other characteristic, it is this framework that imparted the sense of land with*in* and with*out*. The linking, however, of arable with what had been 'meithed and marched' on the ground as assessed land does not mean that assessed arable dated from when land assessments were first imposed. A great deal of arable was likely to be old arable, land that had long been cultivated. What the imposition of land assessments signalled was a change in its status.

It will already be clear from what has been said that we cannot assume that, when land assessments were first devised, they fitted neatly around the holdings or arable that existed at this point. Some would have had a rough-and-ready fit to what was on the ground. The blocky nature of a top-down assessment, coupled with the variegated nature of Highland arable, made such compromises inevitable. The way some assessments embrace multiple touns hints at this. So too, does the presence of dispersed foci of settlement. In the Northern Isles, the conclusion reached by writers such as Thomson was that the early toun was fragmented between different holdings or houses, a conclusion that he saw as applicable to other parts of the Highlands.[131] Gathering together more than one holding under blocks of assessment would have had the effect of binding together the interests of such holdings, with their share of arable, pasture, rents, and so on now having to be negotiated and renegotiated at intervals among all those covered by the assessment. In short, previously detached holdings now leant into each other.

If we assume that assessments burdened on the person or family were being transposed into land assessments by the tenth to eleventh centuries, then there is another aspect to their imposition that needs to be considered. How did they cope with a growth in numbers and the expansion of arable, such as one had during the eleventh to thirteenth centuries? In such circumstances, we can envisage pressure for a revision of assessments but the evidence that we have suggests that, for most of their existence in active use, assessment rating – if not what they denoted – endured unchanged.[132] This rigidity meant that growth had to find a work-around. Three kinds of adjustment were possible. First, as the productive arable core of the toun, we can expect assessed land to have absorbed the initial pressure for expansion through an intensification of output. If infields were not already cultivated intensively, that is, continuously without a break and using all the manure accumulated over winter, then we can expect the caging effect of assessments on growth to have encouraged such a strategy. Second, infield, *sensu stricto*, comprised assessed arable but, in practice, it became equated with all that lay within the head dyke and would have involved meadow and wintering ground. Under pressure for growth, any non-arable within the head dyke would have offered communities scope for expansion. Such expansion, though, carried risks for it was at the expense of meadow and wintering ground, thereby reducing the flow of nutrients needed for such expansion.

Both the strategies just outlined had obvious limitations. In time, growth beyond the head dyke would have been the compelling choice. This third strategy was not a straightforward step, given the implicit boundedness of assessments. The imposition of New Extent in the late fifteenth century and clues, such as the rerating of the davoch from two to four ploughs, suggest that touns may have negotiated around the inertia of their assessment through a revaluation of what assessment units comprised. Yet, whatever the circumstances, we know that touns eventually did push arable outwards beyond their infield assessment for that is precisely what outfield stood for, the land outside what had been the assessed core of the toun. Far from being a sector of great antiquity, outfield can be dated to those phases of growth that post-date the imposition of land assessments. This would put its first emergence at any time from the eleventh to twelfth century onwards. The dearth of data means we cannot fix precisely when but there are clues over its continuing expansion in the post-medieval period. Reference has already been made to Gordon of Straloch's late sixteenth-century comment about arable undergoing a significant geographical expansion.[133] The expansion of arable did not end there for mid-eighteenth-century sources for the eastern Highlands reveal abundant signs of new arable, some labelled as outfield. In Strathavon, almost every davoch had 'new' arable recorded somewhere in its bounds when surveyed in 1762.[134] In some cases, the survey makes it clear that what was colonised was 'improv'd outfield', as at Auchindoun.[135] More usually, we simply find isolated patches of arable on the edges of touns labelled as 'new land' or as 'improvement' as in the touns of Tominlienin and Belno.[136] These patches of new land shaded into the welter of detached intakes and outsets that emerged over the post-medieval period as separate touns. A 1773 survey of the Lordship of Achindown in Mortlach parish, Banffshire, for instance, refers to new intakes such as the 'New Improvement called Crossmaul' and one simply called 'New Tack'.[137]

Of course, outfield was not just land of a different status. It was exploited in a different way. Why communities did not just expand infield but exploited outfield in a different way can be grasped by looking at the pattern of resource use at the point when expansion beyond infield commenced. Nutrient levels in infield depended on the manure that had accumulated over the winter, composted with old thatch, discarded roofing and wall turf, bedding material, and so on, being added in spring plus that provided by the direct dunging by stock when grazing the harvest stubble. At this point, expanding infield at the expense of the remaining hained meadow or grassland would have meant a diminished supply of winter feed and manure spread over a still larger area of arable, a strategy that would have led ultimately to declining yields. The alternative was to use the manure otherwise lost when stock grazed on the hill ground over summer. The manuring of what became outfield depended on tathing, with the cattle being folded over night during summer on that part chosen for cultivation the following spring (Figure 4.8). Naturally, once in cultivation, tathing – being a summer task – was necessarily moved to a new fold. In short, what the formation of outfield did, its innovatory aspect, was to create an all-year-round use of manure, enabling arable to expand by increasing the use of available manure.

While infield–outfield systems of cropping can be documented across the region, there were parts of the Hebrides that practised an outfield-only system. Each year, only a portion, such as a third, of cultivable land was cultivated, the rest left under grass, with each cultivated portion cropped for three, four or more years, then left as much as twice that long under grass. Where intensified, as much as a half of cultivable land might be under cultivation in any one year before being rested under grass. Such flexibility sat uncomfortably with assessments. Two factors may have had a role here. First, though some islands, such

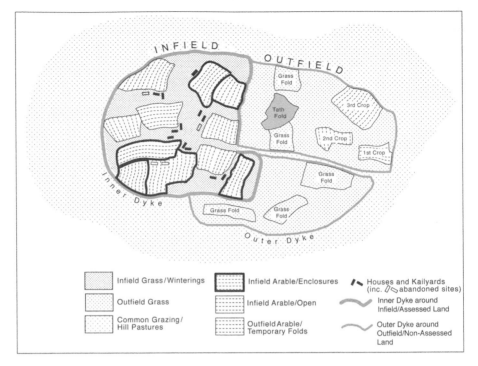

Figure 4.8 Abstract representation of an infield–outfield system.

as the Uists, had some scope for laying out fairly continuous blocks of arable, there were many other Hebridean areas where the ground was too broken or waterlogged for this. These were the areas where estates assessed arable as so many bolls of sowing so the caging effect of assessments was not the same.[138] Second, many Hebridean townships, such as those in the Northern Isles, relied more on seaweed than on animal dung as their primary manure. Seaweed was effective but more short-lived in its effects so would not have been able to sustain the continuous cropping of an infield in the same way as animal dung.

Patently, whether infield or outfield, the key to arable was the need to maintain a flow of nutrients to it. In the mind of Highland communities, arable and manured land were perceived as one and the same thing. Monro drew the connection between the two in the Hebrides, referring to arable always as 'manurit land'[139] while a 1605 division of rendal or runrig land between two landowners at Gravis in Orkney made the same point when it instructed the six honest men charged with dividing land 'with schaft and lyne' so that each had his share of 'the manwrit land'.[140] Cultivated land was not simply about harvesting its natural fertility but about reaping the rewards of what you put in by way of manure. This comes across forcibly from the act passed in 1609 by the barony court for Toyer and Disher on the Breadalbane estate which required that each tenant was to enrich each merkland with thirty-four loads of soil, a transfer seen as for the 'guiding of the land'.[141] The effectiveness of the act is unclear but some tenants were subsequently fined for not carrying out the 'guiding' of their land as instructed.[142] Just how essential it was for land to be 'guided' or manured in some way is emphasised by a 1699 note relating to a new tack for four tenants in the toun of Balnasuim, on the Breadalbane estate. Conveying the idea that its arable was not currently under cultivation, and that it might take two years to bring it back into use or occupation, it simply said that 'the said lands are out of gooding'.[143] The same idea comes across from a 1627 reference to the parish of Nesting, Shetland, where, it was said, 'ilk rowme is constantlie rentallit and payis accordinglie', provided 'it be manured'.[144]

The grazing resources of the typical toun lay intermixed with its arable as well as beyond its head dyke. Regulation of access to these grazings was through the soums of a toun. Precise weightings varied but it was common for two horses to be equated to a soum, one cow to a single soum, and ten sheep to one soum, though some areas equated only five sheep to a soum, and, where kept, ten goats to a soum.[145] In each case, the weighting of soums was taken to include followers. While soums were standardised units, enabling stock to be set against each other in a flexible way, some estates tried to fix their exact composition in terms of exactly how many horses, cows, sheep and goats were to be kept. Logically, each tenant's share of soums was proportionate to his or her share of the toun as a whole. In origin, souming was as much about how much was needed to sustain arable as about what grazings were available. The way some estates maintained an exact relationship between how many bolls could be sown and how many soums kept per unit of land assessment makes this point clear. On Tiree, for example, each *tirunga* as was initially rated as 48 bolls of sowing and 72 soums. By the mid-eighteenth century, as elsewhere in the region, their relationship had weakened as some touns began to expand their stocking levels at the expense of arable.[146]

The movement of stock was a closely regulated part of the toun economy. Touns had to protect arable during the growing season, hain pasture for winterings or hay, provide a grass supply for the support of tathing, ensure stock was taken to the shielings and brought back at the right time, herd stock in the tathfolds overnight, regulate access to harvest stubble and prevent stock damage to woodland. For all these reasons, the seasonal movement of stock was orchestrated, though a common complaint was that touns allowed stock to graze freely in winter not just across their own land but on that of neighbouring touns. The extent of the problem is highlighted by the 1686 act anent winter herding which required touns to herd their stock more closely in winter.[147] Their movements otherwise were normally embodied in barony court regulations. The most notable were those controlling the movement of stock to the shielings in early summer and the return of some, first to the tath-folds and then to graze the harvest stubble. The court books for Menzies and

Rannoch provide a typical example, a 1660 act requiring 'everie tenant within the countrey to goe to the shealls with ther goods, the tenth day of May everie yeir', to 'keep ther that grass for ane monethes space', 'to set out ther folds conform to use and want', and 'to putt thir goods therin for tatheing ther ley land'.[148]

For all their remoteness, shielings were a dynamic part of the Highland landscape. As well as enabling communities to take advantage of the short summer growth of the higher hill pastures, they removed stock from around arable during the growing season and from meadow and wintering ground that needed to be hained. They were also the moment in the calendar for making butter and cheese. There was, however, another function. They functioned as a safety valve, with a fair proportion being upgraded into permanent settlements when the need arose. Upgrading was a logical step for, over time, the nightly concentration of stock around the shieling site created a local store of nutrients, often evident through its residual greening effect. Mather's study of Affraick Forest illustrates the processes involved in such colonisation with, from the late sixteenth century onwards, permanent settlements being established further and further into Glen Strathfarrar at the expense of shielings, reaching Glentilt by the late seventeenth century.[149] A similar upgrading of sites was in progress on the higher grazing grounds of Strathavon from the late seventeenth century onwards. The Gordon estate actively encouraged the process, issuing tacks that required farmers to cultivate part of their shieling ground. In time, the shiels were detached and established as separate farms.[150] To the south, we find further signs of how new 'edge' landscapes were still very much in the making over the early to middle decades of the eighteenth century in the Braes of Mar. In a dispute that flared up in the 1760s over lands in Glens Calder and Corievou, which had previously been shieling ground but part of which had become settled, it was claimed that 'it is most certain that the Cultivation is greatly increased in this Country within the Memory of Man, and many Tacks and Possessions have in that Time been either newly set down, or greatly enlarged'.[151] Further west still, on the rugged ground that bordered Locheil, a survey in 1770 shows some shiels caught in an in-between state, being still classed as shielings but having permanent settlement and arable. Ardnosh, the shieling for Invernally, fell into this category. By the time of the survey, it boasted a small amount of arable that was worked by 'some poor people' who 'live there all the Year manured by ferns and cropped with Oats'.[152]

In an analysis of sites in the southern Highlands, Bil documented other instances of shiels being upgraded into permanent settlements. He cites the example of Laidchrosk which, in 1611, was a shieling of Eister Stuikis. By 1665, it had become a pendicle, a term for a toun created out of another.[153] How elaborate the process could be is illustrated by Dalnatarony in Upper Glengarry. Recorded as a pendicle in 1632, it had probably started out as a shieling. By 1723, it had itself acquired a pendicle, created out of available hill grazing.[154] Speaking generally, Bil makes two points that add materially to the debate. First, while the conversion of shielings into permanent touns is well documented during the seventeenth and eighteenth centuries, the process has much deeper roots. He sees it as a regular process of settlement expansion from the medieval period onwards, with the seasonal use of lower and better sites as shielings being but a stepping stone to a more permanent occupation.[155] Second, he does not see it as a one-way process. The divide between the permanent settlement pattern and shieling was at the interface between what was within the limits of permanent settlement and what was at the edge of these limits, an edge constantly being redrawn with climate change. He raises the possibility that one or two of the shielings that we see being used over the seventeenth and eighteenth centuries may have been permanent settlements at an earlier date. He cites the example of Slowmanowie in Upper Strathearn. Referred to in 1583 and before as if it was a settled farm, 'the lands of Slowmanowie', it appears in 1631 as the 'scheling of Slowmanowie'.[156] To this, we can add the examples of Sallachul in Lochaber cited earlier.

Of course, new settlements were not just restricted to the upgrading of shieling sites. Other forms of colonisation can be recognised. An important process of colonisation in some areas focused on the settlement of former hunting forests. Some, around the eastern edge of the Highlands, like those of Cluny, Birse and Cabrach, were being opened up to settlement by the fifteenth and sixteenth centuries. The process started with their formal deforestation. When the Earl of Huntly was given Cabrach in 1508, he was given the 'power either to turn the lands into arable ground or to keep it as Forrest according to the Forrest Law'.[157] The earl took the first of these options so that, by the end of the century, touns such as Elrick and Greauche had been established on what had been forest land. North of the Great Glen, it was the deforestation of Affraick that led to the progressive push of, first shielings, then permanent settlement into its reserves of land during the fifteenth to seventeenth centuries.[158] Not all forests succumbed to the pressures of colonisation so early. Old established forests, such as Carrick in Argyll, and ones established as late as the fifteenth century, such as Mamlorne in Perthshire, continued in use. Mamlorne was still used as a forest down to the eighteenth century, though part was used as shieling ground by touns in Glen Lyon and a small number of what the Breadalbane estate called its 'Forest' touns, such as Clashgour.[159] By this point, the profitability of stock farming meant that, even if forest land offered limited prospect for arable, it could still yield rent for use by stock farms. This is well shown by a mid-eighteenth-century memorial on the Forest of Glenavon. It began by saying that 'Of Late there were many New Grassings Open'd & the great Charge attends the Management of the Forest nothing can be made by keeping it in the Family's hands'. From its own figures, this encroachment of cattle grazing had started *c*.1700.[160]

I have tried to show that the great change of the medieval and early modern period was the imposition of land assessments and the primary role which they played in shaping the institutional framework around which the farming toun was based. The toun's existence as a community of farmers, the importance of runrig to its layout and even the role played by infield–outfield farming can all be linked in different ways to how land assessments shaped this institutional framework. We still have much to research on these aspects, however, before we can claim a secure understanding of how and when this institutional framework emerged, research that must make greater use of the archaeological potential of the period as well as documentary sources. Arguably, fitting these two types of data together offers the best prospect for genuinely fresh insights on how, why and when the toun developed, fresh insights that draw on different pathways of research. We shall only be able to claim a secure understanding of such landscapes when we appreciate how these different sources fit together.

Notes

1. M. Flinn, J. Gillespie, N. Hill, A. Maxwell, R. Mitchison and T. C. Smout, *Scottish Population History from the Seventeenth Century to the 1930s* (Edinburgh, 1977), p. 111.

2. For example, K. J. Cullen, *Famine in Scotland: The 'Ill Years' of the 1690s* (Edinburgh, 2010), pp. 123–56.

3. P. J. Fowler, *The Farming of Prehistoric Britain* (Cambridge, 1983), p. 34; M. Livi Bacci, *The Population of Europe* (London, 2000), p. 6.

4. See comments of Hunter, 'Pool, Sanday', pp. 192–3; MacKie, 'The broch cultures … part 2', pp. 110–11.

5. M. Parry, *Climate Change, Agriculture and Settlement* (Folkestone, 1978), pp. 95–111.

6. Its effects were, in fact, worse in Europe, see Livi Bacci, *Population of Europe*, p. 81.

7. A. McKerral, *Kintyre in the Seventeenth Century* (Edinburgh, 1948), pp. 76, 78.

8. Cullen, *Famine*, p. 188; see also, Flinn et al., *Scottish Population History*, pp. 164–86.

9. Flinn et al., *Scottish Population History*, p. 9.

10. Ibid., pp. 209–16.

11. Ibid., pp. 12, 235.

12. R. A. Dodgshon, D. D. Gilbertson, and J. P. Grattan, 2000, 'Endemic stress, farming communities and the influence of volcanic eruptions in the Scottish Highlands', in W. G. McGuire, D. R. Griffiths, P. L. Hancock, and I. S. Stewart, (eds) *The Archaeology of Geological Catastrophes*, Geological Society, London Special Publications, no. 71, pp. 267–80.

13. M. Gray, *The Highland Economy 1750–1850* (Edinburgh, 1957), pp. 57–66.

14. Data drawn from Kyd, 1952, and *Census of Scotland*, 1841.

15. M. Macklin et al., 'Human–environmental interactions', p. 117.
16. Dawson, *So Foul and Fair*, pp. 100–1
17. Ibid., pp. 101–27.
18. H. Lamb, 'The Little Ice Age period and the great storms within it', in M. J. Tooley and G. M. Sheal (eds), *The Climatic Scene* (London, 1988), p. 129.
19. R. A. Dodgshon, 'The Little Ice Age in the Scottish Highlands and Islands', *SGM*, 121 (2005), pp. 321–37.
20. http://www.nls.uk/pont./specialist/pont 13.tml; NAS GD44/51/743/7.
21. Lamb, *Little Ice Age*, p. 129.
22. A. G. Dawson et al., 'Complex North Atlantic Oscillation (NAO). Index signal of historic North Atlantic storm-track changes', *The Holocene*, 12 (2002), pp. 363–9; A. G. Dawson et al., 'Late-Holocene North Atlantic climate "seesaws", storminess changes and Greenland ice sheet (GISP2) palaeoclimates', *The Holocene*, 13 (2003), pp. 381–2.
23. McNeill (ed.), *Exchequer Rolls*, pp. 647–8; J. R. N. Macphail (ed.), *Highland Papers*, vol. 1, *SHS*, 2nd series, v (Edinburgh, 1914), pp. 277–85, 311–16.
24. NAS, GD170/629/61; GD112/9/35.
25. J. R. N. Macphail (ed.), *Highland Papers* vol. III, *SHS*, 2nd ser., xx (Edinburgh, 1920), pp. 72–3.
26. McKerral, *Kintyre*, p. 78.
27. Ibid., pp. 74–8.
28. Ibid., p. 76.
29. McKay (ed.), *Walker's Report*, p. 54.
30. DC, MDP, 1/466/22.
31. DC, MDP, 2/487/10; ibid., 2/487/12.
32. DC, MDP, 1/466/22 and NLS, Map of Harris, Bald, 1804–5.
33. McKay (ed.), *Walker's Report*, p. 64.
34. Ibid., p. 64.
35. Ibid., p. 64.
36. NAS, FE, E656/1.
37. NAS, GD403/1/1–2.
38. CDC, LMP, GD221/5914; NAS, RH2/6/24; HULL, DDBM/27/3.
39. McNeill (ed.), *Exchequer Rolls*, pp. 614, 648.
40. IC, AC, vol. 65. Instructions . . . 23 October 1750
41. SRO, RHP 8826/2.
42. Duke of Argyll, *Crofts and Farms in the Hebrides* (Edinburgh, 1883), pp. 6–7.
43. IC, AP, vol. 65; NAS, RHP 8826/1 and 2.
44. B. T. Rumsby and M. Macklin, 'River response to the last neoglacial (the "Little Ice Age") in northern, western and central Europe', in J. Branson, A. G. Brown and K. J. Gregory (eds), *Global Continental Changes: the Context of Palaeohydrology*, Geological Society, special publications, no. 115 (1996), pp. 217–33.
45. For example, NAS, GD112/17/8.

46. Mackay, W. (ed.), *Chronicles of the Frasers, SHS*, xlvii (1995), p. 272.

47. NAS, FE, E745/59, 1771; *OSA*, 1791–99, ix, p. 487.

48. NAS, GD112/5/5/31–3, letter, 27 January 1791.

49. R. J. Adam (ed.), *John Home's Survey of Assynt, SHS*, 3rd series, 52 (Edinburgh, 1960), p. 16; *OSA*, 1791–99, xi, p. 206, and xii, p. 320.

50. NAS, FE, E783/98; V. Wills (ed.), *Reports on the Annexed Estates 1755–56* (Edinburgh, 1973), p. 77; M. M. McArthur (ed.), *Survey of Lochtayside 1769, SHS*, 3rd series, xxvii (Edinburgh, 1936), p. 4.

51. NAS, GD50/135.

52. NAS, GD44/25/2/76.

53. NAS, GD112/112/17/8; *OSA*, 1791–99, viii, p. 568.

54. *OSA*, 1791–99, vi, p. 249.

55. R. C. MacLeod, *The Book of Dunvegan*, Spalding Club (Aberdeen, 1939), ii, p. 85.

56. DC, MDP, 2/488/18; 1/380/28.

57. *OSA*, 1791–99, iv, p. 559.

58. T. Charles-Edwards, *Early Irish and Welsh Kinship* (Oxford, 1993), pp. 71–7.

59. C. Withers, *Gaelic Scotland: The Transformation of a Culture Region* (London, 1988), pp. 77–8; A. I. Macinnes, *Clanship, Commerce and the House of Stuart, 1603–1788* (East Linton, 1996), pp. 5–6.

60. A. Fenton, *The Northern Isles: Orkney and Shetland* (Edinburgh, 1978), pp. 21–3; Thomson, *Orkney*, pp. 311–14.

61. Fenton, *Northern Isles*, p. 22.

62. S. Storer Clouston (ed.), *Records of the Earldom of Orkney, 1299–1614, SHS*, 2nd series, vii (Edinburgh, 1914), p. lii; Fenton, *Northern Isles*, p. 22.

63. Thomson, *Orkney*, pp. 313–14.

64. Barrow, *Kingdom of the Scots*, esp. pp. 373–83; see also, Crawford and Ballin Smith, *The Biggings*, pp. 14–16.

65. McKerral, 'The tacksman', pp. 10–25; E. Cregeen, 'The changing role of the House of Argyll and the Highlands', in I. M. Lewis (ed.), *History and Social Anthropology* (London, 1970), pp. 153–90; Macinnes, *Clanship*, pp. 1–24.

66. Skene, *Celtic Scotland*, iii, pp. 428–40; see also, Thomson, *Orkney*, pp. 211–12.

67. McNeill (ed.), *Exchequer Rolls*, xvii, pp. 625–33.

68. Ibid., pp. 622–5, 643–5, 648–9.

69. R. A. Dodgshon, *From Chiefs to Landlords* (Edinburgh, 1998), pp. 105–7.

70. Ibid., pp. 107–17, 234–7; I. D. Whyte, *Agriculture and Society in Seventeenth-Century Scotland* (Edinburgh, 1979), pp. 228–34.

71. Ibid., pp. 228–42.

72. Ibid., pp.73–6.

73. For example, B. Smith, *Toons and Tenants. Settlement and Society in Shetland 1299–1899* (Lerwick, 2000), pp. 1–15.

74. Ross, 'The dabhach', p. 66.

75. R. A. Dodgshon, 'West Highland and Hebridean settlement prior to crofting and the Clearances: a study in stability or change?', *PSAS*, 123 (1993), pp. 424–5.

76. A. Geddes, 'Conjoint tenants and tacksmen in the isle of Lewis, 1715–26', *Economic History Review*, 2nd series, 1 (1948–49), pp. 54–60.

77. Shaw, *Northern and Western Islands*, pp. 84–5.

78. Ibid., pp. 85–6.

79. R.A. Gailey, 'The evolution of Highland rural settlement', *SS*, 6 (1962), pp. 155–77.

80. Crawford and Ballin Smith, *The Biggins*, pp. 55–6; Smith, *Toons*, pp. 1–15.

81. Thomson, *Orkney*, p. 107.

82. Ibid., p. 115.

83. Ibid., p. 115.

84. Ibid., p. 115.

85. Smith, *Toons*, pp. 3–4.

86. Crawford and Ballin Smith, *The Biggins*, p. 29.

87. Thomson, *Orkney*, pp. 124–5; see also, F-A. Stylegar, 'Township and Gard: a comparative study of some traditional settlement patterns in Southwest Norway and the Northern Isles', *The New Orkney Antiquarian Journal*, 4 (2000), p. 30.

88. A survey is provided by R. A. Dodgshon, 'Changes in Scottish township organization during the medieval and early modern periods', *Geografiska Annaler*, 55 (1977), series B, pp. 51–65.

89. McNeill (ed.), *Exchequer Rolls*, p. 624.

90. Innes, *Origines Parochiales*, 1855, vol. ii, part ii, pp. 406–7; see also ibid., 1854, vol. ii, part i, pp. 3–4, 191.

91. J. Storer Clouston, 'The Orkney Townships', *SHS*, xxvii (1920), p. 37; Storer Clouston, *Records*, p. lxi.

92. Ibid., pp. 347–9.

93. Ibid., p. 174.

94. Thomson, *Orkney*, p. 53.

95. NAS, GD24/1/146.

96. NAS, Inventory of the Title Deeds of the Estate of Easter Moniack, 1796.

97. NAS, GD16/5/126.

98. M. Bangor-Jones, 'Mackenzie families of the Barony of Lochbroom', in J. R. Baldwin (ed.), *Peoples and Settlement in North-west Ross* (Edinburgh, 1994), p. 82.

99. Ibid., p. 85.

100. A. Mitchell (ed.), *Macfarlane's Geographical Collections*, vol. ii, *SHS*, 53 (1907), p. 272.

101. Ibid., ii, p. 272.

102. *Miscellaneous*, Spalding Club, vol. 4, 1849, pp. 289–90.

103. A. Rogers, *Rental Book of Cistercian Abbey of Cupar Angus* (London, 1880), i, p.188. See also, *Miscellaneous*, v, 1852, p. 226.

104. J. Stewart, *The Settlements of Western Perthshire. Land and Society North of the Highland Line 1480–1851* (Edinburgh, 1990), p. 132.

105. W. Macgill, *Old Ross-shire and Scotland* (Inverness, 1909), i, p. 162–3.

106. McKerral, 'Tacksman', 1947, pp. 10–25; Cregeen, 'The changing role . . . Highlands', pp.153–90.

107. Dodgshon, *Chiefs to Landlords*, p. 126.

108. IC, AM, Book 2531, Rentall of Tirie . . . p. 52.

109. Dodgshon, *Chiefs to Landlords*, pp. 125–35.

110. That is, DC, MDP, 2/61.

111. Storer Clouston, *Records*, p. 167; J. E. Donaldson, *Caithness in the Eighteenth Century* (Edinburgh, 1958), p. 118.

112. R. S. Barclay (ed.), *The Court Book of Orkney and Shetland 1612–13* ((Kirkwall, 1962), p. 18.

113. For example, DC, MDP, 2/130; 2/490/14.

114. IC, AP, 2531, Rentall . . . Tirie, 1662.

115. R. A. Dodgshon, 'Scandinavian "solskifte" and the sunwise division of land in Eastern Scotland', *SS*, 19 (1975), pp. 1–14.

116. J. Anderson (ed.), *Calendar of the Laing Charters*, A.D. 500 to 1286 (Edinburgh, 1899), p. 307.

117. Robertson, *Illustrations*, iv, p. 782.

118. Dodgshon, 'Scandinavian "solskifte"', pp. 1–14.

119. Ibid., pp. 3–4, 11–12.

120. Anderson (ed.), *Calendar*, p. 688.

121. DC, MDP, 2/487/19.

122. M. Bangor-Jones, 'Settlement History of Assynt, Sutherland', in Atkinson et al. (eds), *Townships to Farmsteads*, BAR British ser. no. 293 (Oxford, 2000), p. 212.

123. NAS, GD44/51/745/1/1.

124. NAS, RHP2487/5.

125. Thomson, *Orkney*, pp. 206–7.

126. Stylegar, 'Township and Gard', p. 36; G. Rønneseth, 1975, 'Gard und Einfriedigung. Entwicklungsphasen der Agrarlandschaft Jærens', *Geografiska Annaler*, ser. B, special issue no.2 (1975).

127. Stylegar, 'Township and Gard', pp. 34, 36.

128. Macgill, *Old Ross-shire*, i, p. 371.

129. BPP, *Evidence taken by Her Majesty's Commissioners of Inquiry into the Condition of Crofters and Cottars in the Highlands and Islands* (London, 1884), 23, 1404. N.b., all text references based on pagination in Irish University Press BPP reprint, vols 22–5. See also, ibid., 1350: *NSA*, 1834–45, xv, pp. 13, 116–17; Smith, *Toons and Tenants*, p. 40.

130. J. Brand, A *Brief Description of Orkney, Zetland, Pightland Firth and Caithness* (Edinburgh, 1883), p. 225.

131. Thomson, *Orkney*, pp. 124–5; Stylegar, 'Township and Gard', p. 30.

132. Ross, 'The dabhach', p. 66.
133. Mitchell (ed.), *Geographical Collections*, ii, p. 272.
134. NAS, RHP 2488.
135. NAS, RHP 1824.
136. NAS, RHP 1793.
137. NAS, RHP 1824; RHP 1746.
138. Dodgshon, *Chiefs to Landlords*, pp. 167–78.
139. R. W. Munro (ed.), *Monro's Western Isles of Scotland* (Edinburgh, 1961), pp. 64–7.
140. Clouston (ed.), *Records*, p. 181.
141. NAS, GD112/17/4.
142. NAS, GD112/17/6.
143. NAS, GD112/10/10.
144. A. Macdonald(ed), *Reports on the State of Certain Parishes in Scotland*, 1835, Maitland club (Edinburgh, 1835), p.232.
145. NAS, FE, E729/9/1; IC, AM, V65, Turnbull survey, 1768; NAS, GD174/736.
146. Dodgshon, *Chiefs to Landlords*, p. 167.
147. APS, 1686, viii, p. 595.
148. NAS, GD50/136/1, vol. i, 11 and 12 July 1660; NAS, GD112/17/8; NAS, GD80/384/8, February 1723; NAS, GD461/27; A. Bil, *The Shieling 1600–1840. The Case of the Central Scottish Highlands* (Edinburgh, 1990), p. 183.
149. A. S. Mather, 'Pre-1745 land use and conservation in a Highland glen: an example from Glen Strathfarrar, North Inverness-shire', *SGM*, 86 (1970), p. 163–4.
150. V. Gaffney (ed.), *The Lordship of Strathavon*, Third Spalding Club (Aberdeen, 1960), pp. 28–33.
151. J. S. Mitchie (ed.), *The Records of Invercauld*, 1547–1828, Spalding Club (Aberdeen, 1901), p. 13.
152. NAS, GD50/2/2. See also, NAS, GD44/51/743/7 for the example of Sallachull.
153. Bil, *The Shieling*, p.264; Stewart, *Settlements of Western Perthshire*, pp. 82, 109.
154. Ibid., p. 264.
155. Ibid., pp. 263–4.
156. Ibid., p. 257.
157. NAS, GD 44/1/1/47.
158. Mather, 'Pre-1745 land use', pp. 160–9
159. NAS, GD112/59/22/2; ibid., GD112/59/59.
160. NAS, GD44/51/354/1/7.

ON THE EVE OF THE CHANGE:
A LOOK THROUGH THE SURVEYOR'S EYE

One of the linking themes of this book is that there has probably never been a period in its history when the Highland landscapes were not changing in some way. It was a landscape always in the making. This chapter cuts across this theme of continuing change by taking a more cross-sectional view but at a timely moment. It takes advantage of a growth in evidence, apparent by c.1700, but which becomes abundant by the mid-eighteenth century thanks to the compilation of detailed estate surveys, many of which combined written reports with cartographic coverage down to the level of the individual toun. Though compiled at different points between the mid-eighteenth and early nineteenth centuries, the various surveys and reports provide a detailed view of Highland landscape prior to the Clearances and the emergence of crofting townships. Indeed, many were a preparation for change. Through them, we can look out over the traditional landscape. Some would see this as a window on to an old, archaic landscape, whose settlement, landholding and farming take us back into late prehistory. While there may be elements of what we see which have survived from prehistory down into the early modern period, the link between the well-settled landscapes of late prehistory and the equally well-settled landscapes of the late eighteenth century was far from being a straight line, and certainly not a case of descent without modification. There were ways in which, compared to its prehistoric antecedents, it had become a different species of landscape by the mid-eighteenth century.

My discussion of what we see is arranged under three headings. First, I want to sample some of the surveys available for mainland estates, drawing out what they reveal about touns and how they were ordered. Second, I want to do the same for a sample of surveys for the Hebrides and Northern Isles. Finally, I want to take advantage of the richness of data available by this point to take a more systematic, close-focus view of the Highland toun at its point of disappearance.

On the Eve of Change

The Central and Western Highlands

Even by the seventeenth century, some Highland estates had started compiling simple, but systematic, lists of data relating to touns, their occupiers, number of bolls sown and stocking levels. By the mid-eighteenth century, a different sort of list becomes available. Some were intended to take fuller stock of the estate's resources at a time when landowners looked to capitalise on what they had. Others were more obviously designed as a basis for planning change. Still others were summary surveys compiled in preparation for the sale of the estate. Whatever their purpose, these surveys covered most aspects of the estate's resource endowment, describing each toun in terms of its occupiers and major land-use categories (for example, arable, hill ground, woodland). Adding to their value is the fact that many provide us with the first graphic illustration of how touns were laid out on the ground in terms of land use, settlement and boundaries. Altogether, these surveys were compiled over a timespan of seventy to eighty years; nevertheless, they show areas across the region at a consistent point: the eve of change.

The earliest of the eighteenth-century surveys was Sir Alexander Murray of Stanhope's *The Anatomie of the Parish & Barony of Ardnamoruchan and Swinard*, 1723. It listed what was present in terms of arable and stocking but used each toun's assessment in pennylands to measure arable, with no mention of acreage, nor was there any cartographic representation of touns. In a letter dated 1740, he acknowledged what was deficient. Future surveys, he suggested, should depict the layout of farms 'not only by Plans, but by Solids, representing the true Face and Aspects of the County'.[1] There was still a while to wait before Highland estates could be represented by 'solids' but some detailed estate plans were already being produced by the 1740s to 1750s, such as for parts of Islay produced in 1749–51 by Stephen MacDougall.[2] By the 1750s, they were supplemented by General Roy's large-scale military map.[3] Though Roy's survey conveys a sense of how the settled landscape of the toun and the semi-natural landscapes that lay beyond were interdigitated, there is much that it simply glosses over.[4] As the second half of the century unfolded, however, many of the larger estates on the mainland commissioned more detailed toun-by-toun surveys. One of the better documented estates in the southern Highlands is that of the Breadalbane estate. An extensive estate, it embraced areas such as Netherlorn and Glenorchy in Argyll and Strathfillan and Lochtayside in Perthshire. The Lochtayside portion was surveyed in 1769, the north side by John Farquharson and the south side by John McArthur.[5]

As one might expect, given their contrasting aspects, the two sides differed in character. Touns on the north side were generally larger in terms of the number of occupiers present, especially if one adds cottars and crofters. There were also differences as regards arable, with those on the north side averaging just over 45 acres (23 hectares) and those on the south side averaging just over 29 acres (15 hectares).[6] These figures combine both infield and outfield arable, sectors that were clearly defined on the Breadalbane estate though infield on the south side is labelled as croftland.[7] Broken down between the sectors, infield on the north side averaged 26 acres (13 hectares) while, on the south side, what the survey labels as croftland was 17 acres (9 hectares). As regards outfield, the average on the north side was 20 acres (10 hectares) while, on the south side, it was 12 acres (6 hectares).[8] From the data provided, arable appears to have been cropped with oats and bere, mostly at a ratio of four to one. There was also a significant difference between the two sides of the loch as regards meadow, with the north side having an average of just over 4 acres (2 hectares) whereas the Southside had an average of just over 14 acres (7 hectares) per toun. Meadow acreage, though, may not have been the only part of the equation. A list of cut hay available in the townships of Lochtayside,[9] admittedly for over fifty years earlier, shows the amounts involved, computed in terms of the number of *bolts*, were not noticeably greater in touns of the south side compared to those of the north side.

Farquharson's cartographic survey of north Lochtayside appears to have been a reasonably accurate representation, one that can be fitted to the framework of modern Ordnance Survey maps.[10] His depiction of infield, outfield and pasture provides a fine indication of how landscape in the main body of the Highlands was sectored prior to the Clearances. It also shows the discontinuous nature of arable, its surface broken by bogs, waterlogged soil and rock outcrops, and its edges fretted (see Figure 4.5). Such were the opportunity costs attached to patches of workable soils that we even find detached blocks of arable, especially outfield, at a distance from the core areas. The other noteworthy feature conveyed by Farquharson's survey is the amount of grazing that lay within the head dyke, not just across outfield but around infield. By its very nature, outfield would have been mainly under grass except for the portion that was under crop each year. Like meadow, most would have been hained, or saved for when stock were used for tathing.

Among the most detailed of the mid-eighteenth-century surveys are those relating to the Forfeited Estates: these were the estates confiscated after the Forty-five and managed by an overseeing Board of Commissioners.[11] Their coverage of the Struan estate, including baronies such as those of Slisgarrow and Kinloch and of Murlagan, provide a cameo of landscape in the Central Highlands. Despite being less favourable for farming, their touns were socially more crowded than those on Lochtayside. To judge from lists from the mid-seventeenth century onwards, their comparatively larger social scale extends back at least to the mid-seventeenth century.[12] In reputation, parts were unsettled in nature, with some of its touns, especially those on the edge of Rannoch, having a notoriety for harbouring thieves and broken men. When we come to look at their arable, most were modestly endowed, averaging 19 acres (9.6 hectares) per toun across the two baronies. Yet, if their arable was limited, we cannot say the same about the number of stock present. The large areas of hill grazing available to them meant they could, and did, maintain significant numbers. As well as cattle and sheep, among them were significant numbers of goats and horses. The number of goats present was in spite of the estate's attempts to control their numbers by this point. The number of horses present was probably a sign that they were being bred for sale.[13]

The surveyor identified both infield and outfield but his commentary makes it clear that outfields were not actually described as such by the tenants who worked them. It was simply what the surveyor saw as arable of poorer quality. In at least one case, Camghouran, it included arable at its shieling. Altogether, the ten touns of Slisgarrow Barony had between them 133 acres (67 hectares) of what was called infield and 106 acres (54 hectares) of what was deemed outfield. Much of what was infield was actually haugh land, covered with surface stones and, as on Invercomrie, 'many Stoney Baulks' and subjected to repeated damage from the burns that regularly overflowed across it. When we look at the maps that accompany the survey, arable in many touns comprised detached, irregular blocks despite the surveyor's attempt to round off the edges and to airbrush out wet ground and rock outcrops that must have lain across it (Figure 5.1).

In total, the survey records 67 acres (34 hectares) of meadow in the Barony, but it does not appear to have been closely managed as such. Most of it simply represented what the surveyor himself classed as 'cutting grass' rather than land that was actually managed as meadow. Often, as with the 6 acres (over 3 hectares) of meadow on Wester Feinnard, it was 'full of Low brush, Such as Dwarf willow, Water Myrtles, & Rushes &c'. Beyond its head dykes, Slisgarrow also had 30,088 acres (15,284 hectares) of hill pasture, moor and moss, along with 4,471 acres (2,271 hectares) of birch and 3,066 acres (1,558 hectares) of fir. Significant areas of the wood cover were noted on various plans as stunted, presumably an effect of the relatively high numbers of stock, especially goats, that were present and the harvesting of young growth for wattle.[14]

Further north, available surveys enable us to take a transect across the central Highlands, from Badenoch on one side through Rothiemurchus to Strathavon and Glenlivet on the other. When surveyed in 1771, the surveyor of Badenoch found it 'a Barren Country . . . its surface is mountainous & Steril'. Despite this reaction, some of the touns present in Badenoch were some of the largest in the Central Highlands. In the parishes of Laggan and Kingussie, touns like the Mains of Clunie, the Aird of Clunie, Laggan and Noodmore on the Cluny estate possessed over 100 acres (over 51 hectares) of arable, as well as vast amounts of hill pasture and muir. As in the Mains, most was 'cropt

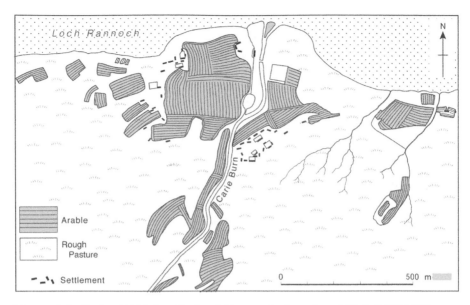

Figure 5.1 Carie, 1756, beside Loch Rannoch, based on NAS, RHP3480.

with Black Oats' but the sowing details show that a small amount of bere was present. These touns present a paradox. Despite the amount of arable present, a general note on the estate reported that 'some of them are necessitated to buy meal . . . two or three months each season'.[15] A part explanation for this might be the fact that returns on seed were said to 'be precarious' owing to the risk-laden climate of the area.

Further east, in areas like Rothiemurchus, Strathavon and Glenlivet, the small touns that had emerged display a different structure of land uses. As elsewhere, arable lay at their core. In some cases, as at Glenmore, this arable was encircled wholly or largely with a belt of pine forest, beyond which lay more open pasture.[16] Glenmore appears as a small free-standing toun, though it was actually divided into two sub-touns, Renachin and Rein Ruick. Around it, but especially further east and south-east, one finds larger, more agglomer-ated toun complexes based around 'daughs and their subdivision into ploughs or lesser parts. While some of these touns making up such 'daughs' were reasonably sized for a toun, others were small by any standards. If we look at the Daugh of Braes, for example, the largest of its nine touns had 40 acres (20 hectares) of arable but others had no more than 5 to 10 acres (2.5 to 5 hectares). Notable about Strathavon touns, such as Braes, is that their 1762 survey makes no reference to outfield, though later surveys do refer to patches of 'new outfield' taken in from the muir.[17] What is also interesting is the extent to which touns used their adjacent woodland for 'under grass' as well as for shelter. Indeed, in terms of the pasture that was private to each toun, the 'under grass' of adjacent woodland was more extensive than their open pasture. In the Daugh of Fodderletter, for instance, there were 8 acres (just over 4 hectares) of open grass and 30 acres (15 hectares) of 'under grass'.[18]

One of the largest surveys carried out for the Forfeited Estates Commissioners was that of the Locheil estate drawn up in 1771.[19] The estate embraced all the land from Loch Eil northwards to the north side of Loch Arkaig. The surveyor saw the chief employment of farmers on the estate as 'attending their Cattle and that is indeed the greatest means of their Support and the fund for paying their Rents'. Like all surveyors, he was looking at it from the viewpoint of what was the most commercial product of the toun, or what sustained rents. From a tenants' perspective, arable was no less important. When we look touns on the estate, most had modest amounts of arable, with very few having more than 20 acres (10 hectares) and many having less than 10 acres (5 hectares) with only a threefold return on seed for their basic crop, oats. The pressure that was being exerted on the arable is highlighted by the significant number of shielings that had arable attached to them at this point, like that of Tonguie, a shieling in Glen Camgory that was used by the tenants of Lagganfern. Given Laggenfern itself only had 10 acres (5 hectares) of arable, it is easy to understand why any opportunity for cropping their shieling ground for arable was seized and why it was lived in all year round. Neither the arable around the main touns nor shielings was labelled in terms of infield or outfield but the surveyor does suggest that folding, the distinguishing husbandry of outfield, was a source of manure for some arable, such as at Invermally where it is made clear that land 'is manured by tathing and cropped for two years and then left in grass for 7 years'.[20] In addition, a number of touns used ferns as manure. Even where ferns were not being added directly as a field manure, that used as bedding would have found its way, via the kailyard compost heap, on to arable. In the seventeenth century, parts of the Locheil area were known for their extensive oak woods and pine forests.[21] The survey, however, reported much of the low hill ground as part covered with the stumps or stools of trees suggesting that, as elsewhere, local communities exerted pressure on their woodland through their demand for wattle and the grazing of their stock.

In the North-west Highlands, John Home's 1774 survey of Assynt is one of the finer surveys. For all its sense of a cultural landscape closely adapted to the infinite variations of Assynt's physical environment, with its mosaic of rock outcrops, lochs, peat and bogs and steeply angled ground, it cannot be regarded as a timeless landscape. In terms of settlement it had farming touns scattered both around its coast and inland, particularly along the trough that runs from Loch Assynt southwards. In his introduction to the published version of Home's survey, Adam thought that 'the broad distinction that emerges is between an arable coastal region already heavily populated and pushing up against the limits of its resources, and a largely pastoral interior'[22] though we should not overstate that contrast when it comes to inhabitants.[23] Nor is the distinction particularly manifest when we look at the differences in the amount of arable present. Coastal touns with conjoint or multiple tenants averaged 56.7 acres (28.8 hectares) of what Home called infield and 45.7 acres (23.2 hectares) of outfield, or what the survey calls 'sheelings'. Those inland touns that carried multiple tenants, meanwhile, averaged 44.2 acres (22.4 hectares) of infield arable and 25.7 acres (13.5 hectares) of arable beyond their infield dyke. Even those inland touns set to tacksmen had 62.9 acres (31.9 hectares) of infield arable and 31.8 acres (16.1 hectares) of arable beyond their infield.[24] For their setting, these represent sizeable acreages but we should also note the surveyor's observation that his depiction of infield arable was simply every-thing with the dykes. It included what was very broken ground. As Home put it in his commentary on Achmelvich, the greater part of its infield 'consists of Rocky Baulks so that only about one half can be keept in tillage'.[25] As with so many areas where arable was cultivated in rigs or lazy-beds, using the spade or foot ploughs, the *cas dhireach* (= straight) and *cas chrom* (= crooked),[26] the baulks or space between the arable rigs was a convenient source of grass. With oat and bere crops said to be capable of growing to the height of an adult, the small arable fields of the coastal touns must have had a striking appearance in late summer. A notable feature of Home's description is his differentiation of arable into infield and what he labels as 'sheelings'. *In toto*, the arable embraced by these 'sheelings' extended to over 600 acres (305 hectares). Some were set at a distance but others were in close proximity to infield, one or two closer to the home settlement than parts of infield (Figure 5.2). The fact that 'sheelings' were tathed before being cropped,[27] and the fact that only a third to a half was under crop in any one year[28] suggest that some were outfields by another name.

The Hebrides

Given the Duke of Argyll's late eighteenth-century references to the 'super-numeraries' or excess population of Tiree, it need not be a surprise to learn that its touns were socially very large units, with each containing on average 6.9 tenants and five cottars or servants and averaging just over forty-nine inhabitants. Walker talked about the touns being 'parcelled out into very small Possessions',[29] a fragmentation which the 1771 report by Alexander Campbell rooted in 'the unrestrained liberty of subset'[30] and which the estate moved to ban by the end of the decade.[31] The 1771 report also noted that most touns, including some sublet through tacksmen as well as those set directly 'in tenen-dry', were farmed in runrig, being divided 'yearly by Lot'.[32]

The map which accompanied Turnbull's 1768 survey shows the layout of touns (Figure 5.3). Taken at face value, it depicts a tidy, even regular, layout of fields as regards both their shape and size. While there are good reasons for seeing Turnbull's mapping of fields as a stylised representation, there are

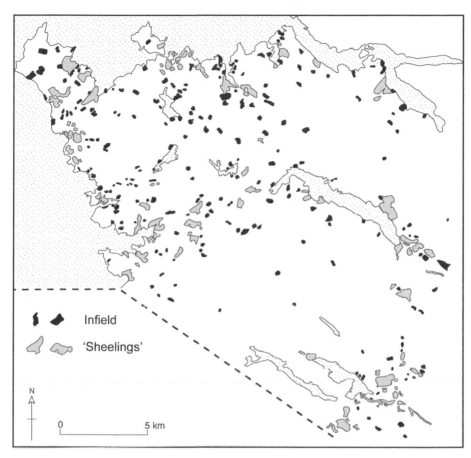

Figure 5.2 Infields and 'sheeling' grounds in Assynt, 1769, redrawn from K. M. Maciver's map, based on Home's survey of Assynt, 1769, in Adam (ed.), *Assynt*, frontispiece.

also reasons for holding back with such a judgement. First, the idea that early communities did not lay out fields in a regular manner has been abandoned, given the evidence we now have for highly regular coaxial fields being laid out at various times during the prehistoric and medieval periods. In fact, traces of regular coaxial fields are present on Coll and the Uists. Second, Tiree was a potentially fertile island, with a high proportion of continuous arable suitable for an orderly layout of fields. Third, the highly formulaic way in which Tiree touns were initially assessed and their rents calculated, with a fixed number of bolls sown per *tirunga*, makes it not inconceivable that this pro rata approach to sowing was carried over into the layout of fields. Fourth, the Tiree landscape was an enclosed not an open landscape. Walker's first survey, *c.*1764, talked about '[A] great part of the Fields are inclosed with Walls of Earth, very broad

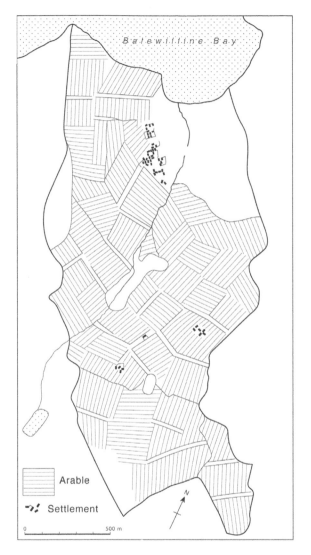

Figure 5.3 Toun of Balewilline, Tiree, 1768–9, showing Turnbull's representation of field layout. Based on NAS, RHP8826.

at the Foundation, five or six Feet high, and covered with Grass from Top to Bottom'.[33] In other words, Turnbull was faced with what must have been one of the most legible field layouts in the Hebrides. Even more telling is what Walker had to say about the nature of fields. What struck him most was 'the Equality of its Fields',[34] which is exactly what strikes us as contrived about Turnbull's depiction of them.

Like Tiree, Skye is one of better Hebridean Islands for eve-of-Clearance survey data. Though not of the scale found in the Outer Isles, the amount of arable present on the threshold of its switch into sheep farms and crofting townships is impressive when added up on a parish-by-parish basis. The parishes that formed Trotternish, for instance, Kilmuir and Kilmaluag, boasted a combined arable of 3,639 acres (1,849 hectares) while Snizort, also on the Macdonald estate, had 2,480 acres (1,260 hectares) in 1810.[35] The extent of pre-Clearance or pre-crofting arable on the MacLeod estate, in the north-west and west of Skye, was no less impressive. An area like that on either side of Loch Beag, stretching from around Ulinish through Struanmore to Bracadale (including Glen Bracadale and Totardor) boasted over 1,500 acres (762 hectares) of arable in 1810.[36] Seen in terms of its average arable acreage per toun, Skye can be positioned midway between the small- to middling-size touns of the mainland and the middling to large touns to be found on islands such as Tiree and the Uists. While there was a broad foundation of touns with less than 100 acres (51 hectares) of arable, most areas, especially around the coastal edges of the northern peninsulas, could boast some touns with 100 to 200 acres (51 to 102 hectares). Ramasaig, an exposed, west-facing toun was typical of these larger touns. Like Lorgill and Ibidale to the south, it possessed over 100 acres (51 hectares) of arable strung out in irregular blocks along gently sloping ground. These were good-sized touns but what was under crop was more a testimony to the capacity of their occupants than to the land itself.

As well as stressing the extent to which arable was worked by the spade or foot plough, late eighteenth-century reports on Skye describe most touns as held in runrig, so that 'no one occupies the same spot He held at the last breaking up'.[37] Of course, such a generalisation tends to downplay the differences between touns set to tacksmen and those set directly to tenants. A 1795 Memorandum on the Macdonald estate was at pains to distinguish between the two forms, noting that, in Sleat, nine of the thirteen touns were set to 'small tenants' while the remaining four were set to single tenants.[38] In not a few cases, the differences between the two forms were only differences of degree. As one Skye commentator put it, some of those who held the tack of a whole toun sublet 'the skirts' to cottars[39] so that such a toun could be as populous as a small runrig toun and part could have the appearance of a runrig toun. We also must allow, however, for the possibility that the 1795 Memorandum distinguished between touns in the hands of 'single tenants' and those in the hands of 'small tenants' because, by this point, the distinction mattered, with some tacksmen now heavily involved in stock farming. When seen in rentals, for instance, we can readily see the potential differences between Trumpan-Beg (or 'Leiterung beg'),[40] which appears from the late seventeenth century onwards as set to a single tacksman, and its sister toun of Trumpan-More which always appears as set to sizeable numbers of multiple tenants: eleven in 1683[41] and twelve in 1724.[42] From what survives on the ground of their premodern layout, the latter was clearly the more crowded toun.

Harris was intimately linked to Skye as part of the MacLeod estate. With its mixture of machair-dominated Atlantic coast, its difficult bays area to the east, its rugged, peat-covered northern interior and its numerous low-lying, offshore Islands, the Harris landscape captures the ecological variations of the region at large. Two summary surveys were produced, one in 1772 and another in 1804–5. What stands out about them is the amount of land shown as cultivated, despite the difficulties of the terrain. In the 1772 survey, the total arable was recorded as 5,047 acres (2,564 hectares) but no attempt was made to differentiate between ploughed or spaded arable in the survey.[43] By the time of the 1804–5 survey, Harris had 1,843 acres (936 hectares) ploughed arable and 5,115 acres (2,598 hectares) of spaded arable mixed with pasture spread across twenty-four touns, so that their average was 76.79 acres (38.8 hectares) of ploughed arable and 213 acres (108 hectares) of spaded arable with pasture.[44] These acreages are probably exaggerated, especially those of spaded land, with the surveyor simply measuring the overall extent, without discounting the areas of rock, lochans, bogs, or grass between the rigs. The surveyor acknowledges the last by categorising some as 'Arable with the Spade or pasture' which he distinguishes from 'Moor and Pasture'.

A detailed comparison between the two surveys is not straightforward owing to the differences in the way arable is presented. Where it is possible, however, such a comparison casts a revealing light on how the landscape was changing. Luskentyre illustrates the direction of this change. It had 197 acres (100 hectares) of arable in the 1772 survey. By 1804–5, this had become 65 acres (33 hectares) of ploughed arable and 397 acres (202 hectares) of spaded arable mixed with pasture. Like other very large touns on the Atlantic coast of Harris, such as Meikle and Little Scarista, Luskentyre benefited from a belt of machair and easily worked soils on its seaward side. Most also had access to an extensive area of wetter, rising hill ground on their landward side, or to the north in the case of Luskentyre. It is across this difficult, broken ground that touns, such as Luskentyre, Borve and Nisibost, developed their extensive acreages of spade with pasture, abundant traces of which survive today (Figure 5.4).

On the opposite side of Harris, facing the Minch, is a line of small, rocky bays known as The Bays area. Though this area is associated with the heavy build-up of settlement following the clearance of touns on the western side of Harris during the early nineteenth century, we can find references to settlement in the Bays' area even before then. These had barely a handkerchief of arable with most, like Manish, having less than 10 acres (5 hectares) that were invariably based on lazy-beds raised out of the broken ground inshore or on the edge of tidal inlets, with fishing as a supplement. Lying across North Harris is the Forest, an extremely rugged area. This was managed as a hunting forest prior to the eighteenth century but, by 1772, its southern coastal area was being actively colonised, with twelve different settlements listed, or fourteen if one includes a mill site and a cultivated shieling, all with a combined arable total-

Figure 5.4 Cultivation rigs on western coastal slopes of South Harris.

ling 195 acres (99 hectares). Typical was Huisnish, with 34 acres (17 hectares) of arable strewn across the small peninsula that formed its territory, and a further 12 acres (6 hectares) at its shieling in the Forest.[45]

A notable feature of Harris is the extent to which the numerous small islands, which lay around its western and south-western coasts, were settled and heavily cultivated. Some were low-lying, machair-rich islands on which heavy seaweed applications were used to enrich soils that were calcium rich but nutrient poor. Others were a mix of sandy soils, poorly drained soils and rock outcrops. Among the first and among the largest of these islands was Berneray. In 1772, it was described simply as having 1,002 acres (509 hectares) of arable. By the time of the more detailed 1804 survey, it was reported as having 1,088 acres (130 hectares) of arable made up of 322 acres (164 hectares) of ploughed land and 766 acres (389 hectares) of spaded arable mixed with pasture. With their exposed, low-lying profile and complete lack of shelter, the scale of cultivation to be found on these islands is witness to the way every niche, every opportunity, was seized upon in the years prior to change, with the MacLeod estate drawing large quantities of grain in rent from them. Islands, such as Taransay and Scarpa or the smaller Killigray and Ensay off the south-west coast, had more mixed ecologies with broken, rugged ground mixed in with soils that offered some opportunities to those willing to use the spade or foot plough. Among these islands, Taransay carried a great deal of arable even by 1772, 430 acres (218 hectares), but, in some ways, the figures for Killigray and Ensay are more telling. Perhaps suggestive of how they had first originated as settled Islands, the 1804–5 survey lumps Killigray with Strong on the mainland of Harris and Ensay with Drymochoind, noting the former as having a combined acreage of 239 acres (121 hectares) of ploughed arable and 319 acres (162 hectares) of spaded arable mixed with grass, and the latter as having an acreage of 273 acres (139 hectares) of ploughed arable and 262 acres (133 hectares) of spaded arable mixed with grass. The earlier 1772 survey, though, makes it clear that a substantial portion of this arable must have been on Killigray and Ensay for their respective acreages then were 298 and 311 acres (151 and 158 hectares). When seen today, there are very few patches on these islands that are not corrugated by cultivation rigs.

The Northern Isles

Traditional landscapes in the Northern Isles were closely defined affairs. In part, we can attribute this to the role played by Udal tenure and the fact that, for part of their history, touns were organised around communities of hereditary landholders whose rights were carefully specified. A sense of their close designation comes across from the description offered by Balfour in his 1859 essay. Each 'TUN or *Townland* with its BOL (Head Boll) or principal farm' was, he wrote, 'enclosed by its TUN GARDE (hill dyke) which separated its GARTH (Infield) from its SEATTUR or HAGI (out pasture or hill)'. Beyond the dyke, 'every enclosure from the Saettur became a Qul (Quoy), which if encircled by an extension of the Tun-garde, became a Tumall, or if again abandoned to pasture, became a Toft'.[46] He doubted whether these later additions, the quoy, tumall and toft, enjoyed the same Udal immunities as the original possessions: the tun, bol and garth. In subsequent divisions of the tun among co-heirs, he suggests that each garth or quoy so created could become the head bull of a new Udal, with its own share of the infield and a proportionate right to the common hagi or saettur.[47] Of course, others have now reworked Balfour's image but its sense of the toun as a set of formalised spaces remains.[48] We can see this from eighteenth and nineteenth-century surveys. The role played by heritors in the Northern Isles meant that quite a number of its touns comprised a form of proprietary runrig. Their removal or division during the eighteenth and nineteenth centuries generated detailed surveys that capture in some detail the structure of touns. An 1846 runrig map of Papa Stour, for instance, shows the exploitation of the island as still based on the former head bull, Da Biggins, and the string of six secondary settlements that had existed back in 1299.[49] The latter probably emerged as houses around the edge of Da Biggins via a family settlement.[50] Each house, like others in Shetland, was organised around *lasts* or sectors in which holdings had been concentrated for greater simplicity. Da Biggins itself comprised nine lasts. At the point when they were first created, we can reasonably expect the layout of each house to have been self-contained rather than intermixed with other houses.[51] Thomson's study of touns such as Laxobigging in Delting parish and Funzie on Fetlar, using early

Figure 5.6
Runrig at
Funzie, Fetlar,
Shetland, 1829.
Redrawn from
ibid., p. 119.

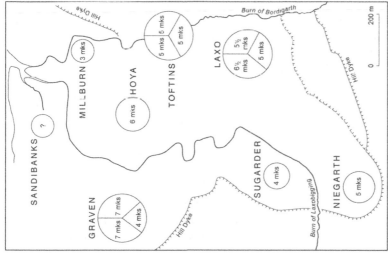

Figure 5.5
Houses at
Laxobigging,
Nesting,
Shetland, 1827,
redrawn from
Thomson,
'Township,
"House" and
tenant holding',
p. 113.

eighteenth- and nineteenth-century surveys, enables us to see how, over time, houses slowly became fused into a single toun layout.[52] He makes his point by setting down the toun of Laxobigging, where a house structure was still visible in its 1827 survey (Figure 5.5), beside what he calls the 'all-toun' runrig system at Funzie (Figure 5.6).

The latter did once have a house system but, by the time of its 1829 survey, it had been overwhelmed by the emergence of this 'all-toun' runrig layout. His study also shows that the kind of designations identified by Balfour were fast disappearing as the layout of touns became reworked as houses were integrated into these larger systems of runrig, with 'the abandonment of features such as *garths*, *punds* and *inner dykes* which had previously created internal systems of division'.[53]

Brand made the point that 'any cornland' in Shetland, 'is ordinarily but a few ridges nigh to the coasts'.[54] Such generalisations need careful handling. No traditional community, especially one with Norse roots, could neglect the uptake of resource from its interior pastures. The presence of items such as cloth and butter wadmel in its traditional rents makes that clear. Even during the medieval period, many secondary settlements pushed the occupied landscape inland. The prime opportunities for arable, however, were, as Brand implied, largely coastal. An inescapably coastal resource such as seaweed provided manure for arable and winter feed for stock. More telling, as population grew over the seventeenth and eighteenth centuries, it was to the sea, rather than just the land, that communities turned to, with more and more cash rents being raised through fishing. This shift into fishing was encouraged by a number of factors. The early success of the Dutch, using small fishing vessels no more than 20 metres in length, called busses, around the Northern Isles, demonstrated that fishing had a commercial, and not just a subsistence, potential. Another factor was the impact of subsistence crises. We can identify a number of crises, such as that of the sixteenth century, associated with King William's lean years in the 1690s and early 1700s, and the famine of the early 1780s. Those who turned to fishing during these crises established routines and investments that were sustained once the crisis receded. For their part, landowners had a vested interest in the encouragement of fishing because it helped to capitalise local resources at a time when cash in hand was still scarce. What is not in doubt is the impact that fishing had on landholding, with a surge in smallholding by the second half of the eighteenth century.[55] For Thomson, this overcrowding may have been a factor in why the house system decayed.

Highland Touns in Close Focus

The more abundant documentation that becomes available from the mid-eighteenth century onwards enables us to give a more precise shape and content to the farming toun. At its core was a community of farmers, one that could comprise a varied mix of tacksmen, tenants-in-chief, subtenants and cottars. Seen through rentals, our view of this community is often partial, with only tacksmen, or those tenants holding directly from the landowner, being listed. Analysed at this level, roughly a half of all touns appear in the hands of single tenants and the other half in the hands of multiple tenants. Where other listings are available, though, we find that many touns, seemingly in the hands of single tenants, were either worked through cottars or were sublet to tenants so that significant numbers were, below the tacksmen or tenants-in-chief, little different from those let directly to multiple tenants.

Runrig was elusive in another way. Runrig layouts were mostly temporary affairs, underpinned by short leases or even annual tenures based on a verbal agreement. Only where heritable tenures were involved can we expect layouts to have had an continuing stability. These were commonplace in the Northern Isles, either based on Udal tenures or on the feu tenures that were exchanged for them by the seventeenth century. They were less common in the main body of the Highlands, though some touns along the eastern edge of the Highlands involved multiple feu tenures while runrig touns, involving wadsets and life-long agreements, would also have enjoyed stability. That said, even where held within a framework of heritable tenures, runrig could still be altered by family settlements, sales or amalgamations so, whatever neatness of order it may have possessed at the outset, would, over time, have been lost or reworked. By contrast, the runrig that prevailed in the vast majority of Highland touns was subject to a regular reallocation of shares between tenants and subtenants at the start of each set. When Little Formestoun, on the eastern edge of the Highlands, was set to two tenants in 1736, for instance, the toun was divided 'att the Sight of Four Judicious Men' and then the two tenants were required to draw lots to see who had which share, of the houses as well as the runrig.[56] Where land was set by tack like this, the runrig order created would have lasted for the duration of the tack, five to nine years being a common period by the mid-eighteenth century. Throughout the tack, tenants would have known their holding as specific strips scattered across the toun. As we move west-wards and north-westwards, though, we find most tenants holding land on a year-to-year basis, usually by verbal agreement or as instant tenants. In these touns, runrig would have been reallocated on a year-to-year basis, so that what they possessed was constantly changing. In fact, given that their runrig lasted only from the turning of the soil or planting of the seed through to harvest, the place of runrig layouts in these landscapes was at best fleeting or ephemeral. It is this yearly flux that so impressed surveyors, a remapping that meant 'no one occupied the same spot He held the last breaking up',[57] a view that Blackadder reiterated in his 1811 survey when he talked about holdings being changed 'from one to another every year'.[58]

Highland cropping schemes were based either on some form of infield–outfield farming, or an outfield-only system. While we can fit most schemes into such categories, the terms 'infield' and 'outfield' themselves were not used everywhere. As noted earlier, croft land was used in the survey of the south side of Loch Tay but infield in the survey of the north side[59] while, elsewhere, wintertown was used in Lorne[60] and intown in Assynt.[61] In some instances, the variants were devised by surveyors, as with Home's interpretation of outfields in Assynt as 'sheelings'. Details of how these sectors were cropped can be gleaned from mid-eighteenth-century surveys, as well as from the *Old Statistical Account*. Where infield or its equivalent forms existed, it was generally divided into three breaks and cropped with two parts under oats and one part under bere.[62] On each break, the bere crop began the infield cycle followed by the two crops of oats. What distinguished infield was its intensive or continuous nature as a cropping sector, though there are signs that a fallow break had been introduced into some infield cropping cycles in parts of the South-west Highlands to create a four break system.[63] In support of its cropping, it received all the manure that had accumulated in byres over the winter, together with a regular flow of old roof thatch, bedding from the byre and, less regularly, turf from house walls, roof foundations and field dykes that had been composted. Where animal manure was insufficient, manures such as seaweed or ferns were used. The first of these required a heavy investment of labour but was short-lived in its effects, 'exhausted with the first crop'.[64] The effect of ferns as a manure is more difficult to gauge but it did have a long-term consequence, encouraging the spread of bracken, as in Knoydart.[65] In addition, infield also received the manure provided by stock when they grazed the harvest stubble over autumn and the grass that invariably existed between the rigs. Needless to say, the way arable was described as 'manurit land' captures this dependence on regular nutrient inputs in a simple phrase.

Those outfields labelled and farmed as such were cropped solely with oats when part of an infield–outfield system, or oats and bere when part of an outfield-only system. Their cropping cycle varied. In principle, most divided it into breaks or tathfolds. After a summer of tathing, each break or fold was then cropped for a number of years before being left under grass until their cropping cycle came round again. The number of breaks employed and the length of their cropping or grass phases varied. In parts of Breadalbane and Netherlorn, we find outfield divided into six breaks, with three years in crop and three under grass in any one year.[66] In Kintyre, it was arranged into ten breaks, with each break cropped for five years after tathing then left in grass for five years.[67] Needless to say, because tathing was concentrated on one break or fold each year, the outfield under arable comprised one break or fold under its first crop after its tathing, another break under its second crop and so on. On the Breadalbane estate at least, a 1706 source suggests that the three folds in cultivation each year may have had distinct names: the tath, the after tath and the third furr.[68]

The key to outfield was its use of summer manure, with the nightly tathing of stock being used to concentrate the manure of stock on a part of the outfield during the summer prior to the onset of its cultivation phase. By the mid- to late eighteenth century, there is no lack of data describing outfield tathing as a husbandry practice. In his 1798 report on Netherlorn, Robertson wrote in his description of Ardluing, that outfield – what was also known locally as 'Tath ground' – was

> divided into Three Divisions. One of these broke up for Oats after what they call tathing their Cattle upon it, and taking two Crops of Oats off, without any more manure, then let it out for Two or Three years carrying a scanty crop of grass till they are over the other Two Divisions by this means the Tathing goes round in a rotation.[69]

To be effective, tathfolds needed to cover that portion of outfield due to be brought back into cultivation the following summer. Their construction varied. Some had their bounding dykes constructed of turf, a technique that added greatly to the scale of turfing and its degradation of pastures. In other cases, wattle fences were used. A revealing description of wattle-based folds is provided by early eighteenth-century data for Glenlyon. It is contained in the fairly abundant documentation generated by a dispute over the use of shieling ground in the forest of Mamlorne. Glenlyon touns maintained what they called home shielings, that is, shielings just outside the head dyke that could be used by stock brought back from the more distant pastures to tath outfield.[70] In their depositions regarding the practice during the 1720s and 1730s, many tenants refer to the number of 'flakes' that were present in the toun, or the number that they had been left by the previous tenant, including some that were described as 'cow flakes' and others as 'sheep flakes'. Thus, John McIntyre, a tenant of a fourth part of Westermore, said 'that it is the Custome to get for a fourth part of That Toun Sixteen Cow fflakes, or five Sheep fflakes'. His co-tenant, John More McCoil McCallum, tenant in another fourth of the toun varied the figures a little, reporting that 'when he Entered to the Sixth part of the sd Town He gott Eight Cow fflakes . . . Depones that he should leave Sixteen Cow fflakes, and Six Sheep fflakes for his present possession'. When we read closer, it becomes clear that the flakes involved were used for tathing, for McCallum went on to say that 'these flakes if he was to fauld all his Cows and Sheep, would be too few for them'.[71] Donald McAlester in Castill added a further point of interest. As well confirming that the available flakes could not contain their whole cattle 'if they foalded them', he went on to say that, as elsewhere in the Highlands, 'they were not in use to ffoald their yeald Cattle'.[72] Clearly, the summer landscape of outfields and their tathfolds in Glen Lyon would have been a complex mosaic of wattle-based hurdles.

As we move westwards and north-westwards on to Tiree, Skye, and the Outer Hebrides, the character of cropping shifts. Outwardly, it appears as an outfield-only system, with land being cropped for a few years before being left under grass for a few years but it did not carry the label of outfield. On North Uist, for example, as in Netherlorn, it was cropped for three years then left for three years.[73] Whilst tathing and turf were the prime manures for outfield in the main body of the Highlands, touns in the far west and north-west supplemented animal dung or replaced it with manures such as seaweed, shell sand, peat and ferns. As noted earlier, seaweed was transient in its effects so, where touns were heavily reliant on it, a grass–arable system was probably needed to enable the land to recover between its arable cropping phases. Emphasising their difference from outfield systems on the mainland, these grass–arable systems of the Hebrides were not cropped solely with oats but also with bere. One should also add that, while we might label these grass–arable systems as outfield-type systems, they were not referred to as outfield systems by contemporaries. Outfield was labelled as such only where juxtaposed with an infield system.

Two further aspects of Highland arable are revealed by the richer sources of eighteenth- and early nineteenth-century surveys. The first is that most arable was thinly sown, a boll per acre (3.11 bolls per hectare) being a standard rate, with some touns sowing as thinly as a boll per two acres (1.56 bolls per hectare). One of the consequences of this was that arable in many touns was as rich in the growth of weeds (for example, corn marigold, wild carrots) as it was in the growing crop.[74] Marshall's report on the Central Highlands provides us with a particularly graphic description of the weeds that affected the face of arable, with wild mustard, marigold, spurrey, corn scabious, and thistle being especially dominant plus 'a tribe of minor weeds'.[75] Of course, for those who looked at the problem through the eyes of the would-be improver, such as Marshall, it was a problem of weed infestation. Duncan Forbes of Culloden's much quoted survey of Tiree in 1737 was penned in this frame of mind, castigating local farmers for the abundance of weeds on their arable. Hardly 'one tenth' was corn, he maintained, 'the rest is all wild carrot, mustard, etc'.[76] Yet there were reasons why communities were not too attentive over weeding out such growth, for some, like silverweed, wild carrot, sorrel, wild spinach, mugwort, ground elder, cow parsley, Scotch parsley, common white blite, burdock, nettles and even the common thistle helped combat the subsistence deficit that followed from a failed or poor harvest.[77] In fact, Tiree communities were said to live on silverweed for a whole month each year, when their meal ran out, though silverweed was as much a weed of the machair as of arable.[78] When we add in to the equation the baulks of grass between arable rigs, the difference between what was 'weed-infested' arable and what was species-rich grassland must sometimes have been relative rather than an absolute.

As detailed maps, such as Farquharson's 1769 map of North Lochtayside, help demonstrate (Figure 4.4), pre-improvement arable was adapted to a highly variegated local environment. There were strategies employed to improve what was available, including adding peat, sand and fresh soil, removing some of the stones present and raising cultivation out of boggy ground via rigs and

Figure 5.7 Cultivation rigs at Europie, Lewis.

lazy-beds, but we still find arable having to fit itself in around rock outcrops, large stones, patches of thin soil, waterlogged soil and bogs. Hence, more often than not, parts of a toun's arable were both broken and fretted in outline. The regular fields of continuous arable that we see in the modern landscape were still to come, a byproduct of later improvement. That said, there would have been a core arable in many touns that was ploughed. As we move westwards and north-westwards, however, especially north of the Great Glen, the overall challenge to cultivation became greater. There were, of course, exceptions to this generalisation. Islands, such as the Uists and Tiree as well as parts of Orkney, had low-lying ground which was suitable for the plough if holding size was sufficient. Overall, though, the further north and north-west, the more likely we are to find communities having to cope with broken, marginal ground. Where combined with a predominance of small holdings, we find an increasingly greater use of handtools, such as the spade and foot plough. They were invariably used in association with lazy-bed cultivation, a technique of raised rig cultivation that involved the gathering up of soil from the land between on to the rig, along with available manures, thereby sacrificing the land between rigs for the sake of deepening the soil on the rig itself (Figure 5.7).

This made the acreage of spaded land difficult to calculate. Surveys, such as that for Harris, acknowledged this by presenting its data as 'spaded arable with grass' but other such tabulations, such as that for Ardnamurchan 'ploughed,

Figure 5.8 Shielings on the northern slopes of Ben Lawers.

1504 ac, spaded 1260 ac' and Sunart 'ploughed, 567 ac, spaded, 803 ac', do not.[79] Whatever the problems of calculating its extent, what is certain is that the complex of spade/*cas dhireach*/*cas chrom* and lazy-bed enabled cultivation to be squeezed between rock outcrops, run up slopes, floated out over water-logged ground and to defy the handicap of thin, stony soils.

Beyond arable and beyond the head dyke of touns lay the hill grazings that were exploited via designated shieling grounds (Figures 5.8 and 5.9). Some lay just beyond the head dyke; these were the home shielings from which stock could be used to tath outfield during summer.

Others grounds lay at a distance. For example, Speyside touns such as Cluny Mains and Aird had access to extensive hill grazings or shieling grounds located around Loch Ericht, 12 miles to the south where they would 'drive up all their Sheep and Cattle in the summer and there make quantities of Cheese & Butter, and in the meantime their pasture nearer home is much hained and saved for the winter season'.[80] As mentioned in the previous chapter, local courts exercised a close control over the dates when stock were to be removed beyond the head dyke in early summer, when they were to be taken to the summer shielings and when cattle, mainly milk cows, were to be brought back for tathing and when the rest of the stock should return.

When it came to stock, a small, but equally important part of the toun landscape, was the meadow. Though having a presence in many touns, there

Figure 5.9 Shieling sites, Duirinish, Skye.

is something enigmatic about its role in the Highland toun economy. Taking into consideration the need for the winter housing of livestock in many parts of the Highlands, and the role played by winter manure in the maintenance of infield arable, we would expect the management of hay meadows and the production of hay for winter feed to have been important. Undoubtedly, there were areas, especially in the Central and Southern Highlands, where managed meadows were very much a part of the toun landscape. The figures cited earlier for the extent of formally designated meadow in touns beside Lochtayside, where most touns had meadow, demonstrate this. We must avoid overstating its significance, however. To judge from other mid-eighteenth-century surveys, formally designated meadow was not present everywhere. Even in the Central Highlands, estates like that of Struan were seemingly deficient though, if we refer to the cartographic surveys, the surveyor does note some patches of grassland within the head dyke of some touns that, in his view, were for 'suitable for hay'.[81] The dearth of formally designated meadow is especially striking when we look at data for the western seaboard and the Hebrides, with surveys either reporting no meadow present or only small amounts. Of course, harvesting hay need not have been confined to cutting it from formally designated meadow. Any hained area of herb-rich pasture, or what some referred to as 'cutting grass', would have sufficed. If, however, we shift the question and ask whether surveys refer to the cutting of hay, as opposed to the presence of meadow, what stands out are the number of reports about the extent to which tenants across the Highlands neglected the provision of winter feed or hay for their stock. For some, the neglect appears to have been total. A tenant of Teachnock, a farm on the Lovat estate in the parish of Kilmorack, for instance, said how, when his grandfather took over the farm in 1711, he 'was the very first whoever cut and win hay on this farm, and by his example the neighbours came by degrees into the same practice'.[82] Such references need to be seen alongside more general comments that also stress the failure of many farmers across the region to cut sufficient hay for winter feed.[83]

The neglect of haymaking meant that many touns suffered from a lack of winter feed for their stock, a deficiency that drew the attention of many observers. It meant that not only did stock reach spring in poor condition but also less manure was accumulated from the byre for arable. To an extent, different parts of the region faced different problems. We can see this in the Hebrides. A number of sources makes it clear that, on islands such as Tiree and the Uists, stock were wintered out of doors down until about the 1720s and 1730s.[84] Quite apart from the fact that milder winter conditions in the Hebrides attached fewer risks to the outdoor wintering of stock, the heavy reliance on the use of seaweed for arable meant that there was less need for the manure that might have accumulated from indoor wintering. This situation changed during the second quarter of the eighteenth century when landowners started to appreciate the value of seaweed for kelp manufacture and began to restrict its use as a fertiliser. Communities responded by making greater use of animal dung, a switch that meant stock were now wintered indoors.[85] On the mainland, and certainly in the Central and Eastern Highlands, the winter byring of stock, overnight if not all day, played a more critical role because of the harsher conditions but it was still not a practice followed everywhere. The extent to which some touns made use of local woodland as a source not just of winter shelter for their stock but also of vital winter grazing, with stock feeding on the so-called 'under grass' and bark, can tell us a great deal about these wintering strategies on the eve of change.[86]

Where the winter byring of stock was practised, we can expect byres to have been present. Some would have been part of a longhouse arrangement, such as the eighteenth-century examples excavated at Easter Raitts in Badenoch.[87] Others, though, were free-standing structures. In Stratherrick, for example, a mid-eighteenth-century commentator reported that cattle there were kept in 'a great many small houses on their farms' during winter,[88] a description that suggests touns had separate byres. Data on how stock were housed show that, whether in separate byres or longhouses, stock was not just herded into them without distinction. In the first place, many of the larger tenants may have had separate byres for their stirks and for sheep, as well as for their cows. Just as it informs us about tathing, the large body of material generated during the early 1740s by the long-standing dispute over grazing rights in the forest of Mamlorne enables us to shed light on how touns there approached the byring of their stock. From the evidence presented, it is clear that tenants in Glen Lyon saw the stocking capacity of their holdings as much in terms of how many cattle 'stalls' and 'flakes' their share of the toun was expected to have as in straightforward stock numbers. While exceptions were acknowledged, the general custom in the area was that each merkland had twenty-four soums or cows plus 'Twenty four stalls and Twenty four flakes' attached to it, and the tenants 'look't upon this as the Rule of Souming'. Yet, while the importance of 'stalls' in the scheme of things would suggest that housing stock was the norm in Glen Lyon, at least one tenant claimed he only housed his stock 'whenever the Storme forc'd him' but added that his available hay often fell short of needs.[89]

Questions over winter feed and the byring of stock are, in part, bound up with dairying. In a region where dairy produce formed a vital part of everyday diet, the extraction of cheese and butter from the toun economy posed problems. When we use the more abundant data of the post-medieval period to look more closely at how herds were organised, two features stand out. First, herds typically were divided into milch cattle (or 'calfit kye') and yeld cattle (or 'farrow kye'), that is, those in milk (and, therefore, those which had calved) and those that were not in milk (and which, therefore, had not calved that year). Second, a number of sources refer to cattle as organised into couples. This concept of couples is fundamental to any understanding of how traditional diary production was organised. In practice, as Alexander Menzies observed in his 1768 survey of Kintyre, communities reared 'one calf to every couple', an observation which he repeated later in his Tour when commenting on Nether Lorn.[90] Given this organisation of herds into couples, it might be assumed that a couple involved a pairing of a milk and a yeld cow, with one calving each year. In fact, as Ninian Jeffrey made clear in his description of dairying in Coigach, local herds had 'one calf for every two milk cows'.[91] This 'one calf for every two milk cows' was because one calf was always killed. It was said, James Loch, a 'practice, universally adopted, of killing every second calf',[92] leaving the remaining calf to be fed on the milk of the 'two calfit kye'. The operation of such a system shaped herd structures. As well as having an equal number of 'new calfit kye' and 'farrow kye', the killing of every second calf meant that each herd also comprised half the number of 'new calfit kye' present in terms of three-, two- and one-year-old calves, with some also distinguishing newborn or half-year calves. This rather rigid herd structure can be seen at both tenant and toun levels. We can see it very clearly through a series of Breadalbane rentals and, later, tacks from the late sixteenth century onwards. The tenant who held the bow toun of Ledour, for instance, held six 'calfit kye', three three-year-old calves, three two-year-olds, three one-year-olds and three newborn, while the tenant of Botuarie Beg held twelve 'calfit kye', twelve 'farrow kye', six three-year-old calves, six two-year-olds, six one-year-olds and six newborn.[93] Though their numbers had thinned by the end of the seventeenth century, examples of such arrangements continued in parts of the Breadalbane estate down to the mid-eighteenth century.[94]

We can link this early modern data back to Bigelow's argument with regard to medieval Shetland. Faced with heavy payments of cheese and butter for rent, but in a context where milk yields were low, he suggested that Norse communities in the Northern Isles maximised the milk available for dairy produce by killing some of the calves produced soon after birth. This meant that more cows were in milk than were needed to suckle the calves left alive so that a surplus of milk was available for cheese and butter. Bigelow's ideas are clearly supported by the evidence for 'couples'. His assumption that surviving calves could be suckled by more than one cow, an aspect of his argument that was strongly rejected by some of his fellow archaeologists, is actually given support by Martin Martin. 'When a Calf is slain', he reported, 'it's a usual Custom to cover another Calf with its Skin, to suck the Cow whose Calf has been slain, or else she gives no Milk.'[95] Admittedly, the slaughter of surplus calves may not have taken place immediately after their birth. Payments of veal recorded in some rentals, such as those for Mull, raise the possibility that some may have been treated as veal calves before being slaughtered.[96] Loch thought that the reason for the practice was 'on account of the want of winter keep',[97] a conclusion that McCormick favoured when interpreting the preponderance of young stock found in late prehistoric bone dumps.[98] The early modern rental data on couples, though, suggest that, notwithstanding its side effect of easing the problems of winter feed, the extraction of extra cheese and butter is likely to have been the prime driver. We see this in the rental agreements for bow touns, with most being burdened with significant payments of cheese and butter. Bowmen in Gairloch, for instance, were required 'to produce for every two cows, one calf, two stones of butter weighing 24 lbs, English, and four stones of cheese'.[99]

When we look at what else comprised the landscape of the typical toun, one of the most valuable resources available to it was that of woodland or forest. The Highlands and Islands could not be described as wooded countryside on the eve of the Clearances. Initial estimates, based on General Roy's military map of 1755, covering the core areas of the Highlands on the mainland, put its woodland cover at 4 per cent. More recent analysis, however, has suggested that Roy's map does not provide a reliable indication of the amount of woodland present, understating some woods and missing out others altogether, and has suggested instead that a cover of 8 to 9 per cent might be a better estimate.[100]

By this point, *c.*1755, the woodland present existed in three forms. First, and the most significant, were the large stands of fir woods that some formerly regarded as the vestiges of the Great Wood of Caledon that once covered the region *in extenso* but that view is now much qualified.[101] On the eve of the Clearances, substantial stands of these fir woods were then, as now, still to be found north and south of the Great Glen, in areas such as: Speyside (including Rothiemurchus, Abernethy and Glenmore); parts of Deeside; Strath Glass and its offshoots (Glens Cannich and Strathfarrar and Affric); Loch Arkaig, Glen Nevis and Ardgour; Rannoch and Glen Orchy; and, to the north, in parts of Wester Ross, such as beside Loch Maree and Achnashellach.[102] At this stage, the pine woods present would not have been pure stands of Scots pine but would have been mixed, with other species, such as birch and aspen, present.[103] Second, many touns had access to more localised blocks of woodland or forest on the lower slopes or valley ground that lay around them. Some of these localised blocks comprised fir but others were a mixed oak–birch woodland. Along the western seaboard, in areas such as Benderloch, Ardnamurchan, Sunart and Glenelg on the mainland and on islands like Mull, stands of the latter amounted to significant reserves. Even in the main body of the Highlands, stands of oak–birch could form a valuable resource, such as that which straggled along Lochtayside.[104] Forming a third type of wood were the patches of thin, open, scrubby woodland made up of a mixture willow, birch, ash, hazel or pine that could be found across the lower hill ground surrounding many touns. Reports on such wood by some mid-eighteenth-century surveys, such as that of Lochaber (1767), reported on the extent to which its growth was generally stunted, with scattered tree stumps or boles present, and fresh, young growth being grazed out by stock or cut for wattle.[105]

Woodland was not an open resource that could be freely exploited by the farming community but was regulated via arrangements defined through local barony courts. In many touns, local birlawmen were held responsible for ensuring that these court regulations were observed but, by the mid-eighteenth century, some estates also employed more specialist forms of supervision.[106] On the Breadalbane estate, a system of local 'wood keepers' or 'wood officers' were used or there were what Watson called 'tenant-foresters', officials appointed from within the toun community as the estate's eyes and ears responsible for upholding the management of the estate's woodland.[107] To the west, the Argyll estate introduced what called its 'wood rangers'.[108] Again, these were members of the farming community with attached holdings. All tenants and their cottars would have had timber needs, ranging from the heavier timbers needed for roof couples, doors and plough beams to the lighter pieces needed for spade handles, fencing, wattle and the like. Given that the harvesting of such timber was closely regulated, the more so by the mid-eighteenth century, we can be sure that most tenants were inclined to conserve and reuse as much of the heavier timber as possible. The most worn-out piece, which had already been through different cycles of use, would still have been regarded as having a use value. Even Lord Breadalbane can be found ordering the retrieval of roof couples from hay barns being pulled down at Portbeg and their storage in expectation of their later reuse as they will last another 'fourtie years'.[109] For toun communities, however, the perceived value of local woodland extended beyond what it could supply by way of timber. Its saplings, leaves, bark and undergrass, as well as its winter shelter, were all strategically valuable for a toun's stock. Its bark, of course, was also widely used for tanning leather.

Initially, tenant rights were defined through customary rights of wood leave that allowed them to harvest what was sufficient for basic needs[110] but, otherwise, to leave all wood 'hanging and standing' at the end of their occupancy.[111] We can document examples of wood leave in operation down to well into the eighteenth century. On the Argyll estate, for example, tenants on Tiree had access to no ready supplies of local timber. Instead, they enjoyed rights of wood leave in the estate's woods beside Loch Sunart in Morvern[112] but were allowed to cut timber there only under the supervision or orders of the local wood ranger.[113] A comparable system of wood leave appears to have operated on Harris. Like Tiree, Harris had negligible supplies of local wood.[114] As a solution, MacLeod of Dunvegan allowed his tenants a right of wood leave in his Glenelg woods.[115] It was a right measured in terms of a given number of boat loads. When plans to sell Harris were first laid out in the 1770s, this right of wood leave was clearly threatened. Among the estate's proposals was that its tenants from Harris be barred 'from cutting Timber of any kind in the Woods of Glenelg or elsewhere belonging to MacLeod' but that he would 'Annually' provide 'such Number of boat loads of Hazel and Birch as they shall appear to have received annually upon an average of twelve years past'.[116] Further east, we find the Earl of Breadalbane's tenants on Lismore being allowed six boat loads of timber out of Kingarloch for 'the support of their houses',[117] while tacks issued in 1753 by Campbell of Glenure, including one for Sallachan in Ardgour, allowed tenants the liberty of timber for their needs, notably house timbers.[118] In neither case was a charge levied.[119]

The greater volume of evidence available for the mainland, though, makes it clear that, while some traditional forms of wood leave survived relatively unchanged into the eighteenth century, at least for favoured tenants if not for everybody, this was not the case everywhere. From when detailed data first become available in the seventeenth century, some aspects of how tenants and cottars exploited local woodland resources were being slowly monetised. We can see this in seventeenth- and early eighteenth-century barony court proceedings that have survived. In them, we find tenants and cottars fined for cutting the wrong type of wood, cutting it in the wrong place, in the wrong way, or to excess. Seventeenth-century court records for the Menzies and Rannoch estates, for instance, show tenants fined for cutting timber in forest land, cutting it too low, cutting hazel, peeling bark, and so on.[120] On the Breadalbane estate, some were even fined for gathering fallen wood[121] while, on the Mackintosh estate, some were fined for cutting it at the wrong time of year.[122]

Surveying the action taken by the courts over specific abuses, the most serious offences were those involving the destruction or clear felling of trees. One of the reasons why muir burning within open woods, or even just close to woods, was so frowned upon, and repeatedly acted upon in court proceedings, was because of its potential for damaging trees or inhibiting the regeneration of trees. A 1702 entry in the barony court proceedings for the Breadalbane estate talks about the 'great damage of our forests and to the great destruction of our woods both firr and oyr timber old & young' caused by muir burning, and talked about how those responsible should be 'severly fyned to the terror of oyrs'.[123] The Gordon estate was equally severe in its response. When the estate freshened up its regulations for Lochaber in 1770, muir burning 'where there is any stool or appearance of growing Timber' was one of a number of practices which, if it destroyed wood, was to be punished by the forfeiture of a person's tack.[124] Reports of woodland being damaged through stock grazing out young growth or stripping bark are fairly common in the eighteenth century. All stock, including sheep and cattle[125], was capable of such damage but goats were particularly destructive.[126] They were very effective in grazing out young growth but their capacity for bark stripping meant that they caused the trees 'to grow crooked or to wither away' thereby reducing the value of the timber.[127] Estates responded to their threat by banning their presence, at least in those touns that were close to woods.[128] Any such problems would have been accentuated by the greater difficulties which especially pine had in regenerating compared to the eastern Highlands.[129]

The harvesting of vast quantities of wattle, especially of alder and hazel, was one of the most damaging practices for woodland when not part of a systematic coppicing cycle. House walls, internal partitions, folds and fences all consumed huge amounts of wattle and did so on a recurrent basis. In other words, its harvesting sprang from ordinary, everyday forms of use. We need only look at the sheer scale of timber consumption created by a single creel or basket house to appreciate how such housing could have stunted tree growth around touns, with one mid-eighteenth-century estimate talking of about two thousand of the 'streightest & best of the young wood' for a single dwelling, and adding that they were 'annually repaired & often renewed from the foundation'.[130] Despite this use as primary material for house building, its impact on woodland regeneration meant that landlords increasingly tried to ban its use for building or for folds. The Duke of Argyll, for instance, banned the use of wattle for housing in 1733[131] but his action did not have the intended effect for a later duke can be found complaining that the practice was still damaging his woods on Mull and in Sunart, not least because of how it created a continuous need for wood. The Breadalbane estate also turned against its use. In fact, it can be found banning the use of what it called ryce techniques as early as 1620, with an act declaring that 'na barmkin or house be biggit wt ryce in tyme coming'.[132] There must have been a wider prohibition on its use because we find tenants elsewhere being fined for using it to build partitions, flakes and folds, such as in 1639 when a tenant in Botuarie Beg was fined for having a load of wattle to build a stall for his cattle and another in Finlarig was for having wattle for his byre and flakes.[133] The timber used for wattle would have been wood such as young hazel but communities also made use of fir for folding to the extent that, in 1728, the estate specifically banned its use for this purpose.[134] Even at this point, they were still grappling with the use of wattle for walls and partitions for, in 1730, their court for Disher and Toyer enacted that tenants were not only to build up their side walls to their full height in stone but also to build all internal walls in their houses, barns and byres in stone. The building of side walls to full height in stone was said to be in the best interest of tenants, given that the present practice of using stone to half height was 'very destructive to ye woods and expensive to themselves now that they are obliged to buy their timber'.[135]

The reference here to tenants being 'now . . . obliged to buy their timber' might be seen as signalling an important shift was taking place regarding how estates viewed their woodland but it was not an all-embracing, overnight shift in policy, nor was it progressive in how it spread. Signs of it are present in the seventeenth century yet, even in the mid- to late eighteenth century, we find some landlords struggling to apply effectively their policy on the pricing of timber used by tenants. Among the earliest signs are those showing how the Breadalbane estate approached the problem. As noted above, its barony court proceedings contain abundant reference to tenants being fined for harvesting timber wrongly or without supervision. The procedure followed was for each tenant or occupier to attend the court and to declare what timber had been used. While some were fined for abusing the regulations over timber use, others were burdened with a 'fine' simply for having cut or taken timber, as if this was simply how the estate applied its pricing policy for tenant's normal needs. It clearly all depended on how the matter was reported by a toun's birlawmen or wood officers. A similar dual policy also comes across from early eighteenth-century barony court proceedings for Arisaig. Entries show tenants and cottars being fined for infringements of the estate's policy on timber use but, alongside such fines, there also existed a price list of what occupiers had to pay for wood harvested legitimately, the prices set according to what it was to be used for, such as roof couples, plough beams, spade handles, wattle, and so on.[136] Slowly, we find tenants paying for items that had previously been at the very heart of wood leave, such as roof timbers or couples. References, across the seventeenth and eighteenth centuries, to tenants removing their house timbers and doors at the end of their tack suggests that such tenants had paid for the timber involved.[137] The shift in policy, though, was uneven. The Breadalbane estate used meetings of its barony court for Disher and Toyer during the 1720s to lay down a formal process by which timber was to be requested, paid for and then cut under supervision by tenants occupying its Lochtayside touns.[138] To judge from the protests of some of its tenants at the time,[139] the idea of charging people for what had been a customary right of wood leave appears to have been newly introduced. Yet later, a 1783 report for the Argyllshire portion of the estate was able to argue that, wherever tenants still had wood leave, the woods suffered because they were careless over what they used. Echoing what had already been agreed for part of, if not all, the estate, the report's author recommended that wood leave should be replaced by a rate per use.[140] That it was not always easy to break with the cake of custom is illustrated by what happened on the Argyll estate at about the same time. By the 1770s, the estate had inserted clauses into its tacks for Tiree specifying how much was to be paid by tenants for the cabers and pan-trees given to them.[141] That it may have been a new, rather than revamped, system of charging is hinted at by the fact that the duke was soon complaining that the 'wood rangers' had not drawn up accounts of the sums paid to them as they were meant to.[142] In fact, the attempt to levy a charge does not appear to have been wholly successful for, by 1789, the duke was still complaining about the large quantities of wood used by tenants on the island, arguing that they only way 'to check this I believe is to insist with the tenants to build stone and lime walls and to buy timber themselves'.[143] Interestingly, when the duke proposed to sell his wood at Achnacross, he wanted to ensure that his tenants had continuing access to timber by inserting a clause requiring his tenants be sold timber at 'reasonable prices'.[144]

These shifts in attitudes towards woodland management were probably brought about by two significant sources of pressure that built up on traditional systems of wood use over the seventeenth to eighteenth centuries. First, there were the bottom-up pressures, as many touns became slowly burdened with more occupants, especially over the eighteenth century. This raised the demand for timber, with more occupants needing roof timbers and wattle. In some cases, it may have driven an expansion of arable at the expense of the more open woodland. In addition to these bottom-up pressures, though, there were also top-down pressures as landowners – many of whom were grappling with problems of indebtedness by the early eighteenth century – looked more fully to capitalise their estate resources, including the capital contained in woodland. To this end, they began to be more protective of their woodland and how it was used in order to maximise its commercial potential. Some of these commercial opportunities (for example, felling of large constructional timber, charcoal burning and commercial bark exploitation) were already being exploited during the seventeenth century but the scale and the extent to which estates pursued them as commercial opportunities grew noticeably during the eighteenth century. Given that they represent a shift of emphasis away from the traditional uses of timber by the farming community, I have left my review of these more commercial forms of timber exploitation to the next chapter.

Notes

1. A. Murray of Stanhope, *True Interest of Great Britain , Ireland and Our Plantatations* (London, 1740), esp. 'Anatomie of Ardnamurchan and Sunard'.
2. M. Storrie, *Islay: Biography of an Island* (Port Ellen, 1981), p. 63.
3. W. Roy, *The Great Map. The Military Survey of Scotland 1747–1755* (Edinburgh, 2007).
4. G. Whittington and A. J. S. Gibson, *The Military Survey of Scotland 1747–1755: a Critique*, HGRG Research Series, 18 (1986).
5. McArthur (ed), *Lochtayside*; S. Boyle, 'Mapping landscapes of the Improvement period: surveys of North Lochtayside 1769 and 2000', *SGM*, 125 (2009), pp. 43–60.
6. McArthur (ed.), *Lochtayside 1769*, p. xlvi.
7. For example, *OSA*, x, 611; IC, AM, v.65, 1771.
8. McArthur (ed.), *Lochtayside*, p. xlvi.
9. NAS, GD112/9/5/18/4.
10. Boyle, 'Mapping landscapes', pp. 43–60.
11. Wills (ed.), *Reports . . . Annexed Estates*; V. Wills (ed.), *Statistics of the Annexed Estates 1755–1756* (Edinburgh, 1973); A. Smith, *Jacobite Estates of the Forty-Five* (Edinburgh, 1982), pp. 1–53.
12. NAS, GD50/136/1; GD50/156, Lists of Inhabitants, 1695–1747.
13. Wills (ed.), *Reports . . . Annexed Estates*, 1973, pp. 30–6; Wills (ed.), *Statistics . . . Annexed Estates*, 1973, pp. 30–9.
14. NAS, FE, E783/98.
15. NAS, FE, E745/59.
16. NAS, RHP 2504.
17. NAS, RHP2488; RHP1824; RHP1793.
18. NAS, RHP2488/3.
19. NAS, GD50/2/3.
20. NAS, GD50/2/3.
21. T. C. Smout, 'Cutting into the pine. Loch Arkaig and Rothiemurchus in the eighteenth century', in T . C. Smout (ed.), *Scottish Woodland History* (Dalkeith, 1997), pp. 114–24.
22. Adam (ed.), *Assynt*, p. xlviii.
23. A. Simms, *Assynt*, Giessener Geographische Schriften, 16 (Giessen, 1969), seite 6a.
24. Adam (ed.), *Assynt*, p. xlvii; Simms, *Assynt*, seiten 5, 8.
25. Adam (ed.), *Assynt*, 1960, p. 3.
26. Ibid., p. 42.
27. Ibid., p. 43.
28. For example, ibid., p. 15.
29. McKay (ed.), *Walker's Report*, p. 184.
30. IC, AP,V65 Remarks . . . 1771.

31. E. Cregeen (ed.), *Argyll Estate Instructions: Mull, Morvern, Tiree, 1771–1805*, *SHS*, 4th ser., 1 (1964), p. 110.
32. IC, AP, V65 Remarks . . . 1771.
33. McKay (ed.), *Walker's Report*, p. 184.
34. Ibid., p. 184.
35. NAS, RHP 5990/1–4.
36. DC, MDP, Plan . . . Jas. Chapman, 1810.
37. NAS, RH2/8/24.
38. HULL, DDBM/27/3.
39. DC, MDP, 4/254, letter, dated 1761.
40. Ibid., 2/485/12.
41. Ibid., 2485/24/3.
42. Ibid., 2/493/1.
43. Ibid., 1/466/22.
44. Ibid., 1/466/22; Ibid., 1/466/24; NLS, Map of Harris, Bald, 1804–5.
45. DC, MDP, 1/466/22; NLS, Map of Harris, Bald, 1804–5.
46. D. Balfour, *Oppressions of the Sixteenth Century in the Islands of Orkney and Zetland*, Maitland Club (Edinburgh, 1859), p. xxxi.
47. Ibid., p. xxxi.
48. Smith, *Toons*, p. 32–3; Fenton, *Northern Isles*, pp. 24–50.
49. Crawford and Ballin Smith, *The Biggins* , cover map; Smith, *Toons and Tenants*, p. 33.
50. Crawford and Ballin Smith, *The Biggins*, p. 29.
51. Ibid., pp. 32–3.
52. W. P. L. Thomson, 'Township, house and tenant holding. The structure of runrig agriculture in Shetland', in V. Turner (ed.), *The Shaping of Shetland* (Lerwick, 1998), pp.107–27.
53. Ibid., p. 108.
54. Brand, *Brief Description . . . Orkney, Zetland*, p. 112.
55. Fenton, *Northern Isles*, pp. 56–7.
56. NAS, GD312/30/11, 1736.
57. NAS, RH2/8/24.
58. CDC, LMP, GD221/116; IC, AP, vol. 65, Remarks . . . Tiry, 1771.
59. McArthur (ed.), *Lochtayside*, pp. 4 et seq., 89 et seq.
60. NAS, GD112/ 12/1/2/14.
61. NLS, 313/3161/7, Tacks, 1765–66.
62. McArthur (ed.), *Lochtayside*, p. 11; Wills, *Reports . . . Annexed Estates*, p. 77.
63. *OSA*, 1791–99, x, 198.
64. Ibid., iv, 558; IC, AM, v65, 1788.
65. NAS, FE, E741.
66. McArthur (ed.), *Lochtayside*, pp. xlvii–xlviii, 8–9; NAS, FF, E729/1: NAS, RHP972/5.
67. NAS, E729/9/1; *OSA*, 1791–99, pp. viii, 198.

68. NAS, GD112/10/10, p.136.
69. NAS, RHP 972/5.
70. NAS, BP, GD112/59/22/2.
71. Ibid.
72. Ibid.
73. NAS, RH2/8/24.
74. McArthur (ed.), *Lochtayside*, p. 25; *OSA*, 1791–99, xv, p. 111.
75. W. Marshall, *General View of the Agriculture of the Central Highlands* (London, 1794), pp. 38–9; *OSA*, 1791–99, viii, p. 568.
76. BPP, *Report of the Commissioners of Inquiry into the Conditions of the Crofters and Cottars in the Highlands and Islands* (London, 1884), Appendix A, pp. 389–92.
77. R. A. Dodgshon, 'Coping with risk: subsistence crises in the Scottish Highlands and Islands', *Rural History*, 15 (2004), pp. 1–25; J. Lightfoot, *Flora Scotica: or a systematic Arrangement, in the Linnean Method, of the Native Plants of Scotland and the Hebrides* (London, 1792), 2 vols.
78. Ibid., i, pp. 268–9.
79. NAS, AF49/1.
80. NAS, FE, E745/59.
81. NAS, FE, E783/98.
82. A. Millar (ed.), *A Selection of Scottish Forfeited Estates Papers 1715;1745*, SHS, lvii (Edinburgh, 1909), p. 105. See also, NAS, E729/1; Mackay (ed.), *Walker's Report*, 43, p. 185.
83. *OSA*, 1791–99, viii, p. 339; xi, p. 424; Grant, *Highland Folk Ways* (London, 1961), pp. 73, 75, 97–8; A. Fenton, *Scottish Country Life* (Edinburgh, 1976), p. 136; J. H. Dixon, *Gairloch in North West Ross-shire*. (Edinburgh, 1886), p. 136.
84. R. A. Dodgshon, 'Strategies of farming in the western Highlands and Islands of Scotland prior to crofting and the clearances', *Economic History Review*, xlvi (1993), pp. 697–9.
85. Ibid., pp. 697–9.
86. For example, NAS, GD112/9/3/3/14; see also, T. C. Smout and F. Watson, 'Exploiting semi-natural woods, 1600–1800', in T. C. Smout (ed), *Scottish Woodland History* (Dalkcith, 1997), p. 91; T. C. Smout, A. R. Macdonald and F. Watson, *A History of the Native Woodlands of Scotland, 1500–1920* (Edinburgh, 2005), p. 119.
87. O. LeLong and J. Wood, 'A township through time: excavation and survey at the deserted settlement of Easter Raitts, Badenoch, 1995–1999', in Atkinson, Banks, and MacGregor (eds), *Townships to Farmsteads*, pp. 44–5
88. Millar (ed.), *Scottish Forfeited Estates*, p. 141
89. NAS, GD112/59/22/2.
90. NAS, FE, E729/2.
91. NAS, FE, E746/166; see also McKay (ed.), *Walker's Report*, p. 65; *OSA*, 111, 1795, p. 372.

92. J. Loch, *An Account of the Improvements on the Estates of the Marquess of Stafford* (London, 1820), pp. 64–5

93. NAS, GD112/9/20. See also, GD112/9/9; GD112/9/16; GD112/10/7; GD112/10/8; R. A. Dodgshon, 'Bones, bows and byres: early dairying in the Scottish Highlands and Islands', in P. Rainbird (ed.), *Monuments in the Landscape* (Stroud, 2008), pp. 165–76.

94. NAS, GD170/431/5, 8A and 8B.

95. M. Martin, *A Description of the Western Islands of Scotland* (London, 1716 ed.), p. 155.

96. Macphail (ed.), *Highland Papers*, vol. I, pp. 277 et. seq.

97. Loch, *Account*, pp. 29, 64–5; Dixon, *Gairloch*, p. 136.

98. F. McCormick, 'Calf slaughter as a response to marginality', in Mills and Coles (eds), *Life on the Edge*, p. 51.

99. Dixon, *Gairloch*, p. 136; NAS, GD305/1/163/119. See also, NAS, GD112/9/9.

100. Smout and Watson, 'Exploiting semi-natural woods', p. 86; Smout *et al.*, *History of the Native Woodlands*, pp. 61, 64, 67.

101. For example, Breeze, 'The great myth of Caledon', in T .C. Smout (ed.), *Scottish Woodland History* (Dalkeith, 1997), pp. 46–50.

102. H. M. Steven and A. Carlisle, *The Native Pinewoods of Scotland* (Edinburgh, 1959), Figs, 2 and 13–37; Smout et al., *History of the Native Woodlands*, p. 4.

103. Ibid., p. 74.

104. M. Stewart, *Loch Tay: Its Woods and its People, The Woodland History of Lochtayside* (Aberfeldy, 2000).

105. NAS, RHP 2494, vol. 1.

106. Smout and Watson, 'Exploiting semi-natural woods', p. 94.

107. F. Watson, 'Rights and responsibilities. Wood management as seen through baron court records', in T. C. Smout (ed.), *Scottish Woodland History*, p. 105.

108. Cregeen (ed.), *Argyll Estate Instructions*, pp. 110–11

109. NAS, GD112/14/13/6/1, 1722.

110. Smout et al., *History of the Native Woodlands*, p. 139; Watson, 'Rights and responsibilities', p. 106.

111. NAS, GD112/16/4/3/44, 1799.

112. IC, AP, Box 2532, Memorial 1706; ibid., Instructions . . . Morvern, 1733; Cregeen (ed.), *Argyll Estate Instructions*, p. 51.

113. IC, AP, Instructions . . . Morvern, 1733.

114. DC, MDP, 1/466/22.

115. DC, MDP, 380/28, Plan of Glenelg; NAS, RHP23075/1.

116. DC, MDP, 1/466/3.

117. NAS, GD112/16/10/2/1.

118. For example, NAS, GD170/431/5; NAS, GD170/431/41.

119. See also, Smout et al., *History of the Native Woodlands*, p. 106.

120. NAS, GD50/136/1, vol. 1.

121. NAS, GD112/17/11, 28 June 1728, 18 February 1730, 13 December 1732.

122. NAS, GD80/384/, 8 December 1740.

123. NAS, GD112/17/9.

124. NAS, GD44/25/2/76; NAS, GD129/2/353, 30 May 1661.

125. For example, NAS, GD112/16/10/2/20; McKay (ed.), *Walker's Report*, pp. 102, 205.

126. For example, NAS, FE, E741; NAS, GD112/16/10/2/24, Report by Dougall McPherson 1795.

127. NAS, GD50/2/1.

128. IC, AP, Bundle 663, Instructions . . . Campbell of Airds, 1733; NAS, E769/72/6; NAS, GD174/827; NAS, GD112/16/10/2/23; Smith, *Jacobite Estates*, p. 82.

129. Smout and Watson, 'Exploiting semi-natural woods', p. 92.

130. NAS,GD44/51/743/7; Smith, *Jacobite Estates*, p. 59.

131. IC, AP, Bundle 663, Instructions . . . Campbell of Airds, 1733.

132. NAS, GD112/17/4.

133. NAS, GD112/17/6.

134. NAS, GD112/17/11.

135. Ibid., 18 February 1730.

136. NAS, GD201/1/227A.

137. IC, AP, Minutes of Proceedings . . . October 1747; NAS, GD201/227A.

138. NAS, GD112/17/11, esp. 15 June 1722 and 28 June 1727.

139. Watson, 'Rights and responsibilities', pp. 106–7.

140. NAS, GD112/16/10/2/13, Report . . . John Campbell, part 2, 1783.

141. Cregeen(ed.), Argyll Estate Instructions, p.108.

142. Ibid., pp. 109–10.

143. Ibid., p.16.

144. Ibid., p. 111.

THE HIGHLAND TOUN THROUGH TIME:
AN INTERPRETATION

> The Highland 'township' . . . has never possessed any corporate existence in the law of Scotland. It has been, as far as the law is concerned, simply a farm or part of a farm, occupied in common or a division by several tenants.[1]

Surveys produced on the eve of the Clearances and the creation of crofting townships can tell us a great deal about how the Highland toun evolved. Yet, even had these surveys not been compiled, this would still represent an opportune moment for those researching the history of touns. This is because many former runrig touns were simply left abandoned when replaced by a sheep farm or shifted to a new crofting site, with little attempt to tidy up the old in the making of the new. What we can still find by way of upstanding remains – dwellings, outbuildings, rigs, dykes and enclosures – add greatly to the potential information available on how touns were ordered and how they might have evolved.

I want to take the different types of data available at this critical juncture and consider what they reveal about how touns may have evolved in the very long term. My discussion will explore first, the documentary and, second, the upstanding field evidence. Then, in a third and concluding section, I shall combine such data with some of the ideas concerning the development of touns which were touched upon in Chapters 2 to 4 and, through their synthesis, I shall offer a provisional outline of what I see as their long-term evolution.

Touns on the Eve of Change: a Backward Glance

Most eighteenth- and nineteenth-century surveys were intended as a stocktake ahead of change, enabling landowners to look forward, but they also enable us to invert that perspective. Four types of evidence are worth highlighting. The first concerns how touns were arranged on the ground. As already noted,

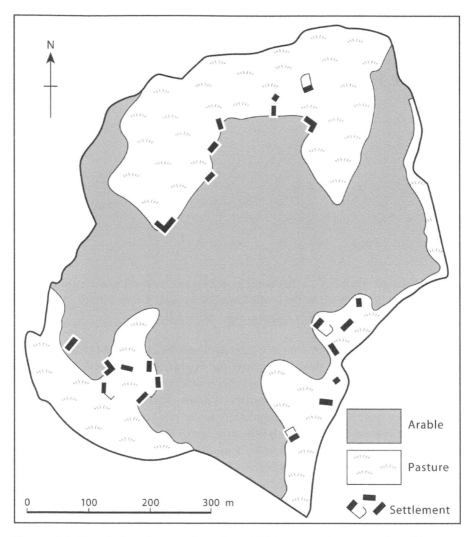

Figure 6.1 Rosal, Strathnaver. Based on 1811 estate plan, reproduced in Fairhurst, 'The surveys . . . Sutherland Clearances', plate 11.

while we can find examples that fit the stereotype of the Highland toun as farm community focused around a small, clustered but amorphous settlement, the cartographic surveys produced on the eve of change show that quite a number of touns were organised around two or more separate clusters of settlement, some being linked to wholly detached blocks of arable within a toun. Rosal in Strathnaver, a toun surveyed in 1807 ahead of its clearance in 1814, provides a good example, with its four or five separate foci of settlement spread out around the edge of its arable (Figure 6.1).

Touns with multiple settlement foci, especially those that also had physically fragmented blocks of arable, raise interesting questions because they
come across as composite touns, meaning that they have the appearance of
having once been separate blocks of arable and settlement that have become
fused at some point. Of course, for reasons mentioned in Chapters 2 to 4,
there were other good reasons why some touns may appear composite. The
internal order of touns was far from being fixed and immutable over the medieval and early modern periods. Tenant numbers increased and decreased. In a
variegated environment such as the Highlands, it need not come as a surprise
if extra numbers were accommodated by adding settlement on a wholly new
site within the toun just as contraction might have been achieved through
the retreat of settlement on to a single site. Similarly, given that they had the
choice, multiple tenants in a toun could arrange themselves into a single runrig
toun, into split touns or even into consolidated holdings. There may even have
been situations where tenants in large touns divided themselves informally
into subgroups, each of which operated a localised system of runrig but which
grazed in common, as Geddes surmised for the touns on Lewis[2] and how Shaw
interpreted her wider documentation of such 'hamlet farms'.[3] In other words,
while, at one level, touns may have formed part of a fixed and unchanging
pattern of landholding, there was another level at which they were responsive
to issues of social scale, the presence of variegated ecologies and so on.

A second type of evidence that can be gleaned from eighteenth-century
surveys is the presence of enclosures. The enclosures being referred to here are
not inner or outer (= head) dykes, or the kailyards, penfolds and other small
stock pounds that were widely present by the early eighteenth century but are
the field enclosures laid out across *parts* of a toun's arable. In some cases, they
come across not as isolated features in the toun landscape but as something
that helped to characterise it at large. To restate a point made in the previous
chapter about Tiree, Walker's first survey, *c*.1764, said that '[A] great part of
the Fields' on Tiree were enclosed. He did not feel they answered the needs
of enclosure as they were 'perpetually crumbling'[4] but that would be a sign
that they were not a recent feature. Robertson's 1769 essay on Orkney made
a similar point, observing that the 'arable fields are in-closed with mud walls
[= earth dykes]' but they were 'irregular & insufficient to protect them from
the incursions of the cattle, sheep & swine that stray thro' the commons'.[5] In
some instances, cartographic surveys went to the trouble of recording these
enclosures. Pre-Clearance and pre-crofting plans for Coll and for the MacLeod
estate on Skye, for instance, both record local areas of enclosure. Those
depicted on Langlands' 1794 map of Coll[6] are difficult to interpret. Both on
Langlands' map and on the ground (Figure 6.2), they come across as composite, embracing enclosures or boundaries of different vintages.

The *c*.1790 map of the MacLeod estate on Skye records field enclosures in a
number of touns along the western side of Waternish, from Unish at the northern

Figure 6.2 Premodern enclosures on Coll.

tip down to sites such as Risagan, Fasach and Trien.[7] Even from such maps, it is clear that enclosures did not embrace all the arable that existed at this point but only parts, suggesting that different parts of arable had different histories.

A third type of evidence concerns how the different spaces within touns were labelled. The standard approach has been to talk generally about the arable of touns divided between an infield and an outfield. In fact, this was not always the case. Only a limited number of Highland areas actually used the terms infield and outfield. Instead, names such as croft land, intown and wintertown were applied to infield, and, where it existed, outfield could be described by terms such as 'sheelings'.

The fourth type of evidence that can be abstracted from eve-of-change sources is that relating to building styles, the materials used and how they varied across the region. Some made use of drystone walling. Others made use of turf walling, either built up over a low stone footing or built up from ground level: as a capping for low drystone wall or as a facing for wattle. Observers from outside the region were especially struck by the use of turf and how it was used in house construction. Robertson's tour of the western seaboard areas in 1768 provides us with a good description of the turf-based houses prevalent in Arisaig and Knoydart.[8] He refers to them as basket houses, or what others called creel houses. These were terms used to describe houses where the turf walls were set against an inner framework of wattle, a frame usually constructed by a technique known as 'stake and rice'. What Robertson had to say about their construction is supported by other sources for the Central Highlands,[9] as well as by mid-eighteenth-century surveys for Barrisdale[10] and Locheil.[11] Even Boswell and Johnson could vouch for their construction for, in their journey through Glenmoriston in 1773, they rested in an inn 'built of thick turf' with 'side-walls' that 'were wainscotted ... with wicker, very neatly plaited'.[12] Settlements also had a range of outbuildings, byres for stock and barns for storage[13] as well as for grain drying. These, too, were generally based on a wattle framework though they may not necessarily have been faced with turf. In his 1771 tour, Robertson talked about cut grain in the central Highlands being carried home immediately after cutting and 'deposited in barns of singular construction. The walls are composed of twigs twisted like basket-work; and the roof is covered with heath, straw turf, or fern.' These drying and storage barns were not small structures but were generally 14 feet (4.27 metres) in width.[14] A 1758 petition relating to a toun in Kilmorack, on the Lovat estate, confirmed this use of wattle for outbuildings but did so by drawing a distinction with the dwelling house. Down to the mid-eighteenth century, it claimed, 'dwelling houses were built of Turff and earth only and their barns with salke and rue'.[15]

Whatever their exact style of construction, traditional dwellings made substantial use of organic or perishable materials, not just for roof cover (that is, heather, straw, ferns or turf) but also for walls or wall capping. This dependence on such resources is particularly well conveyed in an 1829 description by James Loch which speaks of housing being difficult to decipher at a distance because of the way it appeared to merge in with the natural vegetation of the hillsides. Though written when changes in housing were already well under-way, he talked about 'the surprise one feels upon coming suddenly near one of these little towns ... it is not until one is close to them [one can] distinguish them from adjoining mountains'.[16] The use of organic materials had another consequence, though. It was routine for the organic matter used (such as turf, various types of thatch) to be recycled as compost and used as a field manure on a regular basis. Contemporary accounts suggest that roofing thatch was recycled on a short, even annual, cycle while the turf used for walls was replaced on a longer cycle, sometimes as short as three years[17] but, in other cases, every ten to twelve years.[18] Wattle also had a limited life, its need for regular renewal being one of the reasons why, by the early eighteenth century, estates started to ban its use. Overall, the extent and frequency with which the most basic materials used in the traditional house were renewed mean that we need to see such housing as passing through regular cycles of construction and deconstruction. The recycling of wall and roofing turf, together with thatch and turf, as compost for arable played a key role in this process of construction and deconstruction but they were only part of it.

The most valuable parts of the traditional house were its roof couples, that is, the timbers that provided vital roof support. The importance of roof couples to how houses were perceived is made clear by the way in which tacks for the Breadalbane estate defined houses, barns and byres in terms of how many 'cupples' they were expected to comprise, with tenants being required to build or to leave, at their end of their tenure, buildings, including byres, made up of a given number of couples.[19] Where tenants had the right to remove couples because they had paid for them,[20] it would have meant that houses were reduced to their wall footings, or 'doun cassin' as one source puts it, at the point when tenants left the tenancy of a holding.[21] Like the periodic removal of roofing for compost, such 'doun-cassin' would have added to the sense of process that must have surrounded traditional housing. It puts in perspective those references to the eviction of tenants being signalled by the 'doun cassin' of their houses. What may seem a drastic response by an estate was actually what happened generally when tenants reached the end of their occupancy and left a toun.

Given these regular cycles of construction and deconstruction, traditional housing needs to be seen more as a continuing process than as fixed and durable structures. We gain a sense of the cycle from an observation by Loch. Emphasising the fact that little attempt was made to sweep out interiors, and probably overemphasising the point, he claimed that 'when the accumulation of filth rendered the place uninhabitable, another hut was erected in the vicinity of the old one. The old rafters were used in the construction of the new cottage, and that which was abandoned formed a valuable collection of manure for the next crop.'[22] Such a turnover has two implications. First, house-site clusters would have been fairly dynamic affairs, with the houses in occupation moving between different positions and orientations. Second, among the upstanding remains that we see present on former toun sites, we cannot assume that all sites were occupied at the same time.

If we review mid- to late eighteenth-century documentary data on housing, it is clear that the balance between, on the one hand, turf walling and, on the other, the use of drystone or stone and mortar, was changing. Even by the early eighteenth century, traditional housing's consumption of young timber for wattle and mature timber for roof couples brought it into conflict with estate policies that were increasingly protective of what it now viewed as a valuable resource. Likewise, the huge demand for turf generated by its use for walls, roof cover and dykes, especially where cut from green pasture, brought it into conflict with those who saw the growing value now attached to livestock as having a greater claim. This is why we find estates banning the use of wattle and the cutting of green pasture for turf. Bound up with these restrictions, we find estates taking a more direct interest in the nature of housing. As early as 1618, tenants on the Benderloch portion of the Breadalbane estate were being ordered by an act of court to build 'no house no bigging ... Bot onlie with stane'.[23] Elsewhere, the switch away from turf and wattle appears to be more of an eighteenth-century change, as with the 1730 act passed by the Disher and Toyer Court in Lochtayside requiring that all houses, byres and barns being built or repaired 'shall be made with sufficient stone gavills & and any divisions to be made with stone',[24] while another banned the use of 'ryce' for walls and partitions on the estate. Later still, a 1769 order on the Gordon estate required houses to be rebuilt in 'stone & lime or stone & clay or mortar' with 'no feal used'.[25] Others nearby needed no such prompting from estates. The tenant of Dell on the Lovat estate petitioned in 1741 that given his 'dwelling, a creel-house after the ordinary manner of the county, is entirely ruinous and the couple trees and other timber worn out', he was 'desirous of rebuilding the farm-house with stone walls instead of the earthen walls now used'.[26] When we get detailed reports later in the century, some depict local housing as caught between the old and the new, with houses built from stone in the process of replacing creels or basket houses. This was the situation when the Locheil survey was compiled in 1760, with some touns having a mix of the two types. In Eracht, some, newly built of drystone, were still 'turfed on the outside' to provide extra insulation.[27] Changes in building style in the Hebrides can also be detected through documentary sources. References to mid-eighteenth-century housing suggests that Hebridean housing at this point had a smaller footprint but it still made significant use of turf for walls and roofing. Boswell's mid-eighteenth-century comment about turf cabins being 'widespread' on Skye[28] would have been applicable to other parts of the Hebrides though, in many cases, the turf was used to add capping or height to drystone walls and to provide roofing cover. As on the mainland, the growing appreciation of the need to maintain pasture for stock encouraged estates to regulate turf use more closely.[29] By the mid-eighteenth century, tacks for the Lochbuie estate on Mull and MacLeod estate on Skye can be found addressing the problem directly by banning the use of turf for building and stipulating that, thereafter, all housing had to be built of stone.[30]

Highland Touns on the Eve of Change: Reading the Ruins

A prominent feature of Highland landscapes is the way in which decayed houses and dykes of old farming touns are still clearly visible on many sites. Even where touns were replaced *in situ* by new crofting townships, we can find in abundance remains of its pre-crofting forms owing to the way in which the new townships were displaced or put on a different alignment. When we look at what survives, toun sites do not come across as unidimensional land-scapes, fragments of an old landscape that has survived unchanged. What we see amounts to a compounded landscape, one made up of elements that were clearly added at different times.

From this field evidence, I want to look at three particular aspects of this compounded landscape: settlement and housing; arable and the marks of culti-vation; and the distinction between open and enclosed space.

Settlement and Housing

The archaeological potential of the many touns abandoned for sheep or crofts was first demonstrated by Fairhurst's work on the split toun at Lix (East, West and Mid) in Perthshire. This toun was cleared for sheep in the mid-eighteenth century and excavated by Fairhurst in the late 1950s to early 1960s, ahead of its afforestation. The site was found to have a number of stone-based longhouses and outbuildings arranged around a courtyard but, when excavated, their base layers extended back only to c.1700. In fact, Fairhurst found no archaeological trace of settlement dateable to before c.1700 and reasoned that earlier settlement on the site must have been based on perishable materials, such as turf or wicker-frame walling.[31] Finding traces of turf-based housing on any site can be a challenge, especially where communities routinely recycled used turf as compost. The fact that Lix was a split settlement might also have a bearing on the depth and location of particular occupation layers at Lix because splitting usually involved shifts of settlement. More recent work, however, carried out at Lairg in Sutherland[32] and at Kiltyrie on the northern side of Loch Tay[33] have located archaeological traces of former turf-based dwellings, their search made easier by the way in which they were developed over low stone footings. The work at Kiltyrie was especially revealing. It was carried out as part of the Ben Lawers project and focused on a localised scatter of subcircular mounds and rectilinear structures with rounded gable ends positioned above the head dyke, at a height of 320 metres. All were seen by Atkinson as 'ephemeral in nature and constructed of turf . . . set around a timber frame which supported the thatched roof'.[34] The outstanding feature of the excavation was the way it established a real-time depth to settlement. In the case of one site, occupation was spread over three phases, the earliest being between the late eighth and tenth centuries AD, the second between the mid-twelfth and late thirteenth centuries AD, and the third comprising a phase when shieling huts had been constructed over earlier structures, possibly during the fourteenth and fifteenth centuries. In the case of another site, excavation uncovered structures that were dated to the twelfth to fourteenth centuries AD, though traces of earlier occupation were also evident.[35] The Kiltyrie excavations are important because they provide us with a sense of how rural housing in the southern Highlands changed over the medieval and immediate post-medieval period. The larger, more regular, stone-based dwellings, whose foundations still existed at Lix and which can still be found on the lower ground of many Lochtayside touns, were not built until the mid-eighteenth century. As Atkinson makes clear, we need to distinguish such structures in scale, plan and materials used from what existed during the preceding five or six centuries.[36] A similar conclusion was reached by the excavations at Easter Raitts in Badenoch. Four longhouses were excavated but none showed signs of the site being occupied earlier than the seventeenth or eighteenth centuries.[37] As at Lix and Kiltyrie, what existed before probably involved different sites and different styles of dwellings. For many parts of the Central and West Highlands, we can expect dwellings with turf-based walls set against a wattle framework to have been fairly common-place down into the first half of the eighteenth century, though regular changes of stance, and the easy reuse of stone footings (or their absence in the first place) will have made their location elusive on many sites. One way forward may lie in work also carried out as part of the Ben Lawers project which has tried to establish a geochemical soil signature to detect former settlement sites and to distinguish between different functional areas, such as living space, byres, and so on.[38]

At the centre of any discussion about drystone building forms in the Hebrides is surely what field evidence can tell us about the nature and origins of the blackhouse, the area's most emblematic form of housing. For many, it was the classic form of longhouse, one that captures the great antiquity for the practice of housing stock and humans under the same roof.[39] Those that survive at Arnol, on the north-west coast of Lewis (Figures 6.3d, 6.4 and 6.5), have been subjected to particularly close study.

First described in the 1930s,[40] it has more recently been subjected to a house-by-house study by Historic Scotland in a survey that follows on from the detailed study of number 42 (as part of its preservation) and the detailed excavation of number 39.[41] What survives at Arnol illustrates the key features of the blackhouse, at least as it developed in Lewis: thick, low walls (around 1 metre in height) that were mostly double skinned and infilled, most with very angular corners, roof couples that rose from the inner edges of the house walls, occasional wall beds, a byre section set at a lower level to ensure drainage away from the living space, and sometimes the use of a door or removable end wall to take manure from the byre. In some cases, an attached *fosglan*, or porch, on one side and a barn on the other (Figure 6.5) served to elaborate the form of the blackhouse.[42] Indeed, following Thomas,[43] some have linked the blackhouse to an early nineteenth-century type known as a *creaga* which had an even more elaborate form, with different houses conjoined so as to make a small complex, cellular structure.[44] Figure 6.3c shows the example used by Thomas as the basis for his illustration of a *creaga*. Yet, far from captur-ing the antiquity of the blackhouse, its scale and structural differentiation of functional space suggest the influence of estate regulations by *c*.1800 over the design of dwellings while the conjoining of examples – a feature of Thomas's definition – can be attributed to the practice whereby close kin who now shared a holding sometimes built side-by-side dwellings.

We can extend this perspective on the blackhouse by looking at its wider Hebridean context. Like those on Lewis, blackhouses on the Uists and Barra have now been extensively studied.[45] In character, they display some differ-ences from those on Lewis. Their walls were not so low and, measured on their long axis, they were not so large but, otherwise, they had the same feature set,

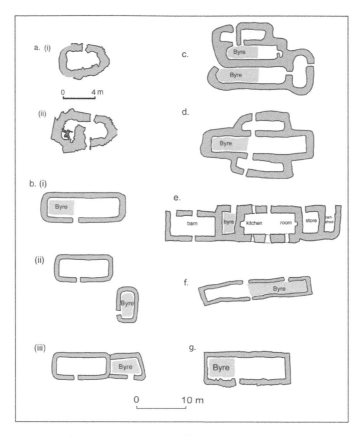

Figure 6.3 Forms of post-medieval dwellings: a. (i) Primary occupation at Druim nan Dearcag, Loch Olabhat, North Uist, after Armit, 'Excavation . . . post-Medieval settlement', *PSAS*, 127 (1997), p. 906; a. (ii) Secondary occupation, Druim nan Dearcag, Loch Olabhat, North Uist, after ibid., p. 906; b. Scaled representations of eighteenth- to nineteenth-century dwellings in the Western Isles based on (i) example from South Uist in which part was later adapted into a byre, as in Parker Pearson, 'Settlement, agriculture and society', p. 420; and (ii) and (iii) based on examples from Barra where byres were added as either free-standing structures or extensions, Branigan and Merrony, *Barra*, p. 218 and Branigan and Merrony, 'Hebridean blackhouse', pp. 1–16; c. Example of *creaga* after Thomas, 'On the primitive dwellings', Figures 4 and 5; d. No. 39 Arnol, after Holden, *Blackhouses of Arnol*, p. 27; e. Dwelling at Ruaig, Tiree, after RCAHMS, *Argyll*, vol. 3: *Mull, Tiree, Coll and Northern Isles*, p. 245; f. Longhouse at Lianach, Perthshire, after J. H. Stewart and M. B. Stewart, 'A highland longhouse: Lianach, Balquhidder, Perthshire', *PSAS*, 188 (1988), p. 303; g. House site, East Lix, Perthshire, after Fairhurst, 'The deserted settlement at Lix', pp. 182–3.

Figure 6.4 Blackhouse, Arnol, Lewis.

Figure 6.5 Blackhouse interior, Arnol, Lewis, showing byre at lower level and presence of *fosglan* and barn.

with access via an entrance on the side (or both sides), roof timbers that rose from the inner edge of their walls, hip-ended roofs, a central hearth and no windows.[46] For some, what mattered, and what classed them as blackhouses alongside their counterparts on Lewis whatever their differences, was the simple fact that they lacked windows. In his study of blackhouses on North Uist, Crawford reached the telling conclusion that no blackhouse in use at that point (the 1960s) was actually built earlier than 1800.[47] We can probably apply this conclusion to all Hebridean blackhouses that survived at this point. In part, this can be explained by the way in which many Hebridean touns underwent a reorganisation into crofting townships over the opening decades of the nineteenth century, with houses being rebuilt on new sites and invariably to new designs. In fact, later, touns, such as Bragar and Arnol, underwent a double reorganisation with an initial switch to a crofting layout, then a further shift to a second layout so as to accommodate extra crofter numbers.[48] The fact that no blackhouse in occupation *c.*1960 had been built before *c.*1800 does not mean that the tradition of building such house types does not take us back further but, equally, it would be a mistake to think its appearance alone suggests it does. We need to take account of what Crawford concluded from his excavations at the Udal on North Uist. Not only were the eighteenth- and pre-eighteenth-century longhouses present at the Udal smaller, and had more rounded corners, than the blackhouses to be found surviving from the nineteenth century but also there was no direct descent in terms of building style from earlier Norse longhouses.[49] The same point, at least as regards the smaller size of pre-eighteenth century longhouses, has also been highlighted by recent field surveys and excavations on Barra and South Uist. On the former, smaller blackhouses, with relatively thinner walls and varying between 7 and 8 metres in length, that date from before the eighteenth century, have been contrasted with the more substantial, double-skinned structures of the late eighteenth and early nineteenth centuries.[50] The excavations and surveys carried out as part of the South Uist project highlight a comparable shift, contrasting the smaller, turf-walled blackhouses built at a site such as Beinn na Mhic Aongheis, just inland from Bornais, South Uist, with the typically more substantial walling of nineteenth-century blackhouses surveyed at Airigh Mhuilinn, barely a mile or so to the south.[51]

This apparent increase in house size can be seen as part of a general shift in house forms that took place across the Long Island over the seventeenth to nineteenth centuries and partly conveyed by comparing the house plans in Figures 6.3a to 6.3b with those in Figures 6.3c to 6.3d, even allowing for claims over the antiquity of Thomas's *creaga*-type blackhouse. Some aspects, however, are obscured by the lack of field evidence from the late medieval and immediate post-medieval period, a gap that may relate as much to a shrinkage of settlements as to a greater use of turf.[52] Armit's excavation of a site at Druim nan Dearcag, beside Loch Olabhat, North Uist, provides some indication as to the nature of basic peasant housing, at least during the sixteenth and seventeenth centuries (Figure 6.3a). The site comprised a cluster of small, oval-shaped or subrectangular structures, five in total. Two were deemed to be houses, two were outbuildings and the fifth, a small enclosure. The two houses measured 4 metres by 2 metres internally and 6 metres by 4 metres externally though one underwent a small extension lengthwise during a secondary phase of occupation. In construction, houses and outbuildings used drystone and turf for walling.[53] There are hints of small, oval or subrectangular structures at other sites, their compact size and shape picked out by low, grassy mounds, as in Tusdale (Minginish) on Skye, and at Harris, on the west side of Rum. Usually, they occur in pairs, one probably a dwelling and the other an outbuilding. Seen in context, they come across as the oldest upstanding remains on pre-Clearance or pre-crofting sites. If we make the comparison with Druim nan Dearcag, they probably represent the field remains of settlement from the early eighteenth century and before. Alongside them, and forming the most prominent form of housing on pre-Clearance and pre-crofting sites, we find larger upstanding structures. Lengthwise, many of the smaller examples range between 8 and 10 metres internally but, on some sites, we can find larger examples that vary between 10 and 12 metres in length. Most are subrectangular in shape, with rounded corners, though some, probably those built just before the Clearances or the spread of crofting, had more clearly angled corners. Many had substantial barns or byres adjoining them, end on or side on, while many also possessed small kailyards attached or free-standing. A site such as the older part of Illeray, North Uist, largely abandoned by the mid-eighteenth century, provides a good illustration of the range of the relatively smaller house structures present, once we get back to the eighteenth century (Figure 6.6).

In origin, the emergence of the larger, more regular dwellings was probably a response to the closer regulation being exercised by estates over the construction of dwellings by the late eighteenth century. Most striking of all is the way in which some housing adopted a more elongated floor plan by incorporating a byre under the same roof as humans, a trend that ultimately produced the large blackhouses, averaging 20 to 25 metres in length, that we see on islands such as Lewis by the mid-nineteenth century. Any comparison between the houses built at Bragar, when it was first laid out as a crofting township *c.*1812, and the larger blackhouses that were built in the mid-nineteenth century, when its crofts were relaid, captures this shift in scale. We can even see the shift in coastal areas on the mainland opposite. A comparable change in housing over the late eighteenth and early nineteenth centuries was also a feature of Assynt, with worked stone replacing gathered stone, clay mortar being used instead of drystone and turf, regular or angled corners replacing more rounded ends, and houses adopting larger ground plans.[54]

It follows from what I have said that we cannot take the antiquity of the blackhouse or other longhouse forms for granted. It may seem an archaic

Figure 6.6 Former dwelling site at Illeray, North Uist.

arrangement but there is no archaeological proof that it had always been the standard for housing in the Hebrides. Admittedly, relatively few medieval house sites have been excavated in the Hebrides. Those of Viking or Norse origin that have been excavated, however, such as those on South Uist, show that even during the so-called Norse period, or post AD 1000, housing stock under the same roof as humans was an exceptional, rather than a regular, feature,[55] a conclusion that suggests Hebridean Norse dwellings were quite different from Norse house forms in the Northern Isles.[56] In fact, drawing on excavations targeted at medieval and early modern sites in the Uists, Parker Pearson et al. concluded that the use of part of the dwelling as a byre, effectively converting it to a longhouse form, appears in a systematic or regular way only during the eighteenth century when approximately one-third of the floor space within dwellings was assigned for such use,[57] a conclusion entirely consistent with the documentary evidence for the Hebrides.[58] These re-emergent longhouses had the same form as those of the Viking period, being elongated, some slightly oval in plan, with rounded corners, but were much smaller in length (8 to 10 metres) than their medieval counterparts. For comparison, a study of all blackhouse forms on Barra, including some that had been built in the eighteenth century, also found no evidence for the byring of stock under the same roof as humans. Instead, when the indoor housing of stock spread, they were mostly housed in frees-tanding byres or byres attached as an extension.[59] Patently, different islands responded in different ways when landlords introduced restrictions on the use of seaweed for manure.[60]

Working the Soil

Traces of pre-Clearance or pre-crofting cultivation survive in many areas but especially along the western seaboard, across the Hebrides and in Shetland. The peak for arable in the Highlands and Islands was probably reached in the decade or so after 1800. By this point, the Clearances were already starting to reduce the arable on farms in the main body of the Highlands. Typologically, we can draw a distinction between broad and narrow rigs. On balance, the former can be linked with the use of a plough and the latter, especially if steep sided, with the spade, *cas dhireach* and *cas chrom*. That said, linking rig form to the implement used is not straightforward. Some broad rigs, for instance, could just as easily have been worked by a group using the spade.[61] What is clear is that the *cas dhireach* or *cas chrom* were pre-eminently suitable for use where ground was broken by rock outcrops, had thin soils or was subject to waterlogging.

In part, the choice between using the plough, as opposed to the spade or foot plough, was bound up with the physical and social ecology of their use. If touns had access to broad, level sites and easily worked soils, such as one had along the machair in South Uist, then we can expect conditions to favour the plough. In contrast, where touns occupied difficult or marginal sites, then we can expect conditions to have favoured the spade, *cas dhireach* or *cas chrom*, with their greater suitability for working difficult, broken ground. In 'the intricacies between the craggs', there was 'no room for the action of a team and plow', a point not lost even on Dr Johnson.[62] Fenton's distribution map of where such implements were used bears out this ecology.[63] We must also not overlook, however, the equal importance of social ecology. The spade, *cas dhireach* and *cas chrom* required more labour than a plough so that we can expect their use only to have been feasible during times of increased population. In return, they provided a yield bonus at a time when it was needed. Of course, times of population increase were also when communities would have been under more pressure to crop difficult and marginal sites, the very sites on which the spade, *cas dhireach* and *cas chrom*, had advantages over the plough. It also follows that, as numbers increased and holding size fell, communities would have been less able to sustain plough horses anyway.[64] Fenton's discussion of early cultivation techniques added a further point. The *cas chrom* or crooked foot plough had an advantage over the *cas dhireach* or straight foot plough in that it gave greater leverage on stony ground, enabling communities to lever up stones and to bring more broken ground into cultivation. Combined with the fact that references to the *cas chrom* do not appear until the late seventeenth century at the earliest, he drew the conclusion that – far

Figure 6.7 Cultivation rigs on slopes overlooking Loch Odhairn, Park, Lewis.

from being a primitive survival – it was probably a development of the *cas dhi-reach* at a time when growing numbers pressed communities in the Hebrides and North-west Highlands to cultivate such ground.[65] As numbers grew and holding size fell, more landholders in these areas would have resorted to the use of spade and foot plough because it used the labour available, coped with the kind of ground that was now to hand, and maximised output. In use, they were effectively creating arable rather than just extending it (Figure 6.7).

The heavily corrugated landscapes produced by them lay behind Macculloch's description of what he saw on his 1811 tour as 'almost Chinese'.[66] We should keep in mind, however, that, where we have detailed data on their use per arable acreage, such as is available for Harris, (1803) and Ardnamurchan (1807), this growing use of the spade and foot plough did not exclude the plough even in these areas.

Was the Highland Toun Always an Open-field System?

The runrig layout of the traditional Highland toun, with its communal regulation of cropping and harvesting during summer and its rights of grazing across the harvest stubble and over all grass within the head dyke during autumn and winter, favoured an open system, one without any internal enclosures. That is why such field systems are called open-field systems. The only enclosure supposedly present was the head dyke, or the bounding dyke of the toun, though it

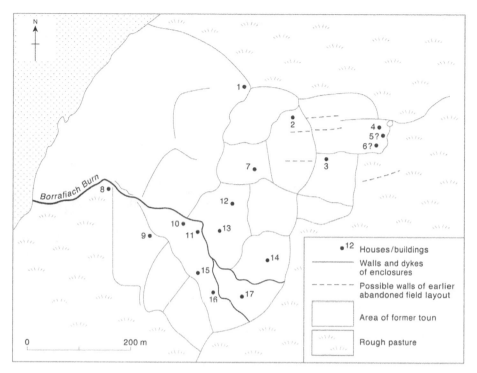

Figure 6.8 Former toun at Borrafiach, Waternish, Skye. Based on Dodgshon, 'West Highland and Hebridean settlement', p. 433.

was accepted that some touns possessed an inner and an outer dyke, one encircling infield and the other outfield. The head dyke was often a substantial affair so that it is not surprising that many have survived into the modern landscape, despite subsequent reorganisation of the toun into a crofting township and its later conversion to a sheep farm. Many touns, though, did not fit this stereotype of being entirely open within the head dyke. Both the field evidence and contemporary reports suggest that some had part of their arable subdivided into enclosures. In some cases, the number of enclosures present amounted to a significant coverage: this was the case at Borrafiach in Waternish, Skye (Figures 6.8 and 6.9).

While those on Skye, especially in the western and northern parts of Trotternish and Waternish,[67] provide particularly fine extended examples of these pre-Clearance enclosures, examples can be found elsewhere in the Hebrides, such as on Coll,[68] on Ulva and opposite, along the north-west and western sides of Mull,[69] on Mingulay and Barra,[70] and on Islay,[71] as well as in parts of the mainland, such as in Assynt,[72] Gairloch and Strathnaver. Inevitably, in what became crowded touns, or those heavily overdeveloped by crofts, enclosures may have been reduced by overdevelopment and stone

Figure 6.9 Enclosures at Borrafiach, Waternish, Skye.

robbing. The sort of local variation we find can be illustrated by the touns occupying the lower part of Tusdale on Skye. The core area of settlement in the lower part of the glen itself is labelled simply as Tusdale or Husdale on the First Edition Ordnance Survey 6-inch map but late seventeenth- and early eighteenth-century rentals suggest it actually comprised two touns, or what the 1683 rental for the MacLeod estate listed as Nuistall Meanish and Nuistall Eicrith.[73] What exists on the ground amounts to a fairly large extent of arable with upstanding remains of settlement at a number of different sites both within the main area of arable as well as around its edges. Some well-preserved enclosures are present but there are clear traces of other less well-preserved dykes. By comparison, the enclosures associated with the two touns positioned on the coastal slopes on either side of the entrance to Tusdale, both small, marginal sites, are reasonably well preserved.

In the context of runrig touns, the presence of such enclosures is anoma-lous. If we are dealing with mature open-field systems, systems that some see as having been developing since the Iron Age, we might reasonably expect no internal divisions of arable, and the only enclosures present to have been kailyards and small stockfolds (that is, penfolds). Of course, one other pos-sibility is that the enclosures present were somehow tied up with outfield cultivation, serving as outfield folds: this was the explanation offered for the pre-improvement enclosures found in Glen Campsie.[74] This is certainly a pos-sibility but, for many of the sites under review here, it faces three problems.

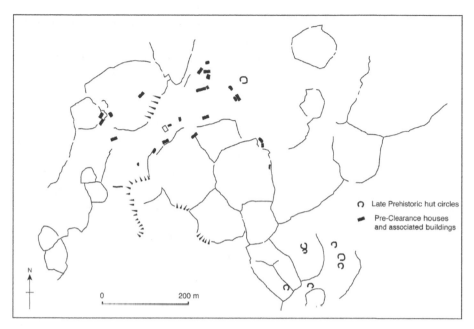

Figure 6.10 Enclosures and settlement in north Kingsburgh, Trotternish, Skye.

First, most descriptions of outfield folding suggest that, typically, most folds were either turf based or wattle/timber based and capable of being moved, reused or recycled. Second, many of the enclosures that we see in pre-Clearance or pre-crofting touns, such as those associated with the touns in the northern part of Kingsburgh on Skye (Figure 6.10), were associated with core arable areas so are likely to have related more to infield and not outfield. Third, some enclosures appear to have been treated as redundant by the time that touns were cleared for sheep or for crofting.

If we seek an alternative explanation, the most logical is to presume that these enclosures represent a residual feature, a carry-over from an earlier form of pre-runrig field layout.[75] The way we find cultivation rigs in Hebridean touns, such as those at Bresikor at the mouth of Glen Tusdale, extending beyond enclosures suggests they certainly did not have much functionality left by the eve of the Clearances. Before exploring this alternative further, we need to look at the wider context in which enclosures are found. Discussion of how touns developed have tended to see the problem in narrow terms, looking simply at what we see recorded as touns in rentals. We need a wider perspective, one that embraces what I have called the 'potential settlement array', that is, all those former sites of settlement or use that are referenced in documentary sources, together with all similar sites for which field evidence of settlement or use during the historic period can be established. Altogether, five different types of site make up this 'potential settlement array'. These sites are not discrete or mutually exclusive but, as types, shade into one another.

The first comprises those touns listed in rentals and surveys as the established units of the landholding. All bore an assessment rating. Some, but not all, also possess signs or traces of enclosures, free-standing or in small clusters, on part of their arable. Outwardly, their most important defining feature is that the majority had some form of head dyke, separating their arable and wintering ground from the hill pastures beyond, a head dyke that effectively made any enclosures within it redundant.

The second type comprises what were secondary or subsidiary touns, located on more marginal sites. Some originated as outsets, or as upgraded former shielings that survived in permanent occupation. This background meant that not all carried an assessment rating. Like Glenerachtie in Struan, they were 'not Valued in Mark lands as the other Farms, being of a later date'.[76] Of course, some may have been attached as daughter settlements to touns that were rated. While the names of some can be recovered from early rentals or from sources such as the First Edition Ordnance Survey maps, the names of others are more elusive, suggesting that they may have been wholly subsumed under a parent settlement. Being on more marginal sites, their arable was usually limited. Further, and this is their defining feature, their arable was enclosed but it was through one or two small enclosures rather than a head dyke bounding the system as a whole. This lack of a head dyke is all the more prominent where the enclosures present are not aggregated into a single cluster but occupy two or more separate clusters. Examples can be found scattered across the hill ground in the coastal areas of Gairloch, such as those that lie to the east of the Gruinard river in an area known as Strathnasalge. The largest, and most complex site, is that of Glenarigolach (or Arigolach), a settlement set down on the ground that rises way from the Allt Creag Odharr, a small burn feeding into the Gruinard river. We find it mentioned in seventeenth- and eighteenth-century sources, usually in tandem with the nearby site of Ridorach.[77] Though both are referred to in some sources as 'pasturages' or 'grazings and sheallings, with one referring explicitly to the 'pasturages thereof called Ridderach and Glenarigolach',[78] the complexity of the site at Glenarigolach, in particular, with its range of house sites, detached clusters of enclosure, cultivation rigs and clearance cairns, suggests that it had claims to being far more than a site of summer settlement (Figure 6.11). The way in which the larger clusters of enclosure are covered by discrete patches of bracken by late summer suggests that ferns were either recycled from bedding or roofing as compost or applied directly as a manure. Defining the site is actually quite difficult because of the number of enclosures that lie at a distance from the core of the site, including a discrete cluster of five that are located 600 to 700 metres to the south-east.

We can find other examples that are very similar to those of Strathnasalge

Figure 6.11 House sites and, above them, some of the higher enclosures at Glenarigolach, Gairloch.

inland from the southern shore of Loch Naver in Sutherland. That of Righcopag is named in the First Edition Ordnance Survey 6-inch sheet but has no explicit mention in early eighteenth-century rentals for the Sutherland estate, though it may have been among the small touns lumped together under the heading of 'Loch Naver'. It is positioned between Lochs Tarbhaidh and Ruigh nan Copag, a mile or so inland from Loch Naver (Figures 6.12 and 6.13), and consists of two separate clusters of enclosure plus some outliers between. Again, there are clear signs of former cultivation in some, with small clearances cairns and rigs lying across them. When surveyed by the RCAHMS in 1965, traces of at least eight buildings were found associated with the northern cluster of enclosures, including a drying kiln, and six with the southern cluster.[79]

The third type of site consists of shieling sites that still functioned as such down into the eighteenth century. These comprised fairly basic dwellings, commonly arranged as a series of small clusters lying across the more sheltered grazing grounds, usually close to a burn. Typically, they lay open, without any sign of associated enclosures. Those to be found on the dry spots that occur randomly across the waterlogged, peaty ground of central Lewis or in isolated parts of the Forest of Harris, illustrate the standard shiel, with two or three turf-based huts raised, where possible, over a stone footing. Exceptionally, we find shiels concentrated in more significant numbers, as with the sizeable clusters to be found on the upper slopes of Ben Lawers and the exceptional cluster

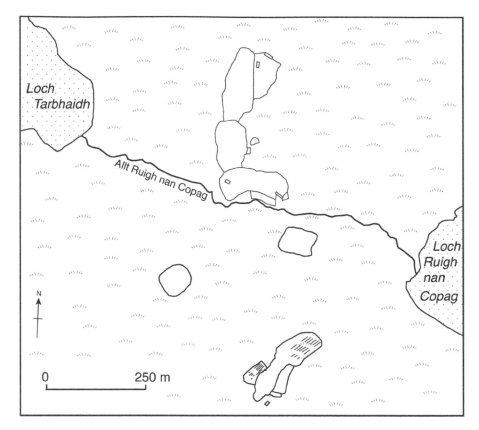

Figure 6.12 Enclosures and settlement at Righcopag, south of Loch Naver, Sutherland.

lying between Ben Forsan and Beinn Chàrnach in Duirinish, Skye (NG152 480).

The fourth type is a variant of this standard shieling but with the shieling organised around one, two or more enclosures. Some of these could have had a shieling-related function, such as the gathering folds around the shiels developed at sites such as Gleann Mór on St Kilda.[80] As shown in Chapter 5, however, many shieling sites served as safety valves for growth, meaning that, when numbers warranted it and climate permitted it, the occupation of shiels was intensified, with arable being created and the site being lived in all year. The large enclosure that surrounds shielings on rising ground towards the upper part of Sleadale in Minginish, Skye (NG 335288) provides a good example. Though classed as a former shieling site on the First Edition Ordnance Survey 6-inch map, the large enclosure associated with the site suggests that it may once have served as the basis for an all-year-round settlement when conditions were more favourable. Less than 2 miles to the south, at the

Figure 6.13 Pre-Clearance enclosures at Righcopag, south of Loch Naver, Sutherland.

head of Glen Caladale, is a site marked on the map as a shieling site. Again, though, it comprises a small cluster of enclosures with clear signs of former cultivation. The shieling site surveyed by Stewart in Monachyle Glen (NN 475215), with its 'complete enclosed field system' and 'rig and furrow patterns in most of the fields' provides a more complex example from the southern Highlands, one which had clearly transitioned into an all-year settlement at some point before reverting back to being a temporary shieling.[81] Despite his potential confusion between what was outfield and what was a shieling, some of the so-called 'sheeling sites recorded in Home's 1769 survey of Assynt fit into the category of shielings which, when climatic conditions were favourable or when need arose, were pressed into use for arable. When we set them down in the landscape, they appear widely spread. Some were actually set at quite a distance from the touns to which they are attached (Figure 5.2). The ambiguity over whether they should have been called outfields, rather than 'sheelings', springs largely from the fact that they were all cultivated, with some even having signs of local settlement as if, at some point, they had served as all-year-round sites. Furthermore, this arable was enclosed within one or two small, irregular enclosures. Collectively, they hint at an all-year-round footprint of occupation that potentially was more extensive than the core touns recorded in Home's survey.

The fifth type of site is represented by a range of small, scattered enclosures that are given no name, status or function in the sources available to us. Examples occur widely across the Northern and North-west Highlands. Even in Assynt, Home's map of 'sheelings' does not include all the small, free-standing, irregular enclosures that exist in the Assynt landscape. In some cases, these more nondescript enclosures were small enough probably to have functioned as stock-control pens. Faced with them in Assynt, Cavers and Hudson thought that the greater proportion 'could be assigned to sheep farming in the post medieval and modern centuries',[82] though one should really separate out the usually very regular enclosures that were built as sheep pens and fanks, following the early nineteenth-century Clearances for sheep, from the irregular forms that predate this point and which, in Cavers and Hudson's words, are more likely to be 'representative of earlier activity'.[83] Some of these irregular enclosures are too large to have functioned as stock-holding pens, however, and some display signs of occupation. Some of those in Strathnasalge, Gairloch, fall into this category. Elsewhere, some of the largest examples are the three curvilinear enclosures that stand detached from each other either beside or across the B871 road (NC 697433), just south of Dalvine Lodge. At least two have signs of what may be former settlement. Though the First Edition OS sheet records them as enclosures, it does not attach a name, nor can we make up for this deficiency by cross-matching them with a name in early listings for the Sutherland estate. What is instructive about such sites is that, if we construct an imaginary head dyke embracing all three, it would form a toun not noticeably different in scale or in the extent of its spread from the toun almost adjacent to them, Rosal.

Whatever their origin and whatever their precise chronology, I want to argue that, when we bring these different types of site together, we can tease out two important points about how the toun may have evolved. First, the human imprint on the landscape appears to have been far wider than that represented by the core touns, invariably assessed touns, that are listed in rentals. Even beyond the sub-touns and outsets, we not only find shieling sites that have clear signs of having, at some point, been used as all-year-round settlements, with enclosures that have clear indications of former cultivation but we also find similar sites that do not even have the status of shielings, either in estate surveys, early estate plans or First Edition Ordnance Survey maps. By way of an explanation, we can bundle these various non-core sites together – the sub-touns, outsets, shielings and the more nondescript sites of former use – and see them as those brought into occupation as demand increased over the early modern period. Some, however, could just as easily be explained as sites that were pioneered into all-year-round use during much earlier phases of expansion. In other words, some may predate the imposition of assessments. As such, they help to build a picture of a more diffuse and more disaggregated pattern of settlement and occupation prior to their imposition, with assessments being focused on those areas where settlement and landholding was already aggregated if not more integrated. Second, and perhaps more important, these lesser sites of occupation, such as

Righcopag, as well as the many shieling sites associated with enclosures, make a simple but vital point. In any pioneer colonisation of a site for all-year-round use, the first step was the separation of arable from pasture, a step accomplished by building an enclosure. As numbers increased, we can expect sites to have acquired a number of holdings based around such enclosures, some accretic, others free-standing. This is exactly what we find at Righcopag and similar sites. Without a head dyke, it can, at best, be seen as at an immature toun stage or even at a pre-toun stage. More than anything else, the head dyke was a statement about touns operating as touns not as individual holdings. Its construction made internal divisions or enclosures redundant. Seen this way, what Righcopag and similar sites may tell us is that the residual enclosures still to be found at more mature toun sites, such as those forming the northern part of Kingsburgh, may represent a similar stage in their growth as touns.

The Highland Toun in Time

While acknowledging that significant gaps still limit our understanding, I want to combine the evidence just presented with that on how assessments may have shaped the development of touns presented in Chapters 3 and 4 and to offer a provisional outline of the how the toun may have developed in the long term.

Before Touns

If we take the early views of writers such as Evans[84] and Uhlig,[85] our starting point for understanding how the farming toun, with its infield–outfield and runrig organisation, would be in the Neolithic, with the very first farmers. In other word, there is no 'before touns' stage of farming. Arguably, the evidence showing the importance of assessment to infield, the fact that outfield developed after infield, and the way in which the temporary character of outfield stemmed wholly from its the use of summer manure, not some basic notion of cropping as Uhlig supposed, undermine such an early chronology. Nevertheless, there are still some who would argue that key traits of the toun, such as infield, were at least present by the late prehistoric period, though such an assumption depends on labelling any area of manured arable as infield. Recently, Parker Pearson has added a new strand to the debate using data from the South Uist project. He highlights the fact that not only was the island's Late Iron Age settlement organised around clusters, that he calls 'villages', but also, if we territorialise the land around them, they show some accordance with the territories of later, eighteenth-century touns. This is a valuable observation but Parker Pearson goes on to make a bolder claim. He suggests that the similarities between the two forms may indicate that what became the 'toun', or what he calls the 'township', was already present by the Late Iron Age in the form of 'proto-townships'.[86] In other words, when it comes to debating the origin of runrig touns, it may have been all over by the Iron Age. Like all arguments that stress a deep-rooted continuity for basic institutions, such an interpretation strips out any sense of a dynamic or movement to the Highland toun across great swathes of its history. To accept such institutional stillness from the Iron Age onwards requires us to see the shifts of population, the equally significant swings in climate, and the far-reaching changes in the sociopolitical organisation of society that occurred over the medieval and early modern periods as leaving the toun untouched. Quite apart from writing out the impact of feudal concepts of tenure on landholding, stressing continuity from the Late Iron Age completely plays down the Norse impact on the island. When others write about how the latter brought about a 'seismic shift in social organisation', a shift 'reflected in settlement forms',[87] then how core institutional forms may have remained unchanged needs to be argued out every inch of the way.

I want to marshal a counterargument to this suggestion that the Iron Age provides us with a *terminus ad quem* for the emergence of 'proto-townships' in two stages. In this section, I want to raise some questions about how such an argument reads the Iron Age data and to put in its place an alternative reading of what the Late Iron Age and early medieval landscape was organised around in terms of settlement and field systems. In the next section, I want to lay out a revised chronology and process of development of how touns developed.

Parker Pearson's case for a similarity between Late Iron Age settlement on South Uist and its eighteenth-century touns rests, in the first instance, on their similarity of form, not their internal character. At no point in his discussion does he show how the very specific institutional character of the mid-eighteenth-century toun and its runrig landholding can be read from the Late Iron Age data, other than through the way clusters of settlement were a feature of both. In fact, at no point does his discussion elaborate on how Late Iron Age settlement organised its landholding or field economy, though the South Uist study as a whole has valuable discussion on what the island's field economy was based on in terms of crops and livestock if not on how its landholding and fields were arranged. Arguably, one cannot make any case for 'proto-townships' by the Late Iron Age until this institutional gap is closed. The territorialisation of settlement is certainly suggestive of some degree of continuity but only at a very generalised level. Indeed, we need to keep in mind how others have interpreted such territorialisation. As noted in Chapter 2, other studies have already drawn attention to how Iron Age settlement in the Uists, on Barra, Lewis and in Shetland, can be linked (using Theissen polygons and so on) to resources of arable as depicted on modern maps, or at least to areas of prime land, with such land seemingly well pie sliced between Iron Age settlements. Far from being seen as a prima facie case for long-term continuity in the detail of landholding forms, most would see it only as a sign of how, when faced with an ecology like the Highlands and Islands, successive farming communities repeatedly made similar choices.[88]

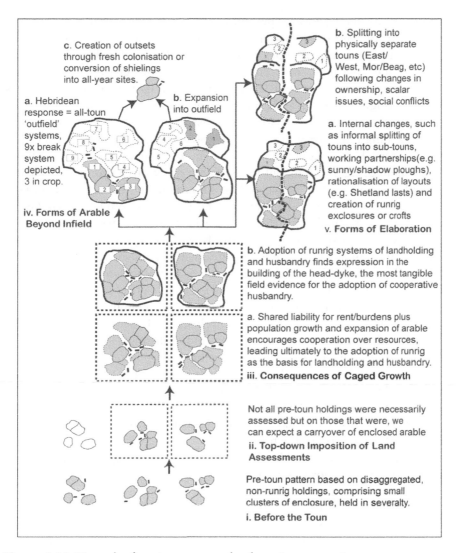

c. Creation of outsets through fresh colonisation or conversion of shielings into all-year sites.

a. Hebridean response = all-toun 'outfield' systems, 9x break system depicted, 3 in crop.

b. Expansion into outfield

iv. Forms of Arable Beyond Infield

b. Splitting into physically separate touns (East/West, Mor/Beag, etc) following changes in ownership, scalar issues, social conflicts

a. Internal changes, such as informal splitting of touns into sub-touns, working partnerships(e.g. sunny/shadow ploughs), rationalisation of layouts (e.g. Shetland lasts) and creation of runrig exclosures or crofts

v. Forms of Elaboration

b. Adoption of runrig systems of landholding and husbandry finds expression in the building of the head-dyke, the most tangible field evidence for the adoption of cooperative husbandry.

a. Shared liability for rent/burdens plus population growth and expansion of arable encourages cooperation over resources, leading ultimately to the adoption of runrig as the basis for landholding and husbandry.

iii. Consequences of Caged Growth

Not all pre-toun holdings were necessarily assessed but on those that were, we can expect a carryover of enclosed arable

ii. Top-down Imposition of Land Assessments

Pre-toun pattern based on disaggregated, non-runrig holdings, comprising small clusters of enclosure, held in severalty.

i. Before the Toun

Figure 6.14 How the farming toun evolved: an interpretation.

Instead, I want to argue that the evidence available for other parts of the Hebrides, as well as the mainland, supports the case for a much later appearance of the farming toun. Its starting point, or what I want to label as the pre-toun stage, is the evidence outlined in the previous section showing that we can find irregular or curvilinear enclosures lying across sections of arable in a significant number of pre-Clearance or pre-crofting touns and settlement distributed across different foci rather than concentrated at a single site within the toun (Figure 6.14, stage i).

Not only do the enclosures present appear both partial and residual when seen in terms of the overall layout but they appear anomalous when set down within a supposedly mature open-field system. When we lift them out of their context, some appear no different in form, scale and character from those present in touns that I would label as proto- or immature touns, such as Righcopag (Figures 6.12 and 13). Sites such as Righcopag can guide us in other ways. Its different blocks of enclosures are associated with different habitation sites, suggesting that part if not all of the spread of settlement between different loci observable in many touns may extend back to a time when landholding was more disaggregated, and focused on a mix of free-standing or loosely clustered enclosures. The variegated ecology of many areas would initially have favoured the take-up of detached, rather than continuous, patches of arable as farmers sought out the more fertile and easily worked pockets of soil first. With any expansion of local needs, we can expect enclosures to have become more numerous on the better sites, with established enclosures being subdivided, new ones being adjoined to them, and others created in the gaps between so as to create a more extended pattern. Yet, even with this closer packing, they would still have been the basis for self-contained holdings, each with its enclosed arable and attached or enclosed settlement. To emphasise the self-contained character of these holdings, and to distinguish them from those enclosures that later became bundled together within a head dyke, I want to use the term *garths* to describe them. In his review of toun structure in the Northern Isles, Balfour equated the term *garth* with infield[89] but I prefer to use the term *garth* for the pre-toun stage because, at this point, these small, irregular enclosures would have been the basis for individual holdings rather than the collective form of the toun.

To suggest that the settled landscape at this pre-toun stage comprised irregular or curvilinear enclosures attached to holdings held in severalty, each with its own attached settlement, does, of course, resemble the kind of landscape that is widely associated with Bronze/Iron Age settlement in the Highlands and Islands. In some cases, such as at Dola, near Lairg, and at Kingsburgh in Trotternish, Skye, we find this late prehistoric settlement actually positioned beside, even within, the enclosures that form part of their eighteenth-century toun layout. Yet, the mere presence of such settlement, even in the quantity evident at Kingsburgh (Figure 6.10), cannot, in itself, be taken as indicating a direct or continuous link between them. Sites such as those at Borrafiach on Skye help make this clear, for the layout of its toun and those quite different field walls which can be linked to the Late Iron Age dun that lies just beyond its head dyke appear as two wholly distinct landscapes. There is a case for a general continuity of site here but what shaped the landholding and settlement of the two layers was probably different. A comparable point is made by looking at the Pitcarmick-type settlement of south-east Perthshire. Set down in an area which boasts many examples of Bronze/Iron Age hut circles and associated field patterns, the Pitcarmick-type farms which emerged over the second half of the first millennium AD ignored what must already have existed on the ground, laying out their fields afresh and adopting a different, more rectilinear style of housing.

The Emergence of Touns

The stage at which we can speak of touns emerging in place of the more disaggregated farms which appear to have existed previously was triggered by the imposition of land assessments (Figure 6.14, stage ii). Obligations, renders and dues which had previously been burdened on the person or family were now territorialised, anchored to specific blocks of land, and made the basis on which land was held. We cannot yet say for certain when this conversion started in the region. In my 1993 discussion of runrig and the origin of the toun, I suggested that a framework of land assessments may not have been fully in place until as late as the thirteenth century.[90] If we use the discussion by Williams and Ross, summarised in Chapter 3, as a guide, it would give us an earlier window for their imposition of between c.900 and 1100.

The patterning of assessments suggest that their imposition took place by gross territorial or district assessments being broken down and emplaced over farms, not by taking what was on the ground by way of land use and building it up into a made-to-measure assessment. It is this downward emplacement that explains the toun-by-toun regularity of assessments in some areas and the almost standard use made of fairly large units of assessment such as the davoch and *tirunga*, or fairly large groupings of assessment such as Five Pennylands. If we work from the assumption that pre-assessment landholding was based around smaller, more disaggregated holdings, each opportunistically picking out the better pockets of soil, we can expect this top-down imposition of large blocks of assessment to have effectively bundled together previously separate holdings under a single rating. We can expect a degree of nip and tuck when it came to fitting one to the other on the ground, some given generous space for expansion by their assessment, others restrained by it. In some cases, the holdings covered may have been localised so that assessments fitted neatly and umbrella-like across them. In other cases, they may have been more geographically spread between pockets of good land, encouraging farms to adopt a split stance (East/West, *Mor/Beag* shares) from the very outset.

We can expect the various holdings brought together by an assessment rating to have become the basis for setting land and for fixing the rent liabilities for those who farmed it. From this point onwards, assessments and the dues and obligations attached to them, together with the extent of arable and the amount sown, bore a relationship to each other. It is from this point onwards that we can speak of a potential for 'touns' emerging as the fundamental unit of set, one that became shared between different landholders who, as a consequence, shared liability for the burdens attached to it as co-portioners. By their top-down nature, however, assessments brought scale to the toun, one that almost guaranteed its communal or multiple occupancy, especially as the countryside became more crowded. By the way in which assessments were used to frame the extent of its arable or sowing capacity, they gave it definition, a layout that needed to be 'meithed and marched' on the ground rather than laid out in an arbitrary way. By the way in which assessments were used to proportion dues and rents between occupiers, they ensured that their multiple possession was based on aliquot shares, or shares that were set proportionately against one another. From their moment of assessment, we can expect previously separate holdings slowly to have leant towards each other, as their occupiers reached agreements over how their collective renders or rent were to be paid through deals over how stock were to be managed, how the cheese, butter or grain was to be chosen, over the cycle of cropping, over how fresh arable was to be divided, and so on. As such deals involved more and more resources, we can expect communities to have opted for an all-toun agreement so that the ultimate outcome of such a coming together is likely to have been the creation of a single runrig toun. That this conversion of holdings into a runrig system, which embraced the entire toun, need not have happened at a single moment is the conclusions that we might draw from what Geddes had to say about touns on Lewis when seen through a detailed rental of 1718.[91] It also comes across, however, from Thomson's analysis of how houses on the edge of runrig touns, such as Laxobigging and Funzie on Fetlar, Shetland, were absorbed into the larger, more incorporated runrig toun at the centre.[92] While houses probably began life as consolidated holdings, many had become divided between different landowners and/or tenants by the time we see them, their layouts taking on the form of mini-runrig touns. In the more fertile, core areas, the crowding of houses, their jostling and competing for the same resources, fostered their integration into larger runrig groupings. The value of Thomson's study is that it shows this process of integration at different stages in different touns, with Laxobigging still having a legible house system but those at Funzie having been integrated via a redivision into a whole-toun runrig scheme. Significantly, he also notes the wider relevance of the process, arguing that Shetland enables us to see it at work because, chronologically, the process unfolded much later there. Other than their freehold basis and later survival, there is no reason to see houses as fundamentally different in character from the farms that may have formed the basis of pre-toun layouts elsewhere in Scotland. For Thomson, their survival and documentation in Shetland helps to specify a process of toun formation via farm or house integration which elsewhere, as with my early work on a similar process of toun formation in the Hebrides, could be assumed only from more circumstantial evidence.[93] Arguably, when a simple, immature site such as Righcopag is set alongside Laxobigging and Funzie, together they provide us with the extremes of a potential trajectory of development along which touns may have unfolded from separate farms or houses to fully integrated runrig touns.

As the number of landholders increased and as deals over joint working matured, we can expect communities to have seen the rationale of replacing the various garth enclosures, which had previously protected their arable from stock, with a single communal head dyke that surrounded all garths (Figure 6.14, stage iiib). Of course, fitting a bounding head dyke around what had previously been small, clustered, or free-standing blocks of enclosed arable would have meant that what lay within the head dyke was not just arable alone but would have included those areas of grass that had to be enclosed in the process of enclosing detached blocks of arable by a single, all-embracing head dyke. Further, as Stylegar observed, head dykes may have embraced grass as well as arable simply because communities also needed to hain or protect what served as meadow and to have wintering grounds.[94] His words are echoed by a 1720s report on how touns were valued, with what lay within the head dyke being 'either Corn Land or meadow and winter grass'.[95] Having grass within the head dyke meant that the meaning of what became infield would have been ambivalent from day one. On the one hand, it was what assessments defined, *sensu stricto*, as arable. On the other, it was what lay within the head dyke, a space which initially may have contained significant resources of grass as well as arable.

The construction of the head dyke was a defining moment in the making of highland touns. To borrow Robertson's descriptive phrase, it became 'a fundamental line' in the Scottish countryside.[96] Quite apart from its separation of what was within and without the toun, it was fundamental because its primary construction announced in a visible way the emergence of the toun as a unit of landholding whose liabilities, as well as whose rights, were shared proportionately between all its occupiers and which, because of that, had formed a unit of farming co-operation. Its power in helping to shape the identity of touns comes across from the local bylaws or tack clauses that routinely bound tenants to maintain their common head dyke in 'a good and sufficient condition',[97] and fined them when they failed. In a late sixteenth-century court enactment, we find the tenants of Lochtayside being required 'to big and make heid dykes, sufficient at Beltyuin and to uphold then and yt thai mak ye said Dykes fra March to March of ye bounding of their malingis grangis . . . fra heid dyke to heid dyke And all boundis and pasturis wtout ye heid dyke to be common amongs them.'[98] There is almost a sense in this extract of a system of head dykes being constructed *de novo*, especially given its powerful tag line declaring all beyond the dyke as common. Once introduced, however, head dykes would have been subject to periodic reconstruction, as the expansion of touns, toun splitting, the need to create a more joined-up system or to create a more substantial system all changed expectations. Breadalbane touns, for instance, were required, as late as 1726, to start afresh, building their head dykes 'straight and upon a new foundation and not to follow the Turnings and roundings of Burns and bankes but where is absolutely necessary'.[99] In fact, when we look at those which have survived in the landscape, they often appear as multiple forms, with different or successive dykes embracing more and more land as demand pushed arable outwards, a succession of forms well conveyed by the RCAHMS map of those in Uig parish, Lewis.[100]

Before leaving this issue of how touns first emerged, I want to consider how such a reading squares with what Parker Pearson argued using data from the South Uist project. In his recent reviews,[101] he specifically sets his ideas on the early appearance of the toun against those that were first expressed by me in 1993[102] and which represent an early attempt to deal with some of the ideas presented here on toun development. His view is that the results of the South Uist project undermine what I wrote in 1993 and, therefore, by extension, undermine what I have written here. Unfortunately, part of his case is based on a misreading of what I wrote in my 1993 papers. He works from the assumption that my argument claimed that the Highland toun, with its nucleated forms and runrig, did not emerge until the post-medieval period,[103] even stating at one point that I date them no earlier than the eighteenth century.[104] In fact, in my 1993 papers and across all my published reviews, I have maintained that Highland runrig touns existed across much of the medieval and early modern periods. When it came to dating their emergence, I proposed in my 1993 papers, as in the discussion above, that the shift from pre-toun to runrig layouts may not have taken place until as late as the thirteenth century AD.[105] Clarifying this point is important not just for my own argument but also because of the bearing it has on that offered by Parker Pearson. The South Uist project generated a major insight into how settlement on the island evolved. At some point between AD 1200 and 1400, it concluded that a reorganisation took place, with most of the island's settlements experiencing a shift from the machair to the edge of the peatlands.[106] Surprisingly, Parker Pearson plays down its significance, describing it as one of the 'minor discontinuities' affecting settlement, a conclusion that preserves intact his reading of an early origin for the toun.[107] Of course, seen in relation to my 1993 papers and what I have argued for here, it takes on a new significance. Given when the shift is thought to have occurred, around AD 1200 to 1400, it is unlikely to have been associated with the arrival of the Vikings, even though some disruption might be expected at the point when they did arrive. Nor can we be certain that the settlement shift occurred during Norse rule for their control over South Uist ended in 1266. With that point made, we may be able to shed light on why the shift occurred by looking again at the introduction of land assessments in the Northern Isles. Detailed comparative studies on the Northern Isles and western Norway have established a good understanding of how Norse settlement and landholding were being configured at this point. On the basis of his work in west Norway, Rønneseth suggested that the shift towards townships (= touns) based on open fields was bound up with a restructuring that took place as late as the period of the Landslov (AD 1274),[108] a restructuring that was probably bound up with what Stylegar has recently called the 'locking' in of the farming landscape to a scheme of 'matriculation', through the imposition of 'bol' units which provided a fiscal (or assessed) framework for occupied land and a shareholding basis for its possession.[109] Stylegar saw Rønneseth's ideas as having direct relevance to the Northern Isles, drawing attention to how Storer Clouston and Thomson both saw the initial pattern of Norse landholding as initially more fragmented and its settlement as more spread out but how, over time, houses or farms became both internally divided into runrig systems and collectively more integrated into what eventually became schemes of 'whole-township' runrig. Just as I argued, using different data in my 1993 papers, Thomson saw this model of how fully integrated touns and eventually all-toun runrig developed as applicable not just to the Northern Isles but equally to other parts of Scotland. Brought together, such arguments offer a plausible context within which we can account for the settlement shift observed on South Uist, around 1200 to 1400, whether we date it to the final years of Norse rule or the century or so after it had ended. Far from being a 'minor discontinuity', it may be the best field evidence we have for the 'major discontinuity' brought about by the imposition of land assessments. Of course, given that South Uist was in Norse hands down to 1266, the imposition of land assessments there, as in the Northern Isles, may have followed a different chronology from elsewhere in the Hebrides or on the mainland.

Touns: What Drove them Apart?

The forces that encouraged the closer integration of touns were matched on occasion by opposite forces that encouraged difference and division. To start with, the crofts which we find in many mainland touns stand out as different because they were excluded from runrig layouts, most being marginal or peripheral to them but some physically enveloped by them. In this proposed outline of toun development, we might see this exclusion as extending back to the very foundation of the toun with some crofts being enclosures or holdings that were kept out of any grand, all-toun reorganisation into runrig. As service holdings (that is, for ploughmen, millers, maltsters, and so on), they were probably separated out so as to maintain their neutrality. More significant for the long-term history of touns were the various levels of splitting that we find at work (Figure 6.14, stage v). At first sight, splitting seems a contradiction of what touns were all about. We need to understand, however, how different phases or circumstances brought different problems. Initially, touns were brought together by the way assessments bound the collective interests of different holdings and farmers into a whole so that they held a share of a whole. Once bounded or caged in this way, their respective interests overlapped and, under these circumstances, an open field system such as runrig offered a way of optimising the management of their resources. With growth, though, we can expect other factors to have come into play that encouraged a different solution. Some instances of splitting touns into smaller touns, for example, were rooted in social tensions, with division being a means of regulating these tensions. In other instances, it was brought about by substantial tenants separating out their arable from that of lesser tenants. Changes in ownership which involved the transfer of part of a toun could also produce examples of splitting. In time, possibly the prime factor was the growing scale of touns, both as regards the number of occupiers involved and as regards the geographical spread of the toun and the attraction of having smaller, more localised working units. Such adjustments could play themselves out at a district level, as with touns in the eastern Grampians which were reorganised into smaller touns during the late sixteenth century.[110]

How Touns Reached Beyond Themselves

The long-term stability of land assessments affected the way in which touns responded to growth. Admittedly, there are clues that some assessments may have been subject to occasional forms of adjustment but not necessarily in the number of land assessment units present.[111] In some instances, it is possible that, while the assessment of touns remained outwardly static, what they represented in terms of sowing or capacity may have been rerated. Viewed over time, though, assessment ratings are best seen as tending towards the inertial, unable to respond in real time to pressures for the expansion of arable. In practice, this would have had a caging effect on touns, constraining their unfettered physical expansion. For this reason, head dykes would have represented something of a stadial line, a line at which the development of touns may have paused for a while. To a degree, some arable expansion would have been possible within their head dyke, given that most would have had reserves of grass within it. In time, though, diminishing returns set in, with any further expansion of arable reducing the grass needed for hay or wintering and, therefore, the supply of available manure. Of course, ultimately, we know from the widespread presence of outfield systems that the limits imposed by a toun's assessment and the physical limits defined by its head dyke were breached (Figure 6.14, stage iv).

The very fact that outfield was different in name is a reason in itself for seeing its emergence as a significant new step for touns, a break out in their development. The same can be read from how its husbandry differed from that of infield. This change in husbandry involved more than just the fresh opportunities offered by land beyond the dyke. Without it, growth would simply have produced a larger infield. Why it did not, why expansion beyond the head dyke involved a different form of husbandry, can be explained by looking at how the different sectors were manured. Infield depended on the animal manure accumulated over winter, the supply of which was limited by the number of stock that could be kept over winter and foddered via stubble grazing, hay or winter grazings. Any sustained expansion of infield arable would have reduced the better pasture available for stock and would eventually have conflicted with the supply of manure needed to maintain an arable expansion. Yet, even with supplements, infield expansion ultimately faced a pinch point. For touns reliant on stock manure, the obvious next best step was to find a way of utilising the manure produced by stock over summer so as to establish an all-year-round use of manure. This use of the manure produced by stock over summer was the basis of outfield. Once developed, some touns constructed a new head dyke, one that became their outer dyke as opposed to the inner dyke (= old head dyke), and now embraced infield and outfield. Of course, the outfield-like systems of the outer Hebrides need to be seen as having a different lineage, one that may lie in the facts that stock was wintered outdoors and that each tath was significantly larger than tathfolds on the mainland, though we must also keep in mind the role played by seaweed for such touns and its temporary effects once applied. Generally, the creation of outfield for many touns was only one part of a wider process of expansion beyond head dykes. Adding to the mix of possibilities were the creation of outsets and the upgrading of shieling sites into permanent touns. In some cases, these were part and parcel of the same process, with early rentals treating them in the same way.

As stressed at the outset, this outline of how touns may have developed is very much a provisional one. As such, it must not be seen as a one size fits all kind of development sequence. Quite apart from interregional variations in the size and ecology of touns, different parts of the region saw different inputs into the equation that may have modified the experience of the touns. There will also have been different chronologies for key influences. We cannot, for instance, assume that the imposition of land assessments in Norse areas shared exactly the same chronology as their imposition in other parts of the region, though their effects would have been the same. Looking forward, we need further work on a number of issues. Quite apart from the challenge of establishing how settlement in the typical Highland toun changed over the late medieval and early modern periods, more work needs to be directed at establishing a development sequence for the different field structures that survive in some of the better preserved pre-Clearance or pre-crofting touns. We also need to establish when head dykes were constructed because, more than any other feature, these signal when touns became a community of farmers co-operating in matters of husbandry. If, overall, the medieval period saw a shift from a more disaggregated form of settlement and landholding to that represented by the institution of the toun, then this would have been a key moment.

Notes

1. BPP, *Report*, 1884, p. 17.
2. Geddes, 'Conjoint tenants', pp. 54–60.
3. F. Shaw, *Northern and Western Islands*, pp. 85–6.
4. McKay (ed.), *Walker's Report*, p. 184.
5. D. M. Henderson and J. H. Dickson (eds), *A Naturalist in the Highlands: James Robertson – His Life and Travels in Scotland 1767–1771* (Edinburgh, 1994), p. 129.
6. NAS, RHP3368.
7. RCAHMS, *Waternish*, pp. 14, 18.
8. A. Mitchell, 'James Robertson's tour through some of the western islands, etc, of Scotland, 1768', *PSAS*, 32 (1897–98), p. 14; Henderson and Dickson (eds), *Naturalist . . . Highlands*, p. 81.
9. For example, Millar (ed.), *Scottish Forfeited Estates*, p. 137.
10. NAS, E788/42.
11. NAS, GD50/2.
12. J. Boswell, *The Journal of the Tour to the Hebrides with Samuel Johnson* (London, 1909 ed.), p. 118.
13. NAS, GD50/2/3.
14. Henderson and Dickson (eds), *Naturalist . . . Highlands*, p. 180.
15. Millar (ed.), *Scottish Forfeited Estates*, p, 105.
16. NLS, SP, 313/1047.
17. *OSA*, 1791–99, viii, pp. 375–6.
18. NAS, E788/42; R. A. Dodgshon, 'West Highland and Hebridean settlement', p. 423.
19. NAS, GD112/17/2, pp. 98, 119–20, 125; see also NAS, GD170/431/41.
20. Mitchie (ed.), *Invercauld*, p. 142.
21. NAS, GD112/17/2.
22. Loch, *Account*, p. 53; Dodgshon, 'West Highland and Hebridean settlement', p. 423.
23. NAS, GD112/17/4.
24. NAS, GD112/17/11.
25. NAS, GD44/25/2/76.
26. Millar (ed.), *Forfeited Estates*, p. 137.
27. NAS, GD50/2/3.
28. Boswell, *Journal*, p. 138.
29. IC, AP, Box 2532, Memorial 1706; Cregeen (ed.), *Argyll Estate Instructions*, 16 and 51; Dodgshon, 'West Highland and Hebridean settlement', pp. 422–4.
30. NAS, GD174/827; DC, MDP, 2/32/4; NAS, GD46/1/278.
31. Fairhurst, 'The deserted settlement at Lix', pp. 160–99. See also, R. Noble, 'Turf-walled houses of the central Highlands', in *Folklife*, 22 (1984), pp. 68–83; Dodgshon, 'West Highland and Hebridean settlement', pp. 421–4.

32. R. P. J. McCullagh and R. Tipping (eds), *The Lairg Project, 1988–96: the Evolution of an Archaeological Landscape in Northern Scotland*, Star (Edinburgh, 1998), pp. 173–93.
33. NTS, Ben Lawers Historic Landscape Project, *Annual Report* 2003–4.
34. Atkinson, 'Settlement Form and Evolution', p. 322.
35. Ibid., pp. 324–5.
36. Ibid., p. 317.
37. LeLong and Wood, 'A township through time', pp. 44–5.
38. P. W. Abrahams, J. Entwistle and R. A. Dodgshon, 'Ben Lawers Historic Landscape project . . . X-ray fluorescence spectrometry', *Journal of Archaeological Method and Theory* 17 (2010), pp. 231–48; C. A. Wilson, D. A. Davidson and M. S. Cresser, 'An evaluation of the site specificity of soil elemental signatures for identifying and interpreting former functional areas', *Journal of Archaeological Science*, 36 (2009), pp. 2327–334.
39. For example, Curwen, 'The Hebrides', pp. 261–89.
40. Ibid., pp. 261–89.
41. A. Fenton, *The Island Blackhouse and Guide to no. 42* (Edinburgh, 1978); T. Holden (with contributions by L. M. Baker), *The Blackhouses of Arnol*, Historic Scotland Research Report (Edinburgh, 2004).
42. J. G. Dunbar, 1971, 'The study of deserted medieval settlement in Scotland: the peasant house', in M. Beresford and J. G. Hurst (eds), *Deserted Medieval Villages* (London, 1971), pp. 238–9.
43. F. W. L. Thomas, 'On the primitive dwellings and hypogea of the Outer Hebrides', *PSAS*, 1866–68 (7), Figures 4 and 5.
44. Holden, *The Blackhouses of Arnol*, pp. 11, 22.
45. I. Crawford, 'Contributions to a history of domestic settlement in North Uist', *SS*, ix (1965), pp. 34–65; K. Braniganand C. Merrony, 2000, 'The abandoned settlements of Crubusdale and Gortien', in K. Branigan and P. Foster, *Barra: Archaeological Research on Ben Tangaval* (Sheffield, 2000), pp. 193–8.
46. K. Branigan and P. Foster, *Barra and the Bishop's Isles* (Stroud, 2002), pp. 117–8, 140–1.
47. Crawford, 'Contributions . . . settlement in North Uist', pp. 34–65.
48. Dodgshon, 'West Highland and Hebridean settlement', pp. 419–38; Holden, *The Blackhouses of Arnol*, p. 6.
49. Crawford and Switsur, 'Sandscaping and C14', pp. 131–3.
50. K. Branigan, 'Human settlement on the Tangaval Peninsula', in K. Branigan and P. Foster, *Barra: Archaeological Research on Ben Tangaval*, pp. 204–6.
51. M. Parker Pearson, 'Settlement, agriculture and society in South Uist before the Clearances', in M. Parker Pearson (ed.), *From Machair to Mountains* (Oxford, 2012), pp. 420–1.
52. Ibid., p. 419.
53. I. Armit, 'Excavations of a post-medieval settlement at Druim nan Dearcag, and related sites around Loch Oabhat, North Uist', *PSAS*, 127 (1997), pp. 899–919.

54. Cavers and Hudson, *Assynt's Hidden Lives*, p. 25. A change from rounded to angular corners is also recorded in J. S. Smith, 'Deserted farms and shealings in the Braemar are of Deeside, Grampian Region', *PSAS*, 116 (1986), p. 447.

55. M. Parker Pearson, N. Sharples and J. Symonds, *South Uist: Archaeology and History of a Hebridean Island* (Stroud, 2004), pp. 125–44.

56. Bigelow, 'Domestic architecture', pp. 23–38.

57. Ibid., p. 179.

58. Dodgshon, 'Strategies of farming', pp. 697–9.

59. Branigan and Foster, *Barra*, pp. 128–31.

60. Dodgshon, 'Strategies of farming', pp. 697–9.

61. A. Fenton, 'Early and traditional cultivating implements in Scotland', *PSAS*, xcvi (1962–63), pp. 302–15.

62. Johnson, *Journey*, p. 79.

63. Fenton, 'Early . . . cultivating implements', p. 286.

64. R. A. Dodgshon, 'Farming practice in the Western Highlands and Islands before Crofting: a study in cultural inertia or opportunity costs?', *Rural History*, 3 (1992), pp. 173–89.

65. Fenton, 'Early . . . cultivating implements', p. 315; Fenton, *Northern Isles*, pp. 285–90.

66. J .R Macculloch, *The Western Highlands and Islands of Scotland* (London, 1824), iii, pp. 94, 118.

67. Dodgshon, 'West Highland and Hebridean settlement', pp. 431, 433–4; RACHMS, *Waternish*, pp. 14–21.

68. R. A. Dodgshon, 'West Highland and Hebridean landscapes: have they a history without runrig?', *JHG*, 19 (1993), pp. 386–9, 392–3. See also the map of the island's irregular enclosures published in RCAHMS, *Eigg*, 2003.

69. Dodgshon, 'West Highland and Hebridean landscapes', p. 392.

70. Branigan and Foster, *Barra*, pp. 125–6.

71. D. H Caldwell, R. McWee and N. A. Ruckley, 'Post-medieval settlement in Islay – some recent research', in Atkinson et al. (eds), *Townships to Farmsteads*, p. 61.

72. Dodgshon, 'West Highland and Hebridean landscapes', pp. 391–2.

73. MacLeod, *Book of Dunvegan*, i, p. 152.

74. RCAHMS, *Well Sheltered Glen* (Edinburgh, 2001), p. 23.

75. Dodgshon, 'West Highland and Hebridean landscapes', pp. 383–98. See also, Caldwell et al., 'Post-medieval settlement in Islay', p. 61.

76. NAS, FE, E783/98.

77. NAS, GD305/1/18/3; GD305//1/18/9; GD305/1/51/1

78. NAS, GD305/1/146/6.

79. RCAHMS, Canmore ID 91293.

80. Harden and Lelong, *Winds of Change*, p. 187.

81. Stewart, *Settlements of Western Perthshire*, pp. 92, 109.

82. Cavers and Hudson, *Assynt's Hidden Lives*, p. 21.

83. Ibid., p. 21.

84. Evans, 'Atlantic Ends', pp. 54–64.

85. H. Uhlig, 'Old hamlets with infield and outfield systems in western and central Europe', *Geografiska Annaler*, 43 (1961), p. 289.

86. M. Parker Pearson, 'The machair survey', in Parker Pearson (ed.), *From Machair to Mountains*, pp. 14, 38–40.

87. Oram and Adderley, 'Innse Gall', p. 131.

88. Hunter, *Fair Isle*, p. 95.

89. Balfour, *Oppressions*, p. xxxi.

90. Dodgshon, 'West Highland and Hebridean landscapes', p. 396.

91. Geddes, 'Conjoint tenants', pp. 54–60.

92. Thomson, 'Township, house and tenant holding', pp. 107–27.

93. For example, Dodgshon, 'West Highland and Hebridean settlement', pp. 419–38 and 'West Highland and Hebridean landscapes', pp. 383–98.

94. Stylegar, 'Township and Gard', p. 36. Also see, Rønneseth, *Gard und Einfriedigung*, pp. 9–263.

95. NAS, GD112/14/13/5/32.

96. I. M. Robertson, 'The head-dyke – A fundamental line in Scottish geography', *SGM*, 65 (1949), pp. 6–19.

97. NAS, GD112/10/1/4/2.

98. NAS, GD112/17/2.

99. NAS, GD112/14/13/5/13.

100. RCAHMS/Historic Scotland, *But the Walls Remained* (Edinburgh, 2002), p. 66; see also Stylegar, 'Township and Gard', pp. 36–8.

101. Parker Pearson, 'The machair survey', pp. 12–73 and 'Settlement, agriculture and society in South Uist', pp. 401–25.

102. Dodgshon, 'West Highland and Hebridean settlement', pp. 419–38; Dodgshon, 'West Highland and Hebridean landscapes', pp. 383–98.

103. Parker Pearson, 'The machair survey', p. 14.

104. Parker Pearson, 'Settlement, agriculture and society in South Uist', p. 419.

105. Dodgshon, 'West Highland and Hebridean settlement', p. 435; Dodgshon, 'West Highland and Hebridean landscapes', p. 396.

106. Parker Pearson, 'Settlement, agriculture and society in South Uist', p. 419.

107. Parker Pearson, 'The machair survey', p. 38.

108. Rønneseth, *Gard und Einfriedigung*, p. 181–3.

109. Stylegar, 'Township and Gard', p. 36.

110. Mitchell (ed.), *Geographical Collections* , ii, p. 272.

111. Smith, *Tenants and Toons*, pp. 1–15; Ross, 'The dabhach', pp. 63–4.

LANDSCAPES OF CHANGE, 1750–c.1815: THE BROADENING ESTATE

The changes in progress by the mid-eighteenth century need to be seen as unfolding along multiple pathways. In addition to the stark and, at times, brutal en masse replacement of farming touns with sheep and the creation of crofting townships of different types (that is, farming-dependent crofts, fisher crofts, service villages) and under different conditions (*in situ*, resettlement), estates also sought to make organisational changes to the management of their estates. Guided by advisers, they sought to make more effective use of available resources (such as forests and woodland, kelp, mineral rights, surplus labour, grain processing), to carry out programmes of enclosure and land improvement, and to encourage better husbandry. As Storrie made clear in her analysis of Islay, when seen systematically, the structural forms of change were not always one-off, singular events but were often combined into simple sequences of change, with land being moved between runrig touns, small farms, crofts, sheep farms, afforested land and, later, deer forests though not in any set order of succession.[1] The Clearances for sheep may provide us with the lead story for what happened, but they comprised only one of a number of changes which were linked and sequenced in ways that provide us with multiple stories about how landscape changed.

The Pressures for Change

Variations in resource endowment played a part in shaping this diversity but the fact that the region was subject to different pressures for change, each with its own chronology and geography, also played a role. Across the eighteenth and nineteenth centuries, five such pressures stand out, two originating from within the region and three from without. First, even by the early eighteenth century, the financial affairs of many Highland landlords were being squeezed between low, irregular rental income and their rising level of lifestyle costs, with many burdened by debts.[2] In the circumstances, it is not surprising that the need to raise estate income provided a powerful agenda for change. Across the region, landlords responded by taking a greater interest in how their estate resources could be capitalised to the full, made to yield a greater rent. Second, the region experienced a phase of population growth which was underway during the closing decades of the eighteenth century and which, as it unfolded, left many areas, especially in the west and north-west, overpopulated, carrying what the Duke of Argyll called 'supernumeraries'.[3] Third, in addition to pressures for change from within the region, there were significant external pressures. Among them, efforts by the authorities to establish social and political stability across it. The perception of those at the centre had long been that the way in which Highland estates were managed was part of the problem. Following the 1715 rebellion, the government's response was to confiscate the estates of those involved but its temporary control over them did not extend beyond that of functioning as a rent collector. It did, though, initiate a road-building programme under the supervision of General Wade (1673–1748). As contemporaries pointed out, these were military roads, designed ostensibly to enable the rapid deployment of troops rather than the marketing of produce.[4] After the Forty-five, the government's response went further, though it took time to deliberate over what its response should be.[5] The estates of those involved in the Forty-five, stretching across the Highlands from Monaltrie to Knoydart and from Arnprior to Coigach, were confiscated by the Annexing Act of 1752 and transferred to the Crown as 'inalienable' property.[6] A Board of Commissioners was established for their management, though its actual hands-on exercise of this management did not begin until 1755. The board saw its remit as using the Forfeited Estates as exemplars for Highland estate improvement, turning the very estates that had harboured opposition into a focus for change, though what that 'improvement' should comprise in a Highland context was much debated, on and off the board.[7] The government's response also saw a further extension to its road-building programme, its efforts now assisted by the board, especially when it came to capital-intensive aspects such as bridge building. The board's support initially favoured what served the estates under their control.[8] As Smith observed, however, 'roads tend to beget roads' and, by the 1770s, not only did the board's support broaden but other estates began the task of building link roads.[9] If, as some argued, change depended above all else on establishing the means of communication, then slowly, over the second half of the century, parts of the region, more south of the Great Glen than north at this stage, were given the prospect of a new orientation. Fourth, by the end of the eighteenth century, the growing indebtedness of Highland landlords led to estates being sold.[10] New money, made in commerce, banking and industry, flooded into the region, bringing with it different values and expectations of how land should be used. Fifth, and the most powerful of all external sources of change, the seventeenth and, especially the eighteenth, centuries saw the growing penetration of market demand from the Lowlands and the south, a demand that initially led to large numbers of cattle being droved out of the region long before there was any explosion of sheep numbers.

Surveying the changes produced by these various factors, we can draw a broad division between those whose chronology was weighted before *c.*1815 and those weighted after. Those occurring before this watershed can be characterised by a rough-and-ready balance between attempts to reform what already existed, including efforts to ensure that many of those present continued to have a foothold on the land, and attempts to explore new ways of exploiting the resources to hand. After *c.*1815, this balance between old and new disappears as many landowners became more and more single-minded in their pursuit of a higher, more stable level of rent, with less regard for the interests of those who had long farmed the land. In short, as one moves more into the nineteenth century, the core objectives that had shaped many change programmes before *c.*1815 – raising the rental income of estates and maintaining a network of well-peopled communities – were no longer seen as so compatible with each other.[11]

We can see this shift in approach before and after 1815 at two levels. On the one hand, we can see it as built up out of the countless on-the-ground decisions made by landlords and farmers as they tried to respond to changing markets. On the other, we can see it as a shift that was prefigured in a more abstract debate about what was the real resource base of the region and how best could it move forward. As Jonsson, especially, has shown, there was a vigorous debate, even during the second half of the eighteenth century, between those commentators, such as John Walker, who thought the region should hold on to its population by ensuring that the smallholder had a place in any schemes of change, and those, such as Adam Smith and James Anderson, who thought that change should maximise the region's market orientation by switching more wholeheartedly into commercial stock production.[12] To a degree, both sides had a case to argue but the debate was always about striking a balance and about whether a portion of the large resident population should be reaccommodated in some way if only because, in the minds of some landlords, they were a resource no less than land, one that could be quarried for military recruitment. What we see over the second half of the eighteenth and early nineteenth centuries is a considered balance between strategies, with many estates finding a place for the smallholder, but what we see from *c.*1815 down to the setting up of the Napier Commission, are schemes that marginalised the smallholder and pursued market strategies in a more single-minded way. The following review of the changes that occurred before *c.*1815 is organised around the different ways in which landowners broadened their estates by making fuller use of the resources available, namely timber, grain processing, labour, kelp, fishing, land improvement and finally pasture, the last being divided into a section on cattle and another on sheep.

The Broadening Estate: the Landlord's Touch

A problem faced by the region in the early eighteenth century was that some of its landholding practices were attuned to older needs, serving local military and political requirements as much as economic and social ones. Some of these practices were addressed by the Statutes of Iona (1609), with those landowners at whom they were targeted being required to raise regular rents from their land. These measures, though, were not a long-term solution to patterns of estate income and expenditure.

Fundamental weaknesses in the toun economy meant that rents remained low and uncertain. Further, the changing lifestyle of Highland landowners, with new houses, new patterns of personal consumption, their growing involvement with Edinburgh and, after 1707, London, came at a price, one that led to further levels of debt. One response was for them to wadset sections of their land, granting it out on long lease in return for the loan of money, often to members of their own family, or *sliochd*, who had been major tenants and tacksmen. The non-return of the loan meant that many wadsets effectively turned into hereditary holdings and served in a modest way to break up estates and to establish a lower tier of small to middling landowners.[13]

For some, another solution lay in ridding their estate of tacksmen. Through their role as middle men, tacksmen enjoyed an advantageous position. The rents paid by them were often fixed by long-term agreements so that they were relatively sluggish in responding to change. By comparison, the rents charged by tacksmen to their under tenants were usually set on a year-by-year basis and were far more price-responsive. For some, the rising levels of conspicuous consumption among tacksmen, including the emergence of the two-storey tacksmen house in the Isles and along the western seaboard, were a manifestation of this advantageous position. The 1733 review of the Macdonald estate captured the thinking of many advisers when it portrayed them as an unnecessary burden on the estate's finances, and recommended leasing land directly to tenants. Duncan Forbes's 1737 report on the sprawling Argyll estate echoed these thoughts. His reference to the tacksman's 'unmerciful exactions' on islands like Coll captures the tenor of his damming analysis.[14] The duke responded by systematically removing tacksmen across his estate, district by district.[15] The outcome was an estate far more directly controlled by the duke and his chamberlains, one across which the landlord's touch – and his profit – was now more direct.

The Broadening Estate: Woodlands

Prior to *c.*1700, the Highland estate economy was spread over a range of resources but, apart from arable, their exploitation was far from optimised. At a time when landlords struggled with growing debts, what was seen as an underexploitation of key resources gave them an obvious way forward. Among the most valuable of such resources was woodland. Long a vital resource for the farming community, its less than optimal use brought farmers into increasing conflict with those landowners who sought to release more of its value for constructional timber, charcoal and tanning bark. Ultimately, what changed attitudes was the growing demand for timber outside the region, a demand already apparent by the seventeenth century, with a regular trade along both the east and west coasts.[16] Pinewoods which offered scope for the removal of felled timber by water, such as those of Speyside and Deeside, or those close to the sea lochs of the west coast, such as Ardgour, were attractive to those who planned to supply boatbuilders on the coast or those further afield who needed good-sized timber, or standards, for house building. Contracts set in the seventeenth century, such as those for the pine forests of Glenmore, Rothiemurchus and Abernethy, were selective, selling only the larger trees and floating them down the Spey. Most contracts for which detailed evidence is available, however, such as the sale of pine from Abernethy fir woods in the 1630s, were rarely completed to everyone's satisfaction, with less wood being removed than expected.[17] In the event, in the Eastern Highlands, there was no great legacy of large-scale clear felling through this seventeenth-century activity. The eighteenth century saw greater amounts of timber removed from Abernethy and, later, from Glenmore and Rothiemurchus.[18] Sizeable amounts of timber were also removed from the fir woods in the Western Highlands, with those in areas such as Loch Arkaig being reduced by felling in the seventeenth century and those in Coigach and Glenorchy being similarly reduced in the eighteenth. The amounts of pine extracted in areas like Glen Etive and Glenorchy may also have been significant enough to make 'an impression' locally.[19] Overall, felling in these western pine woods probably had a greater impact on native pine woods at this point not just because larger areas were felled. As Smout and Watson pointed out, the wetter, more exposed conditions of western areas made natural regeneration of pine woods more difficult after phases of clearance, whether due to felling or fire damage, when compared to clearance events in the Eastern Highlands, so that their long-term impact in denuding the landscape is likely to have been greater in areas such as Ardgour or Glen Etive than in areas such as Rothiemurchus or Abernethy.[20]

Elsewhere, but especially along the western coast, the presence of oak woods was also attractive to those who relied on oak bark for tanning and for those in need of cheaply produced charcoal for iron working. Thinking outside the box, Sir George Hay hinted at what might be possible by establishing an iron smelter, using locally produced charcoal, beside Loch Maree in Wester Ross during the early seventeenth century but his scheme did not last.[21] Iron smelting made a renewed start in the early eighteenth century when smelters were established successively, starting *c.*1725, at: Glen Kinglass, a remote glen feeding westwards into the upper part of Loch Etive in Argyllshire; at Glengarry (Inverness-shire) in 1727; at Abernethy (Inverness-shire) *c.*1730; at Bunawe beside Loch Etive in Argyllshire in 1753;[22] and, lastly, in 1755, at Furnace beside Loch Fyne, also in Argyllshire.[23] In each case, the charcoal used was sourced from woodland within the local area but the iron ore was brought from England. Despite the difficulties of transporting charcoal, most appears to have been made at source, with the majority of the supply sites having stances for preparing it.[24] Inevitably, this put a premium on woodland that could supply charcoal to the smelters using waterborne transport. The smelter at Glen Kinglass, for instance, drew its charcoal from around the upper part of Loch Etive as well as from the woods along Glen Kinglass itself. That at Bunawe, or the Lorn Furnace as it became known, proved the most successful, producing iron down to its closure in 1876.[25] Over its life, it drew its supplies of wood from a wide range of woods across Argyll, all within reach of the coast.[26]

Alongside their use for charcoal, hardwoods, especially oak, were also used to provide tanning bark. When, in 1756, the Lochbuie estate sold part of its wood to Nathaniel Taylor, an agent for the new iron furnace at Bonawe, it allowed him both to work the bark and to make charcoal, the latter by means of stances on the farm of Knock.[27] Likewise, a report of 1786 notes the Breadalbane estate as holding 2,030 acres (1,031 hectares) of wood in Argyll, of which 1,850 acres (940 hectares) were in the hands of the Lorn Furnace Company[28] which Sir Duncan Campbell allowed to coal the oakwood that he had reserved after he had stripped it of its bark.[29] The harvesting of tanning bark was a long-established and lucrative part of woodland exploitation. Following the Irish Cattle Act (1686), the Irish tanning industry responded by tanning more leather in Ireland. Part of its bark needs were supplied by the oak woods of the West Highlands, especially those along the seaboard, with a substantial growth in demand from the late seventeenth century onwards.[30] Irish contractors were to the fore in this trade down to the mid-eighteenth century. They played a less prominent role thereafter but Irish demand still continued to drive the trade into the early nineteenth century. Alongside the large-scale commercial contracts, but less visible, was the barking activity by many ordinary tenants and cottars. The extent of such localised, farm-based activity occasionally surfaces in estate material. By an agreement of 1770, for instance, tenants on Jura were granted the 'privilege of culling woods on the Estate for the use of the ground . . . & Peeling Barks for Tanning their leather'.[31] Increasingly, though, we can expect landlords to have exercised close control, especially where woods were deemed suitable for commercial exploitation or where they were scarce. A 1780 survey of the Lochbuie estate on Mull, for example, restated its regulation that tenants could peel bark only with permission.[32] In some cases, estates imposed a complete ban on tenants peeling bark. The Sutherland estate banned the practice on all small farms in Rogart parish by inserting a clause to that effect in leases issued after 1811. The reason given was that timber had 'been very much destroyed, and in some places, utterly extirpated, by the practice of peeling bark from Trees for the purposes of Tanning leather'.[33]

The growing commercial exploitation of woodland helped to foster two changes in the way in which woodlands were managed. The first was their physical enclosure, ostensibly to prevent stock damage. It was not a new approach[34] but, by the eighteenth century, it had become the standard means by which stock and wood were kept apart wherever estates wanted to conserve wood. In practice, it deprived touns of what had long been a vital source of winter feed and shelter, so there can be no surprise at the hostility that tenants showed towards the process.[35] The extent to which tenants could suffer is shown by the example of the farm of Lower Kinchrackin on the Breadalbane estate, with a 1788 report claiming that the enclosure of more wood before that presently being cut was out of hainage would cause a 'great part of the rent [to] be sunk for a time' because of the loss of winter shelter and under grass.[36] Overall, the extent to which, by the late eighteenth century, the Breadalbane estate pursued a policy of enclosing its woodland by is shown by figures that it compiled in 1786 showing how much of its woodland resources in Argyllshire (exclusive of fir woods) were enclosed by that time, with 2,030 acres (1,031 hectares) enclosed and only 154 acres (78 hectares) still unenclosed.[37] From the point of view of landscape, the greater efforts by estates to enclose wood or wood pasture had the effect of transforming it from one in which the distinction between wood and pasture had been somewhat fuzzy into one in where the difference became sharper. It was sharpened not just by the protection given to woodland which was now enclosed but by what was happening outside these enclosures, with rising numbers of stock, particularly of sheep, after *c.*1750, grazing down open wood pasture.[38]

The enclosure of woodland and the exclusion of stock ultimately paved the way for the second change, or its closer management through more systematic schemes of exploitation. All the indications are that the coppicing schemes, in operation for the extraction of charcoal and bark from oak woods during the seventeenth century and for much of the eighteenth, were poorly implemented, with exploitation, rather than management, being too often in the ascendancy.[39] Available reports make it clear that careless cutting was as destructive as stock damage. The woods along Glen Etive and in Glenorchy, for example, were contracted to charcoal burners who were said to have extensively damaged the woods when setting up their stands for charcoal and when removing timber.[40] Reviewing the problem, Smout and Watson concluded that it was not until later in the century that Highland estates generally began to use carefully specified contracts and systematic coppicing.[41] In best-practice scenarios, this more systematic coppicing involved wood being divided into clear sectors and the frequency of their rotational coppicing being fixed by agreement.[42] Contemporary observers were clearly conscious of this shift to closer management. The *OSA* report for Luss parish is especially informative. Referring to the 700 acres (356 hectares) under oak in the parish, it explains how it was divided into twenty hags or parts, each one cut every twenty years, usually between May and July. It also talks about how, 'formerly', there was little attention paid to how work was performed but 'now care is taken that the trees shall be all cut down and peeled close to the ground'.[43] Once cut, hags were enclosed for a least six years so as to protect regrowth from stock grazing. On the Barcaldine estate in Argyllshire, Lindsay established that Craig Wood, on the north bank of Loch Creran, was subject to a major cutting at intervals of 28, 26.8 and 20.2 years between 1754 and 1830.[44] Coppicing reached a maximum with the increased demands for wood, charcoal and tanning bark during the Napoleonic wars, 1793–1815. With the post-war slump in demand, though, and the use of chemicals for tanning leather, coppicing declined.

The Broadening Estate: Beyond the Harvest

Highland landowners had long exploited the value of their land not just through a rent for what it produced but also through a rent derived from the processing of what it produced, with farmers being thirled to a mill, or compelled to have their grain ground at a particular mill and to pay multures for the privilege. Having a network of mills across an estate yielded an extra source of income. The economics of maintaining a large vertical mill, though, made demands not just on whether a suitable river site was available but also on the scale of crop available locally. Only where output was sufficient to cover the costs involved would landowners have deemed it worthwhile to build such a mill. An estate in the southern Highlands, such as the Breadalbane estate, had such mills, each with its own croft, in each of the commissionaries into which the lands on either side of Loch Tay were divided. But further into the Highlands and particularly as one moved westwards and northwards, such mills became fewer: thus we find vertical mills in areas such as Glenelg and on Lismore,[45] Mull and in parts of Skye but not on Harris or in areas such as Coigach or Knoydart.[46] Instead, grain was prepared using the smaller, so-called horizontal or ladle mill and the hand-operated quern. The scale of the former enabled quite small burns to be pressed into service but its capacity meant it could rarely service more than the needs of a single toun or small group of touns. The quern was the simplest solution, one adapted to the needs of a single family. Widely used across the North and North-west Highlands and the Hebrides, it suited farmers in the more marginal touns, those with no more than a toehold of arable and for whom any saving of multure was a bonus.[47] When pressured to raise their rental income, landowners increased the provision of vertical mills at the expense of the ladle mill and quern, even going to the extent of confiscating querns.[48]

Beyond milling lay another potential stage of grain processing that was split between landowners and tenants: that of malting, brewing and distilling. As with milling, there were official and unofficial dimensions to how they evolved. Across the south and east of the region, most of the large estates had established changehouses, malthouses or brewseats in their core areas, enabling them to exercise control over where tenants had their bere malted as well as over who was responsible for the brewing of ale and its sale. In some areas, we can find quite a high density of official changehouses, malthouses and brewseats. The Lordship of Huntly, for instance, had a network of brewseats, or 'ailhouses', as it called them according to a mid-seventeenth-century rental. In Belly parish alone, ten were listed,[49] suggesting that the thirst of more than its locals was being served. The Breadalbane estate, too, was well provided for, with a number of official malthouses and brewseats, including the malt kilns at Killin and at Taymouth (attached to Rissal Croft) and brewseats at sites such as Stix and Ballemore.[50] Initially, brewing ale was more important than distilling, at least in the main body of the Highlands. Whisky features in seventeenth-century rentals as a payment but only from select touns,[51] but references to distilling are not on the same scale as those to brewseats or alehouses at this time. Youngson thought the balance began to change quickly by the 1770s.[52] Alehouses were increasingly licensed to distil spirits but such official provision did not stop illicit distilling developing in the more remote corners of the larger estates. Starting with an act of 1785 that changed how tax was levied on stills,[53] the government's response was to subject distilling to closer control, culminating in the 1823 act that sought to give encouragement to legal distilling. As a piece of legislation, the act led to a more orderly system of licensing for legal distillers and the active suppression by excise officers of illicit distilling. Its impact helps draw out just how important illicit distilling had become in the Central and Eastern Highlands prior to 1823. '[T]he inhabitants of Glenlivet', wrote the author of Inveraven's *NSA* report, 'almost without an exception, and many also in the parish church district, were more or less engaged in manufacturing and carrying to market smuggled whisky' but that since the 1823 act, this illicit distilling and smuggling had declined under the 'eye of officers of excise'.[54] An 1829 report about Breadalbane estate farms on the south side of Loch Tay captured the same change. It also hints at the extent to which illicit distilling had been practised, for it talked about how 'the vigilance of the excise has of late so much increased that it is impossible for them to dispose of their grain in the usual way of distilling.[55]

As in the main body of the Highlands, the earliest references to the presence of distilling in the Western Highlands and Islands are those provided by the payment of whisky as part of rent. A 1642 rental for Colonsay, for example, notes tenants in eleven out of the sixteen touns as paying 'acqua vitae' as part of their rent.[56] This background level of activity appears to have been expanded sooner than in the Central Highlands. A basic pattern of official changehouses for malting appears to have been established during the first half of the eighteenth century, with those positioned at key transit points, such as ferries, also functioning as inns.[57] The presence of changehouses in areas where good surpluses of bere were produced was understandable. The Breadalbane estate, for instance, established some on Lismore and in parts of Netherlorn,[58] while Islay had as many as seventeen by 1765.[59] The way estates moved to award licences for 'brewing, malting, vending and Distilling all sorts legal spirits and liquors'[60] suggests it was seen as another way in which they could capitalise on their resources. Admittedly, the pattern of official provision comes across as uneven. On Lewis, Walker reported as many as eleven or twelve stills[61] but, on South Uist, '[t]here is little Grain destroyed with Aqua Vitae as there is but one Distillery'.[62] Whether an island had surplus of bere and whether it offered tenants and their landlord a means of raising cash were the determining factors. Jura could boast three official changehouses, including two with malt barns, in its 1764 rental,[63] while an island the size of Skye boasted only four or five stills, with its inhabitants reportedly importing whisky for their own needs.[64] The same was said to be true of South Uist, with spirits consumed by the inhabitants 'being mostly brought from other places'.[65] Such assertions probably have an element of understatement about them, giving the official picture while ignoring what was illicit. This was certainly the case on Skye. A sense of how distilling may have erupted comes across from a comparison between a memorandum produced for the Macdonald estate in the 1730s, that talked about the loss caused by tenants having an 'ignorance of Malting it [= barley] & drawing spirits',[66] and a regulation included in tacks during the second half of the eighteenth century designed 'to put a stop to smuggling and retailing of spirituous liquors on every part of the estate'. This was not aimed simply at imported whisky for it went on to state that no tacksman or tenant 'shall upon any pretence whatever Brew Distill sell or retaile any ale, whisky or other spirituous liquors without a particular licence'.[67] Whether such attempts to regulate the trade worked is doubtful, for a meeting of the Justices of the Peace at Sconser in 1788 was still able to record that the 'retailing of spirits has become a very general nuisance', an observation directed at unlicensed premises.[68] The extent of unofficial distilling is also shown by Barrisdale, on the mainland opposite. No maltmakers or stills were said to exist, yet 'whisky-houses' were everywhere.[69] On Tiree, Walker reported the island as exporting a 'great quantity' in his 1764 survey but does not quantify the number of stills. In fact, few years later, in 1768, Turnbull put the number at fifty,[70] but he does not say if these were simply the official stills. That there were illicit stills and that their operation continued down to at least the end of the century is beyond doubt. Establishing a licensed inn was part of the provision made for Scarinish when it was created in 1802[71] but stocking it was a different matter and the duke's chamberlain complained that 'none could be found in the island willing to undertake distilling in a legal way'.[72] Yet, while illicit distilling did not suit the Duke's interests in 1802, it had for much of the eighteenth century, helping to put cash rents in his pocket. It is probably for this reason that neither Walker nor Turnbull actually censured the use of grain for distilling on Tiree. Kintyre, with its easier access to grain markets, was a different matter. A 1761 comment by the duke himself talked about how farmers in Kintyre 'have got of late into a most pernicious practice of distilling their Barley privately into whisky or acquavitae contrary to Law', and wanted tenants to be warned about the penalties.[73] As a 1790s report made clear, the problem was detecting illicit stills,[74] though licensed stills had shifted the balance by this point.[75]

The Broadening Estate: Domestic Industry

Whereas activities such as fishing and kelp production could supplement subsistence or income in coastal areas, it was only domestic industry that offered the possibility of a supplement to the land for inland areas. At a time when expanding industry was about finding new supplies of cheap domestic labour, the Highlands must have appeared as an ideal frontier as restrictions on the use of hand-intensive techniques, such as the hand quern, freed up female labour during the winter months. In spite of the region having a long domestic tradition of spinning and weaving woollen cloth, the earliest efforts to exploit this pool of cheap labour focused on flax spinning. The first steps were made by the Board of Trustees for the Manufactures which had been established in 1727 specifically to expand linen production, a well-established lowland industry but one whose growth depended on expanding the supplies of yarn. By the 1740s, the idea that the Highlands could be a supplier of such yarn was taken up by the British Linen Bank, founded in 1746, and the Board of Commissioners, the one providing capital and the other providing active encouragement as well as capital in the form of spinning wheels and flax at reduced cost. The Board of Commissioners also sought to encourage the industry by setting up stations during the 1750s and 1760s that comprised warehouses for housing hecklers and scutchers, and spinning schools for teaching Highlanders how to spin flax as opposed to wool, the former being regarded as more difficult than the latter. Stations and schools were set up across the region, with sites at Stornoway, Callander, New Tarbet, besides Loch Carron and Loch Broom and in Glen Morriston. For a time, flax spinning appeared to offer a way forward, with reports that tenants and their children 'are all winter employed in spinning, which draws considerable money back into them yearly'.[76] Efforts, though, were more successful in proximity to established weaving centres such as Crieff, Auchterarder, Perth and Inverness, with private dealers and merchant manufacturers working to provide flax and to collect the finished yarn through local shopkeepers or 'intakers'. Yet, though spinning persisted after the British Linen Bank withdrew from the region, it did not prove a sustained supplement to incomes.[77]

The Broadening Estate: Kelp

Seaweed had long been used as a field manure in the Hebrides and in the Northern Isles, as well as along the western seaboard, but its use for kelp did not begin until the early eighteenth century. Kelp was the calcined ashes of seaweed, a product that had a value in the making of carbonate of soda and iodine. Whereas the seaweed used as manure could just as easily be that gathered as drift on the beach, that is, so-called black wreck and tangle, the best seaweeds for kelp were *Fucus serratus*, *F. digitatus* and *F. nodosus*. These were known to communities as yellow or box wreck, or simply kelp ware. They were rich in soda but, because they grew below the waterline, had to be cut at low tide. Harvested between June and September, when farmers also had labour demands from resources such as peat and hay, it was burnt in coffers, often arranged in a circle on the shore, with 4 tons of seaweed being required to produce a stone (6.4 kilograms) of kelp. During the height of its production in the late eighteenth and early nineteenth centuries, few Hebridean summers would have seen wholly clear skies, as the smoke from kelp burning billowed out along the coast.

The production of kelp reportedly began in Orkney, where reports suggest that it was being made on Stronsay as early as 1719.[78] By the 1730s, it was being made in coastal areas of the Mey estate in Caithness, as well as on North Uist and Tiree. Even during these early years, estates moved to protect their interest by distinguishing what was to be cut for kelp as opposed to the tangle gathered for manure.[79] Where tenants organised kelp burning themselves, it became a charge on their rent, as on Jura, where a rental valuation of the estate for 1770 showed tenants being levied with an extra charge for the 'privilege' of 'cutting wreck fitt for kelp'.[80] A major turning point was reached in 1793 when the outbreak of the Napoleonic Wars led to a block on imports of barilla, an alternative source of soda brought in from Spain. Very quickly, the value of kelp rose dramatically, with a price of £6 per ton in 1790 rising to £20 per ton by the end of the century.[81] Encouraged by this increase in price, many smallholders engaged with kelp making and output rose quickly. On Tiree, output stood at 245 tons in 1790 but this had risen to over 500 tons by 1800. Across the Minch, North Uist produced 400 tons in 1790 but its output had risen to over 1,500 tons by 1810. Landowners such as Lord Macdonald and Clanranald made large profits from kelp during these years.[82] For tenants, it was a mixed blessing. On the one hand, it brought a new source of cash, enabling more to live on the land. In fact, a striking feature of kelp production was

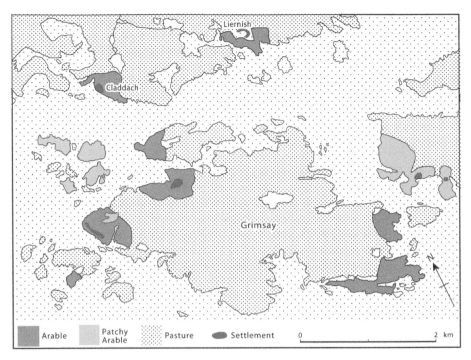

Figure 7.1 Pre-crofting arable around Grimsay, North Uist. Based on Reid's plan of North Uist, 1799, NAS, RHP1306.

the way it squeezed 'a heavier and heavier tenantry on an inelastic system of land'.[83] We must not stress this 'inelasticity' too much, however. It was those touns whose numbers were swollen by kelp production over the late eighteenth and early nineteenth centuries that saw the greatest use of the spade and foot plough to defy these inelasticities. Even those regarded as living by kelp alone, without any formal set of land, are likely to have raised a few rigs for potatoes. The sort of squatter's landscape which kelp manufacture helped to produce is well illustrated by Grimsay. A 1799 map shows a number of small patches of arable scattered across it, over small adjacent islands and the 'mainland' opposite (Figure 7.1), though none are formally referred to in rentals. Even tiny islets were settled and cropped with a few rigs (Figure 7.2). Almost overnight, Grimsay had become 'valuable on account of its kelp shores',[84] though those now settled there would have engaged in fishing as well.

Kelp making continued to provide an income supplement in some coastal areas after 1815 (for example, on the Uists and Tiree) but output never reached the supercharged levels of the war years, 1793 to 1815. Though some landowners, such as the Duke of Argyll, took steps to revive kelp making on islands like Tiree, and achieved some success in doing so using new techniques,[85] profit levels were never as great as in the years 1793–1815.

Figure 7.2 Settled islet beside Grimsay, North Uist.

Its decline left many communities at the tipping point of survival. Admittedly, the potato enabled families to squeeze their resource base but the increased dependence on it brought its own problems given the outbreaks of blight and harvest failures in the years following the end of the wars.[86] Acting in combination, these crises created acute problems of overpopulation for coastal areas.

The Broadening Estate: Fishing

Among eighteenth-century attempts to capitalise on the resources available, fishing was one of the more sustainable. Even before the mid-eighteenth century, it was locally important for subsistence and income. There were, however, regional differences regarding to whom it was important. Around the Northern Isles, fishing had become a major source of livelihood. Coastal communities there had long engaged in fishing. Even in Martin Martin's time, *c.*1700, fishing was seen, at least in Shetland, as being 'the Foundation of both their Trade and Wealth'.[87] For part of summer, it made Lerwick a busy, cosmopolitan port, with Hamburgers, Bremmers and others setting up shop in May and trading items such as linen, muslin, beer, brandy and bread for fish as well as for stockings and mutton.[88] Those Shetlanders who fished shared their waters with the Hollanders whose numerous busses harvested large catches out of the region, enough to regard it as 'their gold mine'.[89] In contrast, few communities along the west coast relied to any degree on fishing prior to 1750. Yes, fishing in the west was important, but it was in the hands of Clydesiders who fished mainly for herring, or what Robertson called this 'heaven-directed treasure'.[90] Their boats would amass in vast numbers in the western sea lochs, such as Lochs Fyne or Hourn, wherever the herring swarmed, with some fishing and some having the task of taking the catch to Glasgow. The movement of herring was irregular but where and when they shoaled, then, as Garnett said of Loch Fyne, it must have seemed as if the loch 'contains one part of water, and two parts of fish'.[91] The involvement of local communities was limited, however. There is evidence for some local specialisation, such as in Loch Fyne and by coastal communities on Bute, but otherwise, most local communities along the west coast, such as those in Knoydart, fished only inshore using small boats and then only to supplement their own subsistence.[92]

Many landlords along the west coast, seeing the visible profit being taken out of the area by Clydesiders, would have been acutely aware of the neglected opportunity. Slowly, the idea that fishing could be used to shoehorn communities on to a smaller footprint of land gained acceptance. The steps taken by the British Society for Extending the Fisheries to develop facilities within the region by establishing fishing stations, together with villages equipped with harbour facilities, warehouses, curing and salting sheds, boat-building yards, barrel-making facilities and small crofts for those involved, were the most notable. Ullapool, Tobermory and Lochbay were all sponsored creations of the British Society for Extending the Fisheries, and designed to foster fishing communities. Conceived around the same time, the late 1780s, and located following discussion with local landlords, they each involved a degree of planning. Largely because of the potential afforded by its site and location, Ullapool was the most immediately successful, growing quickly into a substantial fishing community. Its founding was not the first attempt to establish fishing at Ullapool. Back in 1698, and again in 1712, the Earl of Cromartie attempted to develop commercial fishing there but, though his plans in 1712 got as far as the construction of a curing house, neither attempt came to fruition.[93] Much later, in the 1770s, a Liverpool merchant was allowed by the Board of Commissioners for the Forfeited Estates to establish facilities for curing, salting and storing fish on Isle Martin, a venture that was followed by the establishment of a rival curing station on Tanera in 1784.[94] The British Society's decision to select Ullapool as a site for a fishing village, therefore, cannot be said to have pioneered commercial fishing in the north-west but it was the most ambitious effort. As with its other settlements, it took responsibility for providing the basic facilities needed, including a pier, curing station, salting sheds, and storage warehouse, as well as arranging for the provision of nets, lines and the building of boats. These facilities were laid out on the curve of the bay at Ullapool. Behind, on a more regular grid plan, it laid out crofts of a size that ensured their occupants needed to supplement their income, though it took no responsibility for building the houses. Like Ullapool, Tobermory also possessed a natural harbour but its landward site was more constricted so that, from the outset, it was arranged into lower and upper sites. Despite the constrictions, it grew as rapidly as Ullapool, having a population of three hundred within a few years, with twenty houses of slate and stone along its imposing bay frontage, but a less impressive array of thirty or so huts and cabins on its upper site. Of the three villages founded by the British Society, Lochbay on Skye was the last to be set out (1790) but also the least successful, largely because its crofting structure proved to be too viable in its own terms.[95] In time, though, all the west-coast villages established by the British Society struggled as commercial fishing centres, largely owing to the fickleness of the herring. Notorious for shoaling in different lochs in different years, by the early nineteenth century, the presence of large, dense shoals became sporadic in the west coast lochs.

Alongside these showcase fishing villages, efforts were also made by individual landlords to promote fishing. At first, these efforts amounted to little more than a permissive attitude towards the presence of lesser tenants and subtenants who fished, with many small coastal touns having 'one or two boats' for inshore fishing for white fish, herring or even salmon. In some cases, these impromptu attempts to broaden livelihoods became significant, as with touns in coastal Assynt, such as Baddidarach, Balchladich, Clashmore, Unapool and Inver, with Home's 1774 survey commenting on how important fishing had become for lesser tenants and subtenants on them.[96] The estate encouraged this diversification not just by allowing a build-up of tenant numbers but also by favouring tenants who were 'industrious' at fishing as well as kelp making.[97] Even by this point, they had started to become crowded, busy touns, whose scale of settlement was possible only because of fishing. Yet, other than allowing touns to become crowded in this way, there are no signs that the estate had made any significant investment in fishing facilities.

This reluctance by estates to commit investment to fishing started to change by the late 1780s, no doubt encouraged by the early efforts of the British Society. While some advised the Duke of Argyll that encouragement for local fishing was 'running away with a great deal of money',[98] the duke was one of the first to make an investment, seeing it as helping to put the household economy of his smaller tenants on a more secure footing. That said, the duke's early fisher touns at Scarinish on Tiree, Creich and Bunessan on Mull and Kenmore beside Loch Fyne were modest affairs, with a relatively small number of crofts, underpinned by help with the purchase of boats and lines, though sites such as Scarinish also had harbour facilities. By the beginning of the nineteenth century, we can see a step change, too, in the support provided for fishing in Assynt. When Achnacarnan was set to Donald Macdonald in 1802, it was seen as not for his own 'natural possession' but for subtenants 'to be employed in the fishery or manufactures, and to be provided with boats, nets and lines to fish herring, cod and ling from Cape Wrath to Gairloch. A few years later, in 1811, a report by William Young advised the Sutherland estate of the potential for fishing offered by the safe anchorages to be found in the lochs around Assynt. He saw the ground on the north bank of the Inver as the ideal site for laying out a new fishing village, one that could absorb those displaced by the Clearances and which would make Lochinver 'the *Metropolis* of Assynt'.[99] For landlords, the cost of such settlement lay not just in the laying out of crofts and helping out with boats and lines but in ensuring that facilities existed for the making of essentials, such as boats and barrels, and that curing stations were built for the processing and barrelling of fish. These last were specialised affairs, with those built at Port Ellen and Portnahaven on Islay, Stein in Waternish, Torridon and on Isle Martin, servicing more than local fishing.

Seen over time, the eruption of interest in fishing did not progress smoothly. Those who moved into the newly established crofts were often reluctant to engage in fishing. We can see this at Scarinish on Tiree, where the duke's chamberlain reported in 1802 that 'none of those now settled as crofters in Scarinish seem disposed to begin the fishing under any encouragement'.[100] The problem at Scarinish was twofold, with some crofters being too old and, more importantly, the crofts too large, so that their occupiers had little incentive to fish. This was a lesson learnt quickly by landowners so that, later, new crofts were deliberately kept small to force their occupiers to fish for their subsistence, too. We see this starkly on the Sutherland estate once the estate began to execute its plan to move those displaced by the spread of sheep into fisher touns on the coast. Not all those advising the estate agreed with the scheme. For one, it was 'a kind of heterogeneous jumble', better that 'their whole attention should be directed and kept in its proper sphere of action'.[101] By this point though, just a few years before the end of the Napoleonic Wars, the fisher croft had become an established concept. 'Heterogeneous jumble' or not, it broadened the resources exploited by estates and enabled more people to subsist on less land.

Sources of Change: the Improvers

Just as some thought that a diversification of the rural economy would provide a viable basis of livelihood for the region's population, so also did some see farming improvement as the solution.[102] What they had in mind was the application of Lowland ideas on improvement, with the removal of runrig, the enclosure of the newly consolidated holdings, the adoption of new crops such as turnips and sown grasses, their incorporation into new cropping regimes, and the application of schemes of land improvement which included stone removal, drainage and liming. Such improvements undoubtedly had a place in changing the region's landscape at this stage but their spread was slow and patchy prior to the close of the eighteenth century. Reports from across the region draw out what they see as the hostility of the traditional farmer to change, their refusal to break with the cake of custom. Putting aside the question of how much would-be improvers were inhibited from having a personal strategy of farming improvement within the rigidities of a runrig system, the more serious obstacle for the typical tenant was that of capital. Seen *c.*1750, few would have had the capital, especially when we consider that change brought with it much larger working units than earlier runrig shares. We need only consider the costs of stocking a large sheep farm to appreciate the scalar leap required in working capital. It is not surprising, therefore, if, at this stage of improvement, some of the initiative lay with landowners: they had absolute control over the structure of landholding, could regulate cropping schemes via tack regulations and, most importantly, could provide loans for capital-intensive improvements such as enclosure, draining and liming.

The most fundamental of all the changes, which contemporaries bundled together under the heading of improvement, was the removal of runrig. The wholesale clearance of touns to make way for large-scale, commercial sheep production or their reorganisation into crofting townships undoubtedly formed the prime pathway by which runrig touns were removed but it was not the only pathway followed. Some were removed by a division of shares into matched, consolidated holdings, with each of the latter being commensurate with an earlier runrig share. I want to label these as *commensurate* divisions. They can be contrasted with what I have called *generalised* divisions whereby some runrig holdings were divided into small to medium-sized consolidated farms but with no attempt, in terms of number, size or occupier, to match holdings in the new order exactly to what had been held under the old. Clear-cut examples of commensurate divisions are provided by those runrig touns involving hereditary rights, rights that had to be carried over into the new order. Touns of this type were invariably removed via a formal division under the authority of the 1695 Act Anent Lands Lying Runrig, an act designed solely to facilitate the division of such runrig, enabling one heritor to bring about the division of the whole. Some examples of proprietary runrig existed along the eastern edges of the Highlands[103] but examples were much more common in the Northern Isles. Most of the latter were not divided until the nineteenth and even twentieth centuries but we can find some eighteenth-century examples. Some of these early divisions were anything but straightforward. In some instances, as with the division of Birsay Be-South in Orkney in 1748, we can speak of the phased division of some proprietary runrig. Ostensibly designed to simplify the runrig landholding of heritors, Birsay's division led to the reorganisation of each heritor's land into large, croft-like planks but each plank then became redivided into runrig between a small group of tenants .[104] Elsewhere, Sir Laurence Dundas managed to divide some of his runrig lands in a number of touns during the mid-eighteenth century but not all such plans were implemented. Costs were an obstacle. The tacksmen for the bishop's lands in the parishes of Sandwick, Stromness and St Ola and on the islands of Shapinsay and Holm sought permission in 1762 to have the various lands divided out of runrig. He was given permission to proceed and the surveyor's preliminary advice was to survey the land and to give each heritor his proportion 'conform to the pennyland he possesses'. In the end, the exercise did not go forward because the tacksmen could not reach agreement over who was to pay, but it could not have helped that the surveyor died in the meantime.[105]

On the Scottish mainland, commensurate divisions also affected a few runrig touns occupied solely by tenants. As divisions, these stand out not least because landowners were under no legal obligation to formally divide out the shares of a tenant runrig toun into consolidated holdings that exactly matched former runrig shares once tacks expired. The 1695 act had no bearing on runrig touns held solely by tenants. Many Highland tenants were, in fact, 'instant' tenants or 'tenants at will'. Seventeenth-century Orkney rentals made the same point by distinguishing between 'remoueabill' or 'moweable' tenants on the one hand and 'fewaris' on the other.[106] Landlords exercised the ultimate authority over how holdings were to be divided on the ground and over how many tenants should be involved. The fact that so many Highland runrig touns were simply cleared for sheep makes this authority clear in the starkest way. The only legal process in their clearance would have been the issue of a decreet of removing, whereby tenants were given forty days notice that they were to be removed. Such a process under the authority of the 1652 act, 'concerning the warning of tenants',[107] and the earlier 1546 act, concerning the 'putting and laying of men out of their tacks and steadings',[108] had no specific relevance to runrig touns. It applied where a single tenant was removed from the occupation of a single share within a runrig toun no less than the removal of all tenants in such a toun.

Given this freedom in the matter, therefore, we need to consider why some were prepared to carry out commensurate division of their tenant runrig touns, not least because of the extra costs incurred in surveying runrig layouts beforehand. When the laird of Balnagown wrote to his estate manager, saying that it would be a 'great service' to the tenants to have 'every man's farm . . . set apart and not runrig as they usually are', we can be sure that the interests of the tenant were not the only consideration behind such a reorganisation.[109] Arguably, no single factor was at work. The simplest explanation would be to assume that some were carried out because wadsetters were involved or because sitting tenants had life tacks or claims of kindness or goodwill over their tenure. In some cases, landowners may have faced a situation in which leases were not sequenced to finish at the same time. More likely is that estates might decided a commensurate division preserved the kind of holding sizes best suited to prevailing systems of farming. Around the eastern edges of the Grampians, for instance, interest in cattle rearing meant that many considered small to medium-sized farms as best suited to the area, given the capital needed to stock such a farm. Significantly, it is in this area that we find instances of tenant runrig being subject to a commensurate division. The Gordon estate, especially, provides us with good examples. Much more so than other estates, it commissioned surveys of quite a number of its runrig touns, such as that of Craigwillie and nearby Claymires and Botarie, with detailed plans being drawn up depicting their pre-division runrig status. The fact that there was nothing tidy about their holding layout suggests that it was a layout that had become frozen at some point and had slowly absorbed change.[110] That the estate's intention was to carry runrig holdings over into consolidated holdings in these circumstances is given some support by a 1784 regulation dealing with the estate's touns in Strathavon which declared that 'where tenants had their lands mixed or runrigged with one another, the factor and birleymen could redispose the lands in suitable compact holdings to each tenant'.[111] Though Strathavon is well covered by estate plans, however, there is no sign of the estate commissioning runrig surveys at this point. For this reason, therefore, we cannot rule out the possibility that these divisions were actually generalised rather than commensurate in their execution, with the factor and birleymen simply laying out fresh holdings for tenants but without trying to match the new exactly against the old.

The same uncertainty over precisely what sort of divisions we are dealing with also comes across from the divisions instigated by the Board of Commissioners. The work of Mackillop and, more recently, Jonsson, provides us with valuable insight into the thinking that informed their approach to change. Those with an influence on the formulation of policy were caught between two strategies outlined earlier: the adoption of commercial stock production supported by Lowland ideas on enclosure and new fodder crops, albeit with an accompanying reduction in farm population, or holding on to the region's population through the maintenance of smallholdings, not least because of its potential for regular military recruitment.[112] For the likes of John Walker, a smallholding culture was viable if it was based on a combination of the spade and the potato.[113] Events had a hand in shaping the choice between these pathways. The outbreak of the War of American Independence saw a large and successful recruitment campaign by Highland landowners which, if a surplus was maintained, demonstrated the region's potential for military recruitment. The end of the war, however, saw these social debts called in, with the Board of Commissioners and estates generally feeling obliged to offer land to former soldiers.[114] Against this background, it is not surprising that the Board of Commissioners was not among those who vigorously cleared their land for sheep but set about reaccommodating the main tenants of their runrig touns by laying out a system of small to medium-sized farms and by settling lesser tenants and cottars in smallholdings or crofts. The creation of small to medium-sized farms can be seen clearly in a 1765 report on the parishes of Kirkhill and of Pharnaway and on the Kilmorack portions of the Lovat estate which refers to a surveyor as 'dividing and laying [holdings] out as much together among the present possessors as the situation of the land and houses would admit'.[115] The same sense of those who held under runrig being reaccommodated comes across from changes in the same year on the Struan estate, with the factor reporting that he had 'divided the whole Estate from Runridge into Glebs [= small farms] . . . so that each farm or village can keep and reap at least a third more advantage from them'.[116] As in Strathavon, the factor laid out new consolidated holdings to match how many tenants were present but does not appear to have matched the exact size of their old holdings via a commensurate division. That this was the case is confirmed by plans drawn up for the Struan estate which are labelled as showing runrig but, in reality, they seem only to depict the orientation of cultivation rigs, not the actual possession of runrig strips, the first step towards a commensurate division.

The smallholdings also created by the Board of Commissioners on some of its estates were quite different in character from these small to medium-sized farms. They emerged in a number of ways. Some were created in association with planned villages, such as Kinloch Rannoch, that were laid out by the board over the 1760s and 1770s. Others were established to absorb the lesser tenants and cottars displaced by the settlement of half-pay army officers on holdings in Barrisdale and Coigach at the same time.[117] Others were little more than allotment gardens, a number of which were created to help resettle ordinary soldiers: these were the so-called 'Kings cottagers', such as those laid out at Callander.[118] The limited arable attached to these early smallholdings had a purpose, the expectation being that those settled in them would also be tradesmen or fishermen, with the arable croft as a supplement to, rather than the entire source of, support. Given how questions over whether the region's future lay in maintaining its high levels of population were being debated by other landowners, and not just the board, it is not surprising that other estates adopted such a similar strategy. As early as 1763, for instance, the Earl of Breadalbane petitioned the board for support in resettling ex-soldiers on coastal crofts in Netherlorn, crofts that were directly tied to fishing as their prime form of livelihood.[119] Within a decade, the idea was put to use by the British Society for Extending the Fisheries for the layout of fishing villages such as Ullapool. As a concept, the croft was an old idea, stretching back into medieval times and having a very distinct status. The crofts which began to emerge in the mid-eighteenth century were different. As Mackillop rightly pointed out, those attached to the small planned villages created by the board, or in Barrisdale and Coigach, were a significant step in the evolution of the crofting township, taking the individual form of the medieval croft and turning it into the basis of a collective form or township.[120] These early crofts were small, some no more than 3 acres (1.52 hectares) and most were tied to other forms of livelihood. Those being created by the 1790s and during the early nineteenth century, however, and even those designed as fisher crofts, tended to be much larger, with some averaging 10 to 12 acres (5 to 6 hectares) or more of arable. As the Duke of Argyll observed, fisher crofts of this size provided little incentive for crofters to fish, so we soon find them adjusted downwards in size as a matter of principle.[121] Even non-fishing crofts, though, were slowly reduced in size over time as populations grew and the crofts became split between children.

In some instances, estate documentation makes it clear that generalised divisions were designed only to resettle the main tenants, not necessarily all tenants. We can see this by looking at the Breadalbane estate. The estate set about 'dividing' its touns in Netherlorn in 1785. An initial attempt divided touns so that 'each tenant' had 'his own share' but the estate accepted that 'some small tenants will be thrown destitute of possessions' by the reorganisation so that 'each tenant' could not have meant all tenants.[122] Following their division or reorganisation, some of these Netherlorn touns, such as Ardluing on Luing, were also 'Enclosed & Subdivided wt stone fences' at this point.[123] Whatever its intentions, however, the 1785 division did not entirely succeed in removing runrig. A resurvey, c.1798, found the divided touns had been arranged into separate lots or holdings but that, in a number of cases, such as Ardluing, these lots had slipped back into being 'run-ridge' touns, 'albeit now based around only two tenants' in each lot,[124] and recommended a reorganisation whereby tenants would have 'roughly' the same size of 'lots or divisions as they did before' but now each was to have their share as a separate, consolidated holding.[125]

Generalised divisions also took place on the Argyll estate on Mull and Tiree. In conditions of set for Tiree issued in 1776, the duke had instructed 'that runrig possession of corn farms (arable land) were to be entirely abolished, and every tenant to occupy (by himself or servants, without subletting) a distinct separate possession . . . not below the extent of four mail land'.[126] It accepted that it had a 'great overstock of tenants'[127] but planned to settle as many as possible in holdings that were deemed viable. At the time, the duke considered a 4-mail land unit to be the minimum, enabling 170 out of the 286 tenants who then existed to be accommodated on them,[128] a reduction that meant the divisions were, at best, generalised. Three touns were chosen to pioneer the scheme: Barapole, Kinavar and Kenovar. Barapole and Kinavar were possessed by an Archibald Frackadale, and the initial plan was to divide the 64-mail lands between sixteen tenants, each 'on his own separate farm'.[129] In the event, they were laid out by George Langlands, the surveyor, into only fifteen holdings, with one tenant getting an 8-mail land unit. Likewise, when Langlands came to divide out the 48-mail lands of Kenovar, provision was made for thirteen, not twelve, holdings. In some cases, the duke's tenants were actually encouraged to divide their own farms, with Langlands or the duke's chamberlain checking afterwards whether the division was competent.[130] Yet, even as these early reorganisations were being executed, doubts were expressed over the feasibility of the scheme. In the first place, tenants were reluctant to take up 4-mail land holdings. In the second place, the estate always accepted that it had a problem of excess numbers, its 'supernumeries', and 'that not all of them could be accommodated with such a holding'[131] but it now realised that it had greatly underestimated just how many were squeezed out by such a system. It felt that the situation had been made worse by tenants dividing their holdings between members of their family[132] and by the Passenger Vessels Act of 1803 which had the effect of reducing emigration from the island.[133] Faced with this changing situation, the duke concluded in 1802 that the 'scheme of bringing the farms to 4 mail lands must be given up in so far as not executed effectually, and different farms must be broke down into small crofts to accommodate the people', with those who were formerly tenants now being given only 6 to 10 acres (3 to 5 hectares) of arable, and those who were cottars and tradesmen given 4 acres (2.03 hectares).[134] In other words, for Tiree and Mull, the scheme to create small farms turned out to be simply a stepping stone towards the eventual creation of crofts.

On the Lochtayside portion of the Breadalbane estate, a mix of strategies was employed when it carried out a reorganisation of its touns in 1798. The scheme was part of a wider plan designed 'to divide the Hills from the Lowland on the skirts of the loch'. The former were detached and thrown into 'large pasture or sheep Farms, and the latter entirely into corn, arable or enclosed ones'.[135] The estate proceeded on the basis that the latter, the low-ground farms, 'were [to be] divided into Lots so as each man should have his farm by himself', justifying the inevitable reduction by claiming 'that the remainder [would] have such possessions as they can live upon'.[136] In fact, in its execution, the estate's attempt at the generalised division of its touns adopted a slightly more elaborate plan. Above the road that ran along the north side of the loch, where most former touns had their outfield, it opted to create a number of medium-sized farms of 40 to 50 acres (20 to 25 hectares), while holdings below the road were laid out as 'small lots'.[137] Clearly, with its mix of sheep farms, medium-sized farms and 'small lots', there were contrasting outcomes for landscape here.

The creation of consolidated farms was generally regarded as essential for the improvement of farming practice, one that facilitated enclosure and better schemes of husbandry. The first of these, enclosure, was hardly a new improvement when seen in c.1750. Quite apart from the residual enclosure that still existed within farming touns, what we can label as parkland or pasture-based enclosures had long been used to adorn the estate policies in the Highlands. Some of those around Taymouth, the Earl of Breadalbane's mansion house, for instance, may even have dated back to the late sixteenth century.[138] Such enclosures were part of the way landowners used their policies to control the view both from and towards their main residences. There was also a profitable function to them, however. They helped to provide fattened beef for the house but also enabled landowners to share directly in the rising cattle trade, supplying advanced or fattened cattle to nearby lowland butchers or markets such as Crieff. The importance of estate policies and home farms to early enclosure was made clear by Gailey's study of enclosure in the South-west Highlands. His study stressed the local variations in the way the process unfolded. When seen through Roy's *c.1755* map, the only areas of enclosure were those that formed part of estate policies or home farms, such as those at Stonefield on west Loch Tarbet and Inverneil on Loch Fyneside. The only area that had experienced more general enclosure at this stage was Campbell of Knockbuy's estate. Significantly, Knockbuy was a well-known early improver who engaged heavily in the cattle trade as early as the 1720s. The subsequent spread of enclosure over the second half of the eighteenth and early nineteenth centuries affected some parishes more than others, with those such as Kilninver and Kilmelfort seeing more attempts at enclosure than Knapdale and Glassary.[139]

Over time, we also see enclosure of another kind. By the early decades of the eighteenth century, estates had started to separate out their common grazings. Some agreements targeted the extensive grazings shared between different estates but others moved to assign specific blocks of hill ground to individual touns or clusters of touns within estates. In time, this process of disaggregation was followed by the building of boundary dykes so that the hill ground of each toun or farm became physically defined and bounded on the ground. Patently, this was an important change, enabling the farms or touns affected to maximise their use of hill ground. Yet, ultimately, maximising the use of hill ground also depended on making adequate provision for the wintering of stock. The lack of adequate winter fodder had long been recognised as a fundamental restraint on stocking levels. Because touns could summer far more stock than they could winter, possibly two or three times more according to Robertson,[140] the lack of winter feed acted as a serious bottleneck on the full exploitation of the hills. A solution lay in the use of sown grasses and turnips but its uptake was relatively slow. When Alexander Wight visited the edges of the Highlands in the late 1770s, he came across some farmers planting turnips and sown grasses.[141] Overall, however, their numbers were few and the amounts sown were small. In fact, when we look at cropping regimes even for the 1790s and early 1800s, either through reports in the *Old Statistical Account* or estate documentation, what stands out is not so much the amounts of turnips and grasses being sown but the amount of arable still under oats and barley.

The Broadening Estate: the Changing Value of Pasture

The Rise of the Droving Trade

Observers in the region had long argued that its proper use, its future, lay in the more effective exploitation of its pastures. Initially, however, what they had in mind was a concentration on cattle not sheep. Cattle were seen as offering landowners the prospect of a better, more certain rent.[142] Even in 1783,

Lord Breadalbane could still ask for a list of 'what farms are proper for the purpose of raising and grassing black cattle' as opposed to those proper for grain on the Argyll portion of his estate despite the spread of specialist sheep farms all around him,[143] though he was being advised to reduce the number of black cattle and replace them with sheep on his Loch Tay lands only two years later.[144] The importance attached to cattle was not a sudden transformation. From the early decades of the seventeenth century, there had been a flow of droved cattle out of the region. To start with, the numbers involved were not considerable but they had become sizeable by the early decades of the eighteenth century. There were push and pull factors involved. The increasing demand of landowners for cash rents served as a push factor. For tenants in remoter areas, cattle offered the best prospect of raising this cash because cattle walked to markets. Concept was ahead of reality, however, with estates having to accept cattle in lieu of cash from such areas. It is for this reason that we find landowners themselves engaged in organising droves even in the earliest years of the trade: from Lewis and Assynt in the north to Islay in the south.[145] Alongside the push factors, there was the powerful pull factor of growing market demand from southern markets. Changes in market conditions, such as that which followed the ban on the import of Irish cattle (1686) and later, the growth of demand from the south after the Union (1707), helped to ratchet up the flow of cattle still further

Two kinds of producers were involved in the trade. On the one hand, there were the many lesser tenants for whom releasing even a single animal a year to the drove was a major demand on the farm economy. It did confer a degree of flexibility, however, enabling these lesser tenants to shift part of their rent burden away from arable to grass, leaving more of their crop for subsistence. On the other hand, we find landholders in possession of whole touns whose response to the cattle trade was to place much greater emphasis on cattle production, to the extent that some became specialist cattle producers. Key among these specialist producers were those tacksmen or small landowners who farmed large holdings in hand. Stock lists for parts of Argyll, including Kintyre, suggest that one or two tacksmen may already have started to direct their toun economy towards specialist cattle production by the early seventeenth century.[146] By the early eighteenth century, such specialist cattle producers were to be found in most parts of Argyll. Campbell of Knockbuy was the most notable but he was not the only one. Others can be documented, such as the touns of Glencripesdale, Kinlochteacuis, Laudale and Liddesdale in Morvern, all of which were described in a report of 1770 as operating as specialist cattle farms[147] but are likely to have been dealing as such for a number of decades. A different kind of response can be seen along the southern edge of the Highlands. Here, areas such as Balquhidder saw shieling grounds and other grazings being assembled into separate lets as grazings and wintering grounds by the 1730s, with drovers being among those who managed them, possibly as holding grounds for stock droved down from other parts of the Highlands.[148] The Breadalbane estate, too, had taken steps to carve grazing farms out of part of its shieling ground by the 1730s.[149] Given that it was almost impossible to drove cattle out of the southern Highlands without crossing the Breadalbane estate at some point, it was inevitable that some touns also reached short-term or seasonal deals with drovers over access to grazings,[150] though such deals were not always with the estate's blessing. At one point, in 1726, the estate took the view that the excess drover's stock allowed to graze in areas such as Glenfalloch and Lochdochart had been 'the ruin of the Countrey'.[151] A 1712 list of drovers with tenancies in Glenfalloch, Strathfillan and Disher, twenty-seven drovers in all, gives some idea of the local involvement in the trade, quite apart from the droves that passed through.[152]

The Coming of Commercial Sheep Farming

I have elsewhere said that the first application of sheep to the mountains of the Highlands was like the recovery of an immense area of country from the sea. It is as stupid to object to it as it would be to object to the drainage of the Bedford Level.[153]

Sheep were present in the Highland landscape long before the Clearances began but, as Gray succinctly put it, '[t]here were many sheep but no sheep farms in the Highlands before 1760'.[154] The pioneer of these specialised sheep farms may have been John Campbell, an Ayrshire sheep farmer who had moved northwards after getting into financial difficulties and selling his estate, but precisely where he established specialised sheep farming is unclear. Some have him seeing the potential of commercial sheep farming in the area when he took a tenancy of the inn at Tyndrum which, like other inns during the mid-eighteenth century, had significant holdings attached to it, or those of Auchtertyre and Auchinturin.[155] Others see his first real effort starting when he took a tack of a farm on the Luss estate, beside Loch Lomond in 1747, while still others see his first specialised sheep farm as Glenevoe, beside Loch Lomond, which was 'cleared' of its four runrig tenants and, from Whit Sunday 1756, occupied by Campbell at a significant advance of rent. Almost in the same year, he was reported to have taken a sheep farm, at the head of Loch Fyne, and to have also started large-scale sheep farming there, bringing in cast blackface ewes from the south.[156] Campbell's example was quickly followed by a steady flow of other tenants from Annandale, Nithsdale, Clydesdale and Tweeddale, tenants already skilled in specialised sheep farming and aware of the market, taking over the tacks of former touns in Dunbartonshire, Argyllshire and Perthshire during the 1760s and 1770s. To this extent, the northward march of sheep into the Highlands can be seen as an extension of a process which had already swept across the Southern Uplands, with the Highlands being seen, *c.*1750, as the next best step by market-savvy tenants and sheep masters from the Southern Uplands and Borders. Only in a few areas, such as the higher ground of Angus and Kincardineshire, did the early spread of sheep farming draw on tenants from within the local community.

Because of the way in which the clearances for sheep could be driven by estate-wide policies, we sometimes find touns across a whole district undergoing change more or less at the same time though, sometimes, it could take a decade or so to roll out a policy completely if tacks did not fall for resetting in a synchronised way.

Speaking generally, the majority of sheep farms were created from the wholesale clearance of former runrig touns but this was not the only means by which they were established, especially during the eighteenth century. Instead, some were fashioned out of shieling grounds or hill grazings which had been detached from touns, the latter being left with only their core pasture and arable. Such a strategy can be seen as another way in which landlords tried, in the early years of change, to accommodate new strategies of land exploitation such as commercial sheep farming while, at the same time, trying to maintain the presence of a basic tenantry working on small to medium-sized holdings.

We can see some of the processes involved by looking at the experience of particular estates. The district-based approach is well illustrated by the Breadalbane estate. While it tried to maintain tenant numbers in its reorganisation of areas such as north Lochtayside and Netherlorn, it was happy to see some of its touns in Glenorchy and Glenlochay, as well as some of its former shieling grounds, turned into sheep farms. Many of the touns along Glenorchy, as well as those lying across the southern part of Rannoch, were transformed into sheep farms in the 1780s.[157] A note in 1783 reported that, since the last round of tacks were issued, it had changed so that

> almost every farm in that country having a considerable stock of black cattle and sowing a great number of bolls of grain of different kinds, the whole to a trifle, is now turned under Sheep, and no Tenant of any considerable stock can tell within some hundreds what he has upon his farm.[158]

This observation on what had happened in Glenorchy since the last set was actually by the earl's chamberlain, John Campbell. Though he was well placed, his picture of conditions there was possibly a little overstated for another report had advised a more selective approach, with Arrivean, Clashgour, Blaravon and Drumliart being 'the most proper to be confined to one Tenant each Being extensive sheep farms' but others, such as Achalader, Bochyle, Barchattan and Upper Kinchrackin, were 'well calculated for their present mode of possession as there is of arable land and of land capable of being made Arable', and it was best to allow 'continuing upon them, the number of small tenants necessary to cultivate them regularly'.[159] Elsewhere, the estate took a different tack. Mention has already been made of how the Breadalbane estate was advised to separate out its high-ground from its low-ground farms in the Loch Tay area, putting the former under sheep. More specific proposals were made along these lines with regard to Glenlochay in a letter dated 1791. Its core recommendation was that the extensive (and distant) shieling grounds attached to farms in Glenlochay should be detached and laid into new sheep farms. The letter supported its case by not only suggesting that the touns in the glen were 'without exception the best arable and meadow ground upon the estate' which, with the ample pasture within their head dyke, would still be viable but also pointed to the Braes of Perthshire (as well as to parts of the annexed estates) where shielings had already been disjoined from farms and established as separate sheep farms.[160] A report on farms in Glenlochay records the presence of large sheep flocks by 1810 but it also captures some of the problems, with Kenknock facing 'severe and frequent losses in their sheep stock due to its high and exposed grazings'.[161]

Though some have likened the spread of commercial sheep farming across parts of the southern Highlands and beyond the Great Glen from the 1750s onwards to a 'moving frontier',[162] it was not a progressive diffusion. Some commercial sheep holdings were already to be found north of the Great Glen even by the 1760s, only a few years after the start of the process, with Sir John Lockhart Ross of Balnagowan introducing sheep to the higher ground of his Ross estate in the 1770s, taking farms into his own hands as their tacks fell in and bringing in shepherds to manage the flocks. In the early 1780s, he set the various sheep walks created to a Perthshire tenant.[163] In the decade that followed, touns in other glens and straths which ran eastwards across Ross, including Glen Achary, Strath Oykell, Strathcarron and Strath Rusdale, were cleared.[164]

In the west-central Highlands, the geography of landowning served to produce a locally varied pattern of outcomes. At the outset, estates pursued two parallel policies, creating some sheep farms by detaching extensive blocks of hill or shieling ground and others through the straightforward clearance of touns. We can see the first of these strategies at work in Glen Quoich where a large sheep walk was created in 1782 out of former shielings.[165] Subsequent sheep walks, created across the Glengarry during the 1780s, meanwhile, involved the clearance of touns. To the south, on the Locheil estate, change came later, with only some small-scale clearances taking place along Loch Arkaig before 1800, but the scale of change gathered pace after 1800, with those displaced being resettled in new crofting townships beside Loch Linnhe, at Corpach and Onich. While gradual in its unfolding, the reorganisation of touns eventually left the shores of Loch Arkaig entirely under sheep farms. The imbalance of settlement between the north and south sides, with the former having more touns prior to the Clearances, (even more so if one adds the couple tucked into the lower reaches of Glen Dessary) meant that the disloca-tions caused by clearance for sheep were socially and physically greater on the north side. By comparison, the area beside Loch Eil was subjected to a more mixed strategy of change, with some touns cleared for sheep but others reorganised *in situ* into crofting townships.

Further south, Morvern provides us with one of the more complex landscape histories for the eighteenth and nineteenth centuries. Change had already started to rework its landscape by the late eighteenth century. The *OSA* reported that the parish comprised thirty-two farms, of which seventeen were in the hands of tacksmen, five in the hands of shepherds and the remaining ten set to small tenants.[166] The last were a product of reforms carried out by the Duke of Argyll on his portion of Morvern back in the early 1740s, when he replaced some of his tacksmen-controlled farms with smaller farms set directly to tenants.[167] The switch was not a success and, by the 1750s, some were being regrouped and reset as large units to tacksmen again. Gaskell argues that those reset tacksmen farms were the first to adopt large-scale sheep farming in the 1770s but it was not a switch which involved all farms held by tacksmen as some continued as specialist cattle farms. Using derived measures of their population density, he drew a distinction between two types of tacksmen holdings that turned to sheep. He saw those carrying only a modest resident population as probably organised as specialist sheep farms, using both their low, sheltered ground as well as high-ground grazing; these were to be found in the core areas of Morvern. The second type was distinguished by larger resident populations: these, he argued, were sheep farms which had been developed around the edge of runrig touns, the lesser subtenants and cottars present having access to a small portion of the arable and the right to graze some stock close to the toun but the higher-ground pastures and shieling grounds were now separated out and run as a sheep farm by the tacksman.[168] Gaskell sees this latter type of farm as reflecting the unwillingness of local landowners to allow clearances at this point, with those tacksmen who now saw a future in sheep farming having to work out a modus vivendi between the old and new.[169] The remaining touns, those set directly to tenants, were still organised around runrig arable, with their grazings used for modest numbers of cattle as well as sheep. These small runrig touns had access to sizeable grazings both low hill ground and shielings but, as the local profitability of sheep farming became established, some were cleared so as to release their grazings for sheep. This happened in 1788 on some of the interior touns belonging to the Argyll estate but, at this stage, there was still a concern to accommodate those displaced by resettling them in small farms on a strip of land along the western coast of Morvern, in the so-called Drimnin enclave.[170]

The more the Highland rural economy became drawn into the marketplace, the more it was at the mercy of market fluxes. The self-sufficiency which Adam Smith saw in the traditional Highland economy had its crises but the region's increasing reliance on the marketing of produce brought new problems. Those problems were brought home to it at the end of the Napoleonic Wars in

1815 when, after just over two decades of rising or buoyant prices, prices fell sharply as markets adjusted quickly to a non-war, unblockaded economy. To add to the woes of the region, the immediate post-war years saw old problems revisited on the farm economy, with a number of poor harvests. There was also a new twist to these old problems. The vulnerability which came with a growing dependence on the potato as a source of basic subsistence for many crofters, smallholders and cottars was exposed by outbreaks of potato disease at the end of the war. As we shall see in the next chapter, the conjunction of these various problems shifted conditions and mindsets to the extent that how change unfolded after 1815 also shifted.

Notes

1. M. Storrie, 'Landholdings and settlement evolution in west highland Scotland', *Geografiska Annaler*, 43 (1965), pp. 138–61.
2. D. Watt, '"The laberinth of thir difficulties": the influence of debt on the Highland elite. *c.*1550–1700', *SHR*, 85 (2006), pp. 28–51.
3. Cregeen (ed.), *Argyll Estate Instructions*, pp. 1, 8, 23, 26, 72.
4. A. J. Youngson, *After the Forty-Five: The Economic Impact on the Scottish Highlands* (Edinburgh, 1973), p. 33.
5. Smith, *Jacobite Estates*, (Edinburgh, 1982) pp. 38–53; A. Mackillop, '*More Fruitful than the Soil'. Army, Empire and the Scottish Highlands, 1715–1815* (East Linton, 2000), pp. 78–83.
6. Smith, *Jacobite Estates*, p. 24.
7. Mackillop, '*More Fruitful than the Soil'*, pp. 80–2.
8. Smith, *Jacobite Estates*, pp. 164–82.
9. Ibid., p. 171.
10. Watt, 'The laberinth of thir dfficulties', pp. 28–51; F. McKichan, 'Lord Seaforth and Highland estate management in the First Phase of Clearance (1783–1815), *SHR*, 86 (2007), pp. 50–68.
11. The shift in attitude and policy before and after *c.*1815 is also reviewed by A. I. Macinnes, 'Commercial landlordism and clearance in the Scottish Highlands: the case of Arichonan', in J. Pan-Montojo and K. Pedersen (eds), *Communities in European History* (Pisa, 2007), p. 49.
12. Jonsson, *Enlightenment's Frontier*, pp. 93–146.
13. Shaw, *Northern and Western Islands*, pp. 26–7, 43–6; Macinnes, *Clanship*, p. 144.
14. BPP, *Report*, 1884, Appendix A, pp. 389–92.
15. E. R. Cregeen, 'The tacksmen and their successors. A study of tenurial reorganization in Mull, Morvern and Tiree in the early 18th century', *SS*, 13 (1969), pp. 93–144.
16. Smout et al., *History of the Native Woodlands*, pp. 199–201, 205.
17. Smout and Watson, 'Exploiting semi-natural woodlands', p. 91.

18. Ibid., p. 91.
19. Smout et al., *History of the Native Woodlands*, p. 202.
20. Smout and Watson, 'Exploiting semi-natural woodlands', p. 92.
21. Ibid., p. 93; Smout et al., *History of the Native Woodlands*, pp. 229–31.
22. RCAHMS, *Argyll*, vol. 2 (Edinburgh, 1975), pp. 33–4, 280–92.
23. CANMORE ID23401.
24. RCAHMS, *Argyll*, vol. 2, 1975, p. 281.
25. J. M. Lindsay, 'Charcoal iron smelting and its fuel supply: the example of Lorn furnace, Argyllshire, 1753–1786', *JHG*, 1 (1975), pp. 283–98.
26. Ibid., p. 292.
27. NAS, GD174/737.
28. NAS, GD112/16/10/2/11.
29. NAS, GD112/14/13/9.
30. Smout and Watson, 'Exploiting semi-natural woodlands', p. 93; Smout et al., *History of the Native Woodlands*, p. 250.
31. NAS, GD64/1/86/5.
32. NAS, GD174/827.
33. Adam (ed.), *Assynt*, p. 122.
34. Watson, 'Rights and responsibilities', p. 109; Smout et al., *History of the Native Woodlands*, p. 104.
35. Watson and Smout, 'Exploiting semi-natural woodlands', p. 94; Smout et al., *History of the Native Woodlands*, pp. 103–7.
36. NAS, GD112/9/3/3/14.
37. NAS, GD112/16/10/2/11.
38. Smout et al., *History of the Native Woodlands*, pp. 75, 105, 119.
39. Smout and Watson, 'Exploiting semi-natural woodlands', p. 93.
40. NAS, GD112/16/10/2/1, 1728.
41. Watson and Smout, 'Exploiting semi-natural woodlands', p. 93; Smout et al., *History of the Native Woodlands*, p. 179.
42. For example, NAS, GD112/14/13/9.
43. *OSA*, 1791–99, xvii, pp. 244–5. See also, *OSA*, 1791–99, iii, p. 431; Wills (ed.), *Reports*, p. 11.
44. J. M. Lindsay, 'The commercial use of woodland and coppice management', in M. L. Parry and T. R. Slater (eds), *The Making of the Scottish Countryside* (London, 1980), p. 281.
45. NAS, GD170/420/1/1.
46. NAS, FE, E729/1; NAS, E746/151; NAS, E788/42.
47. Dodgshon, *Chiefs to Landlords*, pp. 116, 217–18.
48. For example, IC, AP, Bundle 663, Instructions ... Morvern, 1733; NAS, GD221/3695/4; NAS, GD221/5037/1; GD201/1/351/12.
49. *Miscellaneous*, Spalding Club, 1849, 4, pp. 261–319.
50. NAS, GD112/9/5/8/13. See also, NAS, GD112/9/5/7/2; GD112/9/5/7/4.
51. NAS, GD112/9/24.

52. Youngson, *After the Forty-Five*, p. 111.
53. Ibid., pp. 113–14.
54. *NSA*, xiii (1834–45), p. 135; see also ibid., p. 73.
55. NAS, GD112/14/3/2/32.
56. IC, AP, NE11, Rentall . . . sett, 1642.
57. Shaw, *Northern and Western Islands*, p. 136.
58. Ibid., p. 136.
59. Smith, *Islay*, p. 483.
60. For example, DC, MDP, 2/2/20.
61. McKay (ed.), *Walker's Report*, p. 42.
62. Ibid., p. 77.
63. NAS, GD64/1/86/5.
64. Ibid., p. 207.
65. Ibid., p. 77.
66. NAS, GD221/3695/2.
67. NAS, GD92/50B.
68. NAS, GD403/40/1–2.
69. Smith, *Jacobite Estates*, p. 199. See also, Wills (ed.), *Reports . . . Annexed Estates*, pp. 39, 48.
70. NAS, RHP, 8826/2.
71. Cregeen (ed.), *Argyll Estate Instructions*, p. 6.
72. Ibid., p. 61.
73. IC, AP, Bundle 2530, Instruction . . . Kintyre, 1761.
74. Youngson, *After the Forty-Five*, p. 115.
75. E. Cregeen and A. Martin (eds), *Kintyre Instructions. The Fifth Duke of Argyll's Instructions to His Kintyre Chamberlain 1785–1805* (Glasgow, 2011), p. 65.
76. Millar (ed.), *Forfeited Estates*, pp. 252–3.
77. A. J. Durie, 'Linen-spinning in the north of Scotland, 1746–1773', *Northern Scotland*, 2 (1974–5), pp. 13–36.
78. BPP, *Report*, 1884, Appendix A, p. 271.
79. Gray, *Highland Economy*, pp. 124–37; Dodgshon, 'Strategies of farming', pp. 697–9; *OSA*, 1791–99, xiii, p. 330; GD46/17, vol. 80; Cregeen and Martin (eds), *Kintyre Instructions*, p. 52.
80. NAS, GD64/1/86/5.
81. Macculloch, *Western Highlands and Islands*, ii, p. 152.
82. Gray, *Highland Economy*, p. 135.
83. Ibid., p. 128.
84. *OSA*, 1791–99, xiii, p. 304.
85. Argyll, *Crofts and Farms*, pp. 45–6.
86. T. M. Devine, *The Great Highland Famine* (Edinburgh, 1995), Chapters 1 and 2.
87. Martin, *Western Islands*, p. 384.
88. Ibid., pp. 385–6.
89. Henderson and Dickson (eds), *Naturalist . . . Highlands*, p. 143.

90. Ibid., p. 143.
91. T. Garnett, *Observations on a Tour Through the Highlands and Part of the Western Isles of Scotland* (London, 1810), i, p.94.
92. NAS, FE, E788/42.
93. E. Richards and M. Clough, *Cromartie: Highland Life 1650–1914* (Aberdeen, 1994), p. 229.
94. J. Munro, 'Ullapool and the British Fisheries Society', in J. R. Baldwin (ed.), *Peoples and Settlement in North-West Ross* (Edinburgh, 1994), pp. 247–9.
95. DC, MDP, 1/384/6.
96. Adam (ed.), *Home's Survey*, pp. xlviii, 9, 10–11, 28–9, 49.
97. NLS, SP, 313/1000.
98. Cregeen (ed.), *Argyll Estate Instructions*, p. 157.
99. R. J. Adam (ed.), *Papers on Sutherland Estate Management 1802–1816*, *SHS*, 4th series, 8 (Edinburgh, 1972), i, pp. 128–9.
100. Cregeen (ed.) *Argyll Estate Instructions*, p. 60.
101. Adam (ed.), *Sutherland Estate Management*, i, p. 27.
102. Mackillop, *'More Fruitful than the Soil'*, p. 80.
103. For example, NAS, RHP1473; ibid., RHP146.
104. Thomson, *Orkney*, pp. 334–5.
105. A. Peterkin, *Rentals of the Ancient Earldom and Bishoprick of Orkney* (Edinburgh, 1820), Appendix, pp. 108–9.
106. Ibid., pp. 82, 90.
107. APS, ii, p. 494.
108. APS, ii, p. 476.
109. Macgill, *Old Ross-shire*, vol. i, p. 177.
110. G. Kay, 'The landscape of Improvement: A case study of agricultural change in north-east Scotland', *SGM*, 78 (1962), p. 101.
111. Gaffney (ed.), *Strathavon*, p. 181.
112. Mackillop, *'More Fruitful than the Soil'*, pp. 82–6; Jonsson, *Enlightenment's Frontier*, pp. 93–146.
113. Ibid., pp. 31–8. See also, Mackillop, *'More Fruitful than the Soil'*, pp. 89–90.
114. Ibid., pp. 88–94.
115. Millar (ed.), *Forfeited Estates*, p. 71.
116. Ibid., p. 236.
117. Mackillop, *'More Fruitful than the Soil'*, pp. 94–8.
118. Ibid., p. 90; Jonsson, *Enlightenment's Frontier*, p. 32. A plan of those laid out at Benebeg (Bennybeg) and Concraig, Perthshire, is reproduced in J. B. Caird, 'The creation of crofts and new settlement patterns in the Highlands and Islands of Scotland', *SGM*, 103 (1987), p. 68.
119. Mackillop, *'More Fruitful than the Soil'*, p. 97.
120. Ibid., p. 99.
121. Cregeen (ed.), *Argyll Estate Instructions*, pp. 64, 199.
122. NAS, GD112/1/2/14.

123. NAS, GD112/9/3/3/14.
124. NAS, RHP972/5.
125. NAS, RHP972/5. For the creation of similar farm lots on Mull and Ulva, see NAS, GD174/244.
126. Argyll, *Crofts and Farms*, p. 8; Cregeen (ed.), *Argyll Estate Instructions*, pp. 55, 196.
127. Ibid., p. 58.
128. Ibid., p. 59.
129. Ibid., p. 51.
130. Ibid., p. 197.
131. Ibid., p. 59.
132. Ibid., p. 199.
133. Ibid., p. 71.
134. Ibid., p. 73.
135. NAS, GD112/16/13/10/2.
136. NAS, GD112/12/1/2/37.
137. NAS, GD112/16/4/2/17 and 22.
138. T. C. Smout, 'Pre-improvement fields in upland Scotland: the case of Loch Tayside', *Landscape History*, 18 (1996), pp. 47–55.
139. R. A. Gailey, 'Agrarian improvement and the development of enclosure in the south-west Highlands of Scotland', *SHR*, 134 (1963), pp. 105–25.
140. Henderson and Dickson (eds), *Naturalist . . . Highlands*, p. 161.
141. A. Wight, *Present State of Husbandry in Scotland* (Edinburgh, 1778–84), III, part ii, p. 718.
142. NAS, FE, E788/42.
143. NAS, GD112/14/12/7/8.
144. NAS GD112/12/1/5.
145. NAS, GD313/918; DC, MDP, 112/9/1/3/45; Gregory and Skene (eds), *Collectanea*, 1847, i, p. 153; Innes (ed.), *Origines Parochiales*, pp. 351–2.
146. IC, AP, N.E. 11 vol. 1543–1610 Rentall . . . Inchaild, 1609; ibid., Bundle 746, Lands . . . Kintyr c.1636.
147. P. Gaskell, *Morvern Transformed. A Highland Parish in the Nineteenth Century* (Cambridge, 1968), p. 16.
148. Stewart, *Settlements of Western Perthshire*, pp. 106–7.
149. NAS, GD112/16/25.
150. NAS, GD112/59/59.
151. NAS, GD112/14/12/5/14.
152. NAS, GD112/17/1/10/11.
153. Argyll, *Crofts and Farms*, p. 36.
154. Gray, *Highland Economy*, p. 38.
155. NAS, GD112/16/25, p. 70; GD112/54/9.
156. Watson, 'The rise and development of the sheep industry in the Highlands and North of Scotland', *THASS*, 5th series, xliv (1932), pp. 6–8.

157. NAS, GD112/16/5/2/21 and 30.
158. NAS, GD112/14/12/7/8.
159. NAS, GD112/16/5/2/30.
160. NAS, GD112/16/5/5/31–3.
161. NAS, GD112/16/13/4/9.
162. Gray, *Highland Economy*, pp. 86–7.
163. Richards, *Leviathan of Wealth*, pp. 185–6.
164. D. Turnock, *Patterns of Highland Development* (London, 1970), p. 25.
165. Ibid., p. 25.
166. *OSA*, 1791–99, x, p. 266.
167. Gaskell, *Morvern Transformed*, p. 7.
168. Ibid., pp. 7–8, 16.
169. Ibid., p. 7.
170. Turnock, *Patterns of Highland Development*, p. 182.

LANDSCAPES OF SHEEP, DEER AND CROFTS: CHANGE AFTER c.1815

The history of the economical transformation which a great portion of the Highlands and Islands has during the past century undergone ... is written in indelible characters on the surface of the soil.[1]

As a point in the history of the Highlands and Islands, c.1815 is not a neat and tidy divide. Each of the different pressures acting on the landscape had a slightly different phasing but, if we could take an average point between them at which the direction of change appeared to shift, then it would be the end of the Napoleonic Wars. The high prices received for farm products, which had been such a feature of the wars, collapsed. So too, did the price of kelp. Quite apart from their impact on smallholders, the financial problems created by these falling prices exerted strain on the already stretched finances of many traditional landowners. Many estates were sold into new hands, with the nineteenth century seeing the arrival of a new kind of landowner, someone who had made money in banking, industry or commerce, and who brought a different set of values and expectations. Whereas a high proportion of change prior to 1815 had tried to balance the need for more productive systems of farming with a concern for those who packed the farming touns,[2] change after 1815 had much less regard for making any such accommodation with the past. Change now became draconian and uncompromising.

Sheep and the 'Ruin'ation of the Highland Landscape

The decades following the end of the Napoleonic Wars saw the final spread of commercial farming to all corners of the region. Sutherland was the county most affected by this nineteenth-century surge, and in no county have the consequences of their spread been more closely scrutinised. A few specialist sheep 'rooms' already existed in the county[3] but the spread of the 'Great Sheep', or Cheviot, the breed increasingly favoured by sheep farmers after 1800, marked a wholly new phase, with people cleared for them across large swathes of countryside. The Sutherland estate was at the heart of this transformation. The estate had been inherited by Elizabeth, the Countess of Sutherland. Her parents had died in 1766, when she was only one year old, and her right of eventual succession to the estate was established in 1771.[4] At this point, the Sutherland estate, like other Highland estates, would not have been particularly profitable. In 1785, however, the countess married Lord Stafford, the heir to the Leveson-Gower estates in Yorkshire, Staffordshire and Shropshire, estates that generated profits from mining as well as from farming. He inherited the Leveson-Gower estates from his father, along with the title of the Second Marquis of Stafford, in 1803, the same year he inherited the profits of the Bridgewater Canal from his uncle.[5] Now equipped with the means, he and the countess began the transformation of the Sutherland estate. On the one hand, they expanded it, acquiring neighbouring estates, including the Reay estate, when they became available. By the time of the Marquis's death in 1833, the estate had grown from *c.*800,000 acres (323,756 hectares) to around 1.5 million acres (625,000 hectares). On the other hand, they gradually developed a large-scale programme of estate reorganisation, one that brought the calculating values of a different world to bear on the remote straths and hills of Sutherland.

The reorganisation followed a coherent plan, one designed to produce 'two parallel economies', with 'a land-extensive pastoral economy of the cleared interior and a labour-intensive coastal economy',[6] and a planned shift of population from the former to the latter. Nor was it a case of only touns in the smaller straths, such as Frithe, Skinsdale or Seilge, being cleared. The larger straths, such as Strath Naver. were equally transformed. In place of what had been a well-settled landscape of runrig touns, the estate created a network of vast sheep walks across the interior of Sutherland, controlled by a new breed of flock masters and their shepherds, displacing the previous occupants to new settlements on the coast. The process of conversion began in 1807 when parts of the estate around Farr and Lairg were cleared. It continued in 1809 with the clearance of tenants from touns around Dornoch, Rogart, Loth, Clyne and Golspie.[7] The same year also saw the clearance of the extensive area of exposed hill and wet moss that lay between Lairg and Loch Naver, including the pasture around Ben Klibreck, to form a vast sheep holding.[8] This was not a densely settled area. Its core formed the headwaters of the Brora and Helmsdale rivers and what settlement existed was squeezed into its narrow straths or occupied the pockets of drier soil that lay on lower, more open ground, such as the string of very small touns, like Gearnsary, that once lay between Badanloch and Loch Choire. Writing about it some years later, Loch argued that developing sheep walks on some of the highest and wettest ground of the estate was something of an experiment, an attempt to see whether sheep coped with its mix of hill ground and low-lying wet moss before then clearing touns in the larger straths so as to provide the wintering ground needed.[9] It was also significant, though, because it formed the first of Sutherland's super farms. Set to two sheep masters from Northumberland, Atkinson and Marshall, it encompassed over 100,000 acres (40,470 hectares) of land once they had added the wintering farms of Ardinduich and Letterbeg which lay south of Lairg.[10]

In fact, their holding formed only one of a number of large, extended sheep walks created over the early nineteenth century. One of the largest was that pieced together by Patrick Sellar, one of the duke's agents, acting with a self-interest that surely played a part in his role as a driving force behind the Sutherland clearances. His poor supervision of the clearances in 1814 led to his trial for murder, oppression and arson. Tried in 1816 but cleared in a single day by a jury the members of which knew their place, Sellar at least stood down as the duke's agent.[11] It left the estate undaunted, however, for it continued to carry out clearances, including the large-scale clearance of touns in the parishes of Kildonan, Laird and Rogart, displacing over three thousand people in the process. Elsewhere, Donald and William Mackay accumulated a comparable cluster of farms which were spread across 150,000 acres (60,704 hectares) of Caithness and Sutherland, including 70,000 acres (29,168 hectares) at Melness on the north coast. By the 1830s, the interior of Sutherland had been accumulated into thirty to forty farms, with flocks that ranged up to fifteen thousand sheep. Whether seen in terms of their sheer physical extent or flock size, these sheep enterprises were far larger than anything we find in the southern Highlands. The speed with which they were developed is measured by the way in which sheep numbers in the county rose from fifteen thousand to 130,000 between 1811 and 1820, with 118,000 of those present in 1820 being the larger Cheviot breed as opposed to the Blackface.[12] Of course, there was a debit side to this build-up, one conveyed by an 1832 listing which noted the touns on the Sutherland estate that no longer existed in the Farr and Strathy districts, with the majority, fifty-one out of seventy-nine, having disappeared, plus a further four out of the five listed for Tongue parish.[13] Most of those that had disappeared lay along Strathnaver and on either side of Loch Naver, though there were also some lying across the small valleys south of Strathy together with a handful towards the north coast. The extent to which this transformed landscape, emptying settlements, is also demonstrated by a list of the resident shepherds employed to manage the various sheep-farming enterprises. Atkinson and Marshall, for example, employed sixteen shepherds, each one based in a former toun. Morton and Culley, meanwhile, employed ten, again each based in what had been a separate toun.[14] As well as dealing with extensive and remote areas of responsibility, the sense of isolation for these shepherds must have been accentuated all the more by the toun ruins that lay around them. 'Of all the solitudes,' said Macculloch, the geologist, when confronted by such abandoned touns in Sutherland on his 1819–21 tours, 'there is no solitude like that of ruins,' arguing that their very particular impact was because it was 'the solitude of art, not of nature'.[15]

Assynt also saw clearances, but the displacement of people from its interior touns to the coast was less pronounced simply because fewer touns were involved. Besides, fishing and kelp-making had already encouraged the build-up of sizeable numbers in coastal touns. For many of these coastal touns, their transition to crofting townships, some in 1812, others in 1818, was an *in situ* reorganisation. In the process, some grew further in size, absorbing not just natural increase but those displaced from inland touns, with quite a number cast into townships of over twenty lots (for example, Elpine, twenty-three lots, Drumbeg, twenty-three lots and Knockan, twenty-seven lots).[16] A listing a few years later, in 1826, suggests that some ended up with still more tenants but others fewer than planned. Uncharitably, the estate still treated them as tenants 'at will'.[17] The contrast between coast and interior was lessened in other ways. Quite a number of interior touns had been held by tacksmen since at least the early eighteenth century with some farmed in hand by them, primarily for cattle. Once the demand for sheep penetrated into the area and began to affect decisions over how touns should be exploited, the switch to sheep would have been straightforward. Quite a lot, however, were still transitioned into sheep farms via a clearance. We can glimpse the process through the affairs of a sheep farmer called Charles Clarke. He already held Achmore on the north bank of Loch Assynt and, in 1819, lobbied the marquis for the tack of Little Assynt, Cromont, Coulin, Camore and Unapool as an extension to his existing lands when he heard that they were 'to be Set for a sheep walk'. At the time, Unapool, Coulin and Camore were held by fairly large communities of conjoint tenants. In his letter, Clarke offered to relieve the marquis 'of removing tenants, from any claims for meliorations and [he would] allow the outgoing tenants to carry away the timber of their houses', though removing the tenants actually turned out be far from simple because of the haggling involved.[18] Clarke's efforts to clear these touns can be approached from another perspective, that of the Napier Commission. At its session in Lochinver, it took evidence from a seventy-five-year-old cottar from Balchladich, Kenneth Campbell. Campbell submitted a list of those touns which, based on his recollections, had been cleared from Assynt. Though he was only a teenager when it happened, almost sixty years earlier, he listed an astonishing forty-eight touns as having been cleared. His list confirms that Unapool, Coulin and Camore were cleared but a number of others on his list were possibly sub-touns rather than touns.[19] The list has value in another way for it shows the sheep farmers who took over the cleared touns. Clarke was responsible for nine, with the largest flock masters, George Gunn and Kenneth McKenzie, controlling eighteen and fifteen former touns respectively.

In comparison to the stark simplicity about change in Sutherland, change in the West Highlands, beyond the Great Glen, followed more varied pathways. Part of the reason for this was the way in which its geography of landowner- ship was so extensively rewritten during the nineteenth century. The scale of turnover can be illustrated by Morvern where, between 1813 and 1838 'every single property in Morvern changed hands, and by 1844 there was scarcely a proprietor left who had any traditional or lengthy association with parish, or (in most cases) with anywhere else in the Highlands either'.[20] In the process, estates also became fragmented, releasing an even greater potential for policy differences between them.

Turnock's early work on the West Highlands adds other revealing insights. In his analysis of Lochaber, he concluded that 'almost without exception estates which remained intact throughout this time are all in the Great Glen area, where change was spread over a long period, while sale and subdivision of landholdings are more typical of the more revolutionary trends further west'.[21] Examples of the 'revolutionary trends', mostly tied in with changes of ownership, are provided by Knoydart and South Morar. In Knoydart, the demands of sheep farming worked westwards from Glengarry. The first large, commercial flocks were introduced on grazings around Loch Quoich on the edge of Glengarry and Knoydart in 1782, taking over what had previously been exploited as shieling ground.[22] As elsewhere, this use of shieling ground for sheep farming stressed the ability of local toun communities to maintain their stock. Of course, just as low-ground communities needed their summer grazing, so also did flock masters around Loch Quoich need the shelter of low ground. It is not surprising, therefore, that the closing decades of the eighteenth century saw touns on the lower ground of Glengarry being dispos- sessed. By the second quarter of the nineteenth century, the westward stream of dispossessed led to an accumulation of even more people around the coast of Knoydart, an area where touns were poorly endowed to begin with.[23] There are few signs that the estate made any effort to resettle those who filled out these touns, though it was not really an area suitable for crofting townships. With the post-war collapse of kelp prices and the lower prices for cattle and wool in the 1820s and 1830s, most estates saw the growing accumulation of numbers in poorly endowed coastal communities as now being more of a problem than a source of profit. The potato famine of the 1840s further exposed their precariousness of livelihood everywhere.[24] Whereas most West Highland estates explored mixed strategies, with local areas assigned to sheep farms and others assigned to crofting communities, the Knoydart estate opted for a more singular strategy. In 1853, the estate carried out one of the more notorious acts of wholesale clearance, removing the bulk of the smallholders, cottars and squatters in one fell swoop, arranging for the 'forced' passage of many to Canada.

When we turn to islands such as Skye and those of the Outer Hebrides, we find their landscapes altering via equally complex pathways of change.

When John Blackadder surveyed Skye and North Uist in 1799, ahead of the Macdonald estate's plans for change, he found the vast majority of touns still farmed in runrig, either held by conjoint tenants directly or sublet from tacksmen, with only a small number farmed in hand by tacksmen.[25] Changes, such as the increasing importance of cattle droving for cash rents, the adoption of the potato, the development of kelp-making and the growth in population, happened largely within the framework of the traditional Hebridean runrig toun. By the time local estates came to change this framework, touns already faced acute problems, with too many people living off a minimum of arable, helped by the potato and the foot plough. Initially, landowners, such as Lord Macdonald and MacLeod of Dunvegan, were reluctant to dispossess their numerous lesser tenants on Skye, so that the first attempts at change on both estates from *c*.1810 onwards followed more of an eighteenth- rather than nineteenth-century strategy, with quite a number of runrig touns being recast into crofting townships. Those that switched to sheep at this point were those that had long been farmed in hand by tacksmen and which already had a well-established livestock-oriented economy focused on cattle. By the start of 1840s, as the *NSA* report for Duirinish put it, sheep had become the 'sole stock of tacksmen'.[26] Easily the most notable examples were the two very large sheep runs created across Minginish, one centred around Rhu-Dunan, or what became Glen Brittle farm and in the hands of M'Askill, and the other somewhat briefly in the hands of a Dr M'Lean and focused on Talisker, with the former extending north to Loch Eynort and the latter south to it. In evidence presented to the Napier Commission, a shoemaker and cottar from Ferrinlea, John M'Askill, said that M'Lean cleared the touns that lay along the shore of Loch Eynort, stretching round into Tusdale.[27] We can expect tenants and cottars to have disappeared with these clearances but some cottars, such as those regularly used by M'Lean to trench his pastures, remained.[28] By the 1830s and '40s, even those most responsible for the welfare of crofting communities now saw crofts and the lotting system as being 'ruinous to the country'[29] and no longer the solution but the problem. Once the fragility of the smallholder economy had been exposed by the 1840s famine, estates responded by seeking a nineteenth-century solution but it was not runrig touns that some now cleared to make way for an expansion of sheep farming but the crofting townships that had initially replaced them. The mid-nineteenth-century clearance of townships such as Suishnish and Boreraig, on the coast overlooking Loch Eishort, are well documented but they were only two among a number of Skye townships on which the Macdonald and MacLeod estates had second thoughts. Other Hebridean landowners, especially those who had recently bought into the problem, also seized upon the aftermath of the 1840s potato famine to act in the same way. It is for this reason that the mid-nineteenth century, especially the years either side of 1850, saw some of the most comprehensive clearances for sheep in the Hebrides, with islands such as Barra, Mingulay and parts of the Uists witnessing wholesale clearances. In contrast to earlier clearances, many carried out during this phase were about the emptying of landscapes, with no provision for reorganising touns or creating crofting townships on new sites, though some were given passage to North America or to Australia.

The mid-nineteenth century probably saw specialised sheep farming at its maximum geographical extent, though some of the gains, in terms of its spread, were already being offset by the conversion of some sheep farms into deer forests. The impact of sheep, though, is about more than just their spread at the expense of people. Following the end of the Napoleonic Wars, prices for crop and stock fell. The price for wool declined further once cheaper imports were available from Australia but sheep and lamb prices recovered and surged strongly by mid-century, ushering in a decade or so of high profits for sheep farmers. We can detect four particular adjustments in response to this renewed profitability. The first involved a substantial increase in sheep numbers. Unlike the surge in the early decades of the nineteenth century, which was brought about by the spread of sheep into the far north of the Highlands, this third-quarter rise saw an increase in stocking levels on existing farms. Second, with the upturn in sheep prices and the reduced demand from the south for cattle owing to the outbreak of rinderpest, many farms now became more exclusively sheep farms, with the folds of cattle that had continued to be kept on many farms now disappearing. Third, there were changes in the composition of flocks, with sheep being sold at a younger age. During the early part of the century, most sheep were sold at four to five years old but, by the 1860s and 1870s, it was common for stock to be sold on at three years.[30] By the 1880s, though, the boom in sheep farming was over. The collapse of prices during the 1880s, coming as it did on the back of rising costs, made sheep farming less profitable. As sheep prices fell, many sheep farmers were caught out by the high level of their rents. Indeed, landlords used the fact that they had increased rents for sheep farmers, while holding down crofting rents, as part of their defence when the Napier Commission came to collect evidence.[31] Fourth, by mid-century, continuing shifts in prices and rents had affected how landlords viewed their earlier attempts at restructuring touns and farms. In comparison with those created in the far north, the sheep farms created in the southern Highlands during the early phases of the Clearances were comparatively small units. As the economics of stock production changed, some landlords revisited the problem, amalgamating their smaller, less competitive farms. Though some mergers were designed to create a more balanced holding, with the right mix of wintering and hill ground, others were simply intended to create larger working units, as with the amalgamation of Strathmashie, Dalachully, Shirrabeg, Shirramore, Garnamore and Garnebeg in Badenoch, all farms previously held by 'gentleman farmers', into a single farm unit.[32]

Leisure and Hunting Estates: Old Pastimes, New Forms

The pursuit of game in the Highlands and Islands is as old as human settlement itself. Glimpsed through the earliest documents, hunting animals of the chase, snaring or shooting game birds and catching game fish, were socially exclusive, a carefully guarded privilege of the Crown or lords. Areas of legal forest in which rights of hunting were reserved probably existed throughout the region but those that come into view by the thirteenth century lay primarily around the southern and eastern edges of the region. To the east, lay reserves such as Lennoch and Rothiemurchus. To the south-east and south were Atholl, Clunie, Alyth and Strathearn.[33] Over the next century or so, we find references to reserves elsewhere in the Highlands, including examples such as Cluanie, Lochaber, Rannoch and Mamlorne.[34] Few references exist to traditional hunting forests in the Hebrides but there were examples, such as that on Harris.

By their very nature, hunting reserves were subject to forest laws. Where they survived in use, something of their character and management is disclosed by documents detailing aspects of their management. A charter of 1584, for instance, appointed Donald Farquharson as Keeper of the King's Forest in Braemar, Cromar and Strathdee. He was bound to 'to caus hayne [= preserve] the said Wodis, Forestis and Mureis; and to serche, seik, tak and apprehend all and whatsumevir personis hauntand or repairand thairin with bowis, culveringis, nettis or ony uther instrument meit and convenient for the destruction of the deir and murefowlis'.[35] Unlike other areas in which hunting reserves were established by the Anglo-Normans, and which practised a form of hunting based on the running or scenting of hounds, the prime form of hunting in the eastern Highlands as in other parts of the Highlands was by means of the drive, using greyhounds provided by surrounding touns. Even in the seventeenth century, tenancy agreements could bind tenants to attend the hunt with dogs, 'eight followers from each davoch of land, with their dogs and hounds' in the case of a 1632 agreement for land in the Braes of Mar.[36]

By the eighteenth century, most surviving areas of hunting forest were now controlled by local estates rather than by the Crown.[37] Their day-to-day management was in the hands of foresters who, usually, were tenants of a local holding, such as Duncan McNab whose tenancy of Feinchrochar in 1722 was linked to the fact that he was forester of Mamlorne.[38] We need only look at local court papers to appreciate how carefully estates still guarded the right to hunt game but, equally, how some in the farm community were just as open to breaching such rules. The court records for Rannoch are especially rich in references to the poaching of deer: this was as much because the farming touns which fringed Rannoch were inhabited by unruly groups as because Rannoch was well stocked with deer. In 1684, a number of tenants were charged not just with killing but with 'murdering' a hundred deer 'that came out of the forest in tyme of the great storm last to shelter themselves and to feed on the strath of the country'.[39] Part of the problem for landowners was that many forests were also used as grazing areas for cattle and sheep so that ordinary farmers could have a right to be in them. Landowners were clearly aware of how this could be a cloak for poaching, some responding with regulations that barred any farmer from carrying a gun or even being accompanied by a dog within the bounds of a forest.[40]

The eighteenth century can be seen as something of a watershed for deer forests. Some were still actively used as hunting reserves but other pressures were now acting on them, with more opened up for stock grazing. A forest such as Mamlorne in Breadalbane had long been opened to use for grazing by the so-called forest farms while touns in Glen Lyon had also established the right, a much contested right, to use parts for shieling grounds.[41] A slightly different arrangement prevailed in the Forest of Coigach, the estate sometimes letting it as a deer forest but 'when not' let, the rent was 'proportioned on to possessions adjacent to the Forest'.[42] In fact, by the eighteenth century, the rising profitability of cattle production had caused some forest land to be treated as a frontier for its expansion, as was the case in the encroachments in Glenavon mentioned in Chapter 4. By the time the Gordon estate compiled a memorial in 1753, there were already 620 cattle grazing in it. The recommendation made was that the forest and its grazings should be divided among the farms in Strathavon. In fact, within a few decades, the estate was taking a different approach. By the 1780s, its view was that touns had more grazings than they needed. What had been forest land was now detached and, along with shieling ground, used to endow wholly new farms. When a new factor took over in 1787, he explained away the new farms names that had appeared in the rental, such as Inverchgar, Terrulan and Balbain, as the result of the improvement of former grazings and shielings.[43]

Yet, once the century had turned, the demand for hunting or shooting reserves recovered. As the new century unfolded, a new kind of sportsman appeared more on the scene. Wealthy businessmen and industrialists from the south now came to the Highlands in increasing numbers to stalk, shoot and fish. Some began to buy estates as growing financial pressures forced many traditional landowners to put their land on the market. Others rented shooting rights. As they did so, game reserves were given a reinvigorated role within the Highlands. Old established areas of hunting were revived or recovered from sheep farms and put under a closer, more active management while extensive new reserves were established. The traditional view of the hunting estate is that they became fashionable again only after the collapse of the profitability of sheep farming during the late 1870s and early 1880s. Orr's work has shown this to be a flawed view. Of the 1.997 million acres (808,975 hectares) reported by the Napier Commission in 1883 to be already under deer forests, a significant proportion had been converted from sheep pasture over the preceding half-century or so, not just during the years of the depression in sheep farming immediately before 1883. Twenty-eight deer forests were formed by 1839 and a further fourteen in the 1840s alone.[44] The scale of this growth is illustrated by the Mar Forest in Aberdeenshire. Only 10,000 acres (4,047 hectares) in 1810, it had expanded to 60,000 acres (24,282 hectares) by 1839 and to 100,000 acres (40,469 hectares) by 1872.[45] Clearly, forests and sheep were not locked into a simple inverse relationship with each other, with forests contracting when sheep farming boomed and vice versa. The fashion for hunting forests was driven just as much by the demands of wealthy businessmen and landowners from the south as by movements in the prices of wool and mutton.[46]

A map of deer forests provided by the Napier Commission for 1883 enables us to take stock of hunting forests at this point. From it (Figure 8.1), we can identify a scatter of reserves or forests in the southern Highlands, such as Blackmount and Mamlorne, and, in the northern Highlands, those such as Langwell, Reay and Ben Armin, and some on Harris/Lewis, such as Amhuinnsuidh, Morsgail and Luskentyre, but most were clustered in two extensive blocks of land. The first of these was in the Grampians and consisted of a fairly compact cluster of estates around Rothiemurchus, including Inveravon, Invercauld, Mar and Athol. The second formed a larger and complex cluster of high-ground estates to the west of Fort William and Inverness, and stretching from around Loch Arkaig northwards as far as Rhisdorroch in Sutherland. The Napier Commission list was compiled at a point when the demand for the purchase or leasing of hunting reserves faced less competition. During the opening years of the 1880s alone, as many as twenty-three new forests were established.[47] Their expansion continued after 1900, reaching a peak in 1912, when 3.584 million acres (1.45 million hectares) were devoted to them.[48]

Crofting Landscapes: Repackaging the Old

As they developed during the changes of the eighteenth and nineteenth centuries, Highland landscapes became more organised around a stark contrast. As well as seeing the emergence of very large farms or enterprises, they also saw the top-down creation of crofting townships and smallholdings. For all their composed, planned appearance, the way crofting townships unfolded was not always simple. Quite a number had prolonged, complex births, acquiring their eventual form over time not overnight. Far from being landscapes of deliberation and decision, they come across as formed from a mix of second thoughts and expediency.

There were good reasons why some do not fit into a standardised mould. To start with, there were differences between landlords over their willingness to preserve a smallholder presence on their estates. Even during the early decades of the nineteenth century, there were still some who felt a responsibil-

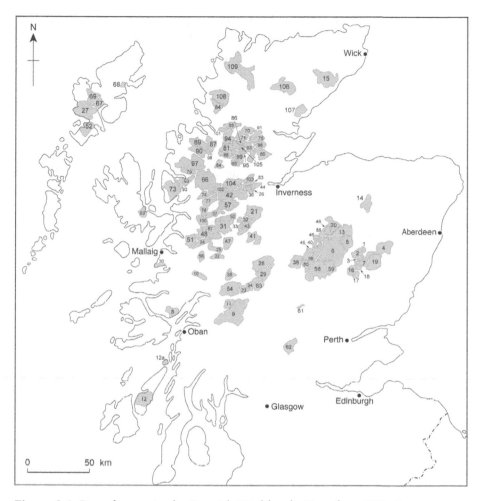

Figure 8.1 Deer forests in the Scottish Highlands. Based on BPP, *Report*, 1884, Appendix C.

ity to lesser tenants and smallholders even though their estate finances did not give them much leeway in the matter. By contrast, some of those who bought into the Highlands at this point brought a different set of expectations, happily accepting the sunken costs associated with those areas of their estate devoted to leisure and pleasure but expecting a secure rent yield and capital return from the rest. As the Duke of Argyll discovered when addressing these issues during the 1790s and early 1800s, part of the problem was in dealing with the chang-ing viability of crofts once they had been created. Surging populations were a prime source of variability, especially once crofters responded by subdividing their lots between children so that arable per family fell away dramatically. Rising numbers posed even more of a threat once supplementary forms of income, such as kelp making, failed and once fish catches declined away during the post-war years.[49] Of all the factors that could undermine the smallholding economy of the region though, none was potentially more devastating than the growing dependence on the potato for subsistence. The poor harvests which can be documented over the first half of the nineteenth century (for example, 1812, 1816), culminating in the catastrophic crop failure of 1845, exposed the structural weakness of crofting, a weakness all the greater once holding size had been compromised by subdivision and alternative forms of income had diminished.[50] Further, each successive crisis acted to build up even greater arrears of rent that were to proved irrecoverable for estates already hard pressed. Without question, some of these problems were already apparent to some landowners before the wars had ended and would have become obvious to all in the years immediately after. Yet, despite this, more crofting townships were created in the three or four decades after about 1810 than before.

Some of the greatest challenges were those posed by the large runrig touns that survived on Skye and in the Outer Hebrides. These were the islands which engaged most with kelp and which, for that reason, had seen some of the greatest gains in population but which, as the wars ended, faced the most acute problems. As if on cue, this was when estates across Skye, North Uist, Lewis and Harris, almost in step, began the reorganisation of their runrig touns, converting many into crofting townships. When seen later, however, from the vantage point of the mid- to late nineteenth century, it comes across only as the starting point for a range of continuing alterations as crofting townships were squeezed, displaced, reformulated or erased to fit a changing set of estate needs.

The Hebrides

If we look first at what happened on Skye, the Macdonald estate commissioned John Blackadder, in 1799/1800 and again in 1811, to survey its property, stretching from Sleat in the south to Trotternish in the north and embracing the eastern side of the island. The former survey described the condition of each toun before change, detailing what was under runrig and what was not. The later 1811 survey, meanwhile, was carried out with the express intention of abolishing the runrig touns, making recommendations about how they could be lotted as crofts. After prefacing its recommendation with the statement that the 'greatest part of this princely estate being in club farms', it determined 'to set every man apart upon his own share' by replacing them with crofting townships.[51] The intention of resettling tenants each 'upon his own share' can be compared with the statement by Lord Macdonald in 1802 that he was against 'dismissing large numbers of his tenants'[52] but, within a few years, or in 1806, he was already adjusting his position, suggesting that, when a farm becomes vacant, it should be offered to the remaining tenants if they are inclined to 'occupy more than they formerly possessed'.[53] Blackadder's scheme shows flexibility over what constituted a croft, with not only modest variations in croft size within townships but a sizeable range of variation between townships. Thus, Erisco in the far north had eight crofts, 4.8 acres (2.4 hectares) each, while Drumuie, a few miles north-west of Portree, had six crofts ranging between 11 and 13 acres (5.6 and 6.6 hectares). When Suishnish and Boreraig, two touns overlooking Loch Eishort, were first cast into crofts, the former was laid out into twelve crofts of 6 to 8 acres (3.04 to 4.06 hectares) and the latter into eight of just over 6 acres (3.04 hectares) each.[54]

As it turned out, the lots proposed by Blackadder were not the last word on either the distribution or the layout of such townships. Looking across the next sixty to seventy years, down to the setting up of the Napier Commission, we can identify five different ways in which townships underwent further change, by: being internally subdivided by crofters or estates; being added to or extended by estates; being cleared for sheep or deer; being relocated, usually on to poorer ground; and having their common grazings reduced. Amongst these changes, the subdivision of crofts, officially or unofficially, was the most widespread, reflecting the fact that population continued to increase down to the 1840s. Greaulin, on the western side of Trotternish provides an example. When divided into a crofting township in 1811, it was laid out as six crofts, a number that matched the number of multiple tenants who earlier had held it in runrig. Their number was later almost doubled, with all but one croft being divided between two tenants by 1851.[55] The eleven tenancies listed in 1851, though, were soon cleared to make way for sheep, the same fate befalling the crofting township established at Linicro next door to Greaulin.[56] The subdivision of crofts evident at Greaulin was repeated elsewhere. Just north of Greaulin, Dellista was laid out by Blackadder into six crofts. In size, ranging between 25 and 28 acres (12.7 to just over 14.2 hectares), they were hardly deserving of the name 'croft'.[57] The 1851 rental for the estate shows the township was still structurally divided into six lots but they were now held by ten different tenants, with two held by two tenants and one by three.[58] At Boreraig, the 1851 survey shows the number of crofts to have increased only modestly from those recommended by Blackadder, or from eight to ten, but, when we look more closely at the data, most were now occupied or shared by more than one tenant, some even by three tenants.[59]

Though crofters themselves initiated some subdivision of crofts, it was not the only means by which townships became more crowded. As already noted, the middle decades of the nineteenth century saw a renewed wave of clearances on Skye as landowners tried to maximise what they perceived as more regular rents with a renewed effort to shift more into sheep farming. Some of those removed emigrated but some were resettled, with established crofting townships being crowded out by new additions. At the same time, landowners, their agents and tacksmen became skilled at clipping the rights of crofters bit by bit, reducing the amount of hill grazing here or restricting what they could stock there. The Napier Commission was given plenty of anecdotal evidence by those whose livelihood had been eroded in these various ways, such as those cleared from the townships around Drynoch, including Somerdale, Craigbreauc, Bendhu, Meadale, Colbost and Glen Bracadale, so as to form an extensive sheep farm based on Drynoch. Those displaced who did not emigrate were crowded on to Coillure or into townships to the north, in Glendale.[60] Some of the those removed from townships in Glens Ose, Colbost and M'Askill were also resettled in Glendale townships.[61] Arguably, we can better appreciate why Glendale became such a centre of reaction in the years leading up to the Napier Commission by understanding how it was used to absorb those displaced from other parts of the McLeod estate, slowly building up a culture of resentment through a background of displacement and over-crowding. Thus, Fasach started out as a toun with twelve tenants *c.*1820 but, when John Tolmie became principal tacksman, he divided it out into sixteen crofts while, at the same time, detaching hill grazings from it, although these were later reattached.[62] Subsequently, MacLeod of Dunvegan himself was said to have reallotted the township into twenty crofts, many of which were subse-quently subdivided into halves so that, by the time of the Napier Commission, it contained thirty-one crofters and six cottars but significantly more rent than it had been burdened with *c.*1820.

Among those who filled the ranks of the crofters in Glendale were some who had been displaced from clearances at nearby Ramasaig and Lorgill.[63] Those involved could not have been surprised at their eviction to Fasach for it was the second time that crofters had been evicted from these touns to make room for sheep,[64] a unique but sad record. In fact, the experience of Ramasaig encapsulates much of what happened on Skye over the eighteenth to nineteenth centuries. It formed the most north-westerly of a small string of touns devel-oped within a series of short, shallow glens along the south-west-facing coast of Durinish, the others being Lorgill, Ibidale, Ollisdale and Lorgasdale, each of which once had sizeable populations and sizeable amounts of arable to match. Seen through rentals of the early eighteenth century, Ramasaig was rated at 3d land and appears in the hands of a tacksman.[65] By the mid-eighteenth century, it was being to set equally to two tenants, a half each. By 1789, the number of tenants listed in rentals jumped to nine,[66] then ten by 1792.[67] In all probability, part of this increase stemmed from the estate setting land directly to all occupiers present (tacksmen and undertenants) rather than simply to the tacksman. In his *Book of Dunvegan*, Canon MacLeod thought that the appearance of listed tenants in this way signalled the fact the toun had now become a crofting township.[68] Bypassing tacksmen and setting land directly to tenants, though, was commonplace on Highland estates by this point and cannot in itself be taken to indicate the presence of crofts. We are on firmer ground with a report drawn up in 1811, a year when quite a number of crofts were laid out on the estate. The report on who held what identified townships that had 'lots or crofts' at this point, such as Roag and Vatten with eighteen lots between them. It listed the occupiers of Ramasaig as each paying the same rent, or £10 per annum, but they are not yet referred to as 'lots' or 'crofters'.[69] Their formation, though, must have been imminent at this point. Within a decade, the estate changed its policy and cleared Ramasaig for sheep, ordering those present at Ramasaig 'to be flitted to Trien and have half of it'.[70] Hardly another decade had passed before the estate reset it as a crofting township. The newly reformed township at Ramasaig had twelve crofts but two were shared between three crofters and three between two crofters.[71] Lorgill was resettled at the same time but, within five years, the sheep farmer who held the adjacent farm of Dibidale – also by chance, the factor overseeing Lorgill – pressed for part of it to be added to his farm so that crofters were left solely on the western bank of the Lorgill, the five displaced crofters moving to Ramasaig which now had twenty-one crofters.[72] Finally, by the mid-1870s, the remaining crofters at Lorgill, together with all twenty-one at Ramsaig, were cleared so that their land could be added to the sheep farm based on Dibidale.[73] Switched from a runrig toun into a crofting township, then cleared twice for a sheep farm, with a second interlude as a crofting township between, the story of Ramasaig clearly embodies more than its fair share of what many Hebridean landscapes were about at this point.

The surveyor of Lord Macdonald's Skye estate, John Blackadder, was also commissioned to survey North Uist in 1813 and to make recommendations there about how its runrig townships might be 'crofted in the same manner as that of Skye'.[74] All but six of its touns were held in runrig. Following Blackadder's survey, most were converted to crofting townships in 1814, a blanket change made all the easier because, apart from the six held on lease by tacksmen, the rest were held at will. When we look at the island through a survey of 1830, after the reorganisation had been implemented, some former touns, such as those of Vallay, Greninish, Scolpaig, Kilpheder and Ballone on the north-west coast of North Uist, appear set to single tenants. Most of the rest, however, were now in the hands of crofters, part of a new landscape of ladder crofts. As noted earlier, the islands of Kirkibost and Baleshare on the west side of North Uist were both prone to severe sand blows. Following Blackadder's recommendations, Kirkibost island continued in the hands of one tenant but the runrig touns on the adjacent island of Baleshare and the mainland opposite became large crofting townships, with Baleshare divided into twenty-three lots, Illeray into fourteen and Claddach Kirkibost, on the mainland opposite, also into fourteen.[75] Whereas the crofts at Baleshare and Claddach Kirkibost were positioned over their former runrig arable, those at Illeray, the toun most affected by sand blow, were repositioned even further away from its former settlement focus (Figure 8.2).

The history of crofting at the northern end of North Uist was well

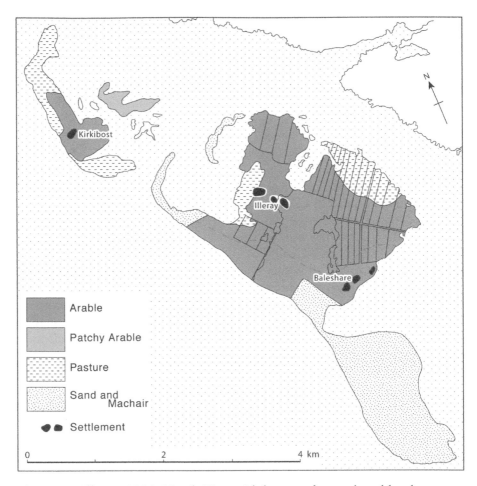

Figure 8.2 Illeray, 1799, North Uist, with later crofts overlayed by the estate, NAS, RHP 1306.

summarised by Moisley. The touns of Knockline, Balemore and Knockintorran were recast into crofting townships in 1814 but, being held by tacksmen, Balranald, Paiblesgarry and Kyles Paible were left unchanged by Blackadder. When the sitting tacksman left Kyles Paible in 1852, however, it was reset to tenants, not as a crofting township but as a runrig toun, one that persisted into the twentieth century. Balranald and Paiblesgarry, meanwhile, contin-ued as single-tenant holdings until 1921 when they were reorganised by the Department of Agriculture into crofting townships, a late origin that explains why they were endowed with the largest average arable per croft among this group of townships.[76] At the southern end of North Uist, Grimsay was not part of Blackadder's scheme even though scattered settlement was shown as present on Grimsay and the surrounding islets in the 1799 map. Yet, when we look at later sources, such as the First Edition Ordnance Survey 6-inch map, crofts had emerged. In short, what began as squatter settlements, exploiting kelp and seafood as much as arable, had become formalised within the estate's structure.

The two or three decades following 1810 were also momentous for Lewis and Harris. On the former, the Seaforth estate announced its intention of dividing touns out of runrig in its new 'articles of set' issued in 1795. Declaring that 'whereas the possessions in Runrig or in common of arable Lands and Lands capable of being cultivated is a practice of a pernicious tendency', it announced that 'it is agreed that the Several farms let to the Tacksmen and tenants Shall be Divided'.[77] Despite this early statement of intention, there was no immediate action. Chapman's plan of Lewis, surveyed in 1807, still shows the layout of touns as they were before crofting. In fact, they were not converted into crofting townships until the second and third decades of the nineteenth century (Figure 8.3).

Even then, it was not a smooth process. If we look at the large crofting townships created along its north-west coast, such as North Bragar and Arnol, crofts were initially laid out running back from 'streets' that ran down towards the shore, where their original settlement had been focused. At a later date, however, *c.*1850, after Lewis had been acquired by the Matheson family, this orientation was changed, with the line of the crofts now running down towards the shore from new 'streets' aligned at right angles to them and tangentially to the former toun's lazy-beds (Figure 8.4).

Figure 8.3 Ladder crofts at Borghastan, Lewis.

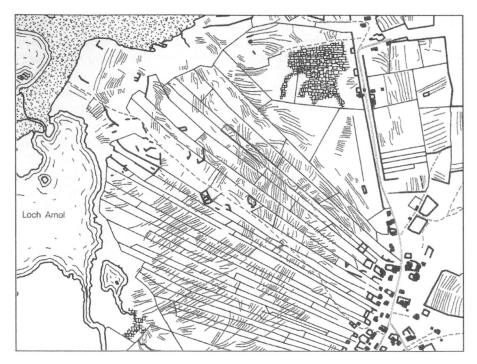

Loch Arnol

Figure 8.4 Crofts and pre-croft rigs at Arnol, Lewis. Abstracted from a map photogrammetrically plotted for the author by M. Weir, Geography Department, University of Aberystwyth.

The changes at Bragar and Arnol were part of a more general reorganisation of crofting townships on Lewis carried out between 1849 and 1851. It was prompted partly by the continued increase in population. As a later report on the island put it, 'on a considerable number of them [crofts], it was found that two or three families had since settled, who either held the lot in common, or had it divided amongst them'.[78] The laying out of crofts afresh sought to address this problem by giving 'a separate lot . . . to each cottar.[79] The outcome was an increase in the number of crofts present. In the case of Arnol, for instance, what started as a township with nineteen crofts became one with forty-five.[80] As at Bragar, we can see the reorientation of the township as designed to accommodate these extra numbers. When the Seaforth estate planned the first crofting townships, it issued guidelines that required crofters to build their houses 'in a straight line with the adjoining houses', on a floor 14 inches (35.5 centimetres) above the surrounding ground, and with separate entrances for their byre and living quarters.[81] The First Edition Ordnance Survey 12-inch map shows that, by 1878, houses at townships such as Bragar and Arnol were aligned beside their new main 'street' axes, though not all were on the same orientation within their crofts. The switch of township layout *c.*1850 also saw the adoption of the larger blackhouses as represented by Number 69 Arnol.

The increasing number of inhabitants in townships like Arnol over the first half of the nineteenth century was not driven solely by natural increase.[82] The displacement of crofters and cottars by sheep or deer elsewhere on Lewis and their resettlement were just as potent sources of expansion. The extent of such disruption is well illustrated by Balallan or Baile Ailein, a township on the north bank of Loch Erisort. In his presentation to the commission, a crofter there talked about the township being 'badly ruled'. His list of nine complaints makes it difficult not to agree with him. 'Forty four years ago', or in the mid-1840s, the estate factor took away part of the township's hill pasture and attached it to the sporting grounds of Aline. Then, sixteen extra crofters were added to it following their displacement from the Park district, on the other side of the loch, followed by the creation of the new township of Arivruach on part of what remained of its hill pasture. When a new factor, John Mackenzie, took over the running of the estate, he redivided Balallan into sixty-four crofts, an expansion that was followed by rent increases. A further change of factor saw the addition of new hill grazings and new places designated for peat cutting but the scale of the problems facing the township are underlined by the estate's announcement, just before the setting up of the Napier Commission, that it proposed to remove all those without a formal set, thirty-two in all. Clearly, Balallan was a part of Lewis that saw successive, continuing changes to its crofting landscape.[83]

Balallan's problem was one of location. Across the loch, it faced an area that saw the greatest disruption on Lewis during the nineteenth century: the Park district. The nature and extent of this disruption were mapped by the Geographical Field Studies Group. An edited version of the map can be seen in Figure 8.5.

Originally, the bulk of Park was forest land. As with the forest district of Harris, there appears to have been fairly widespread encroachment by settlement over the seventeenth to eighteenth centuries. That which emerged in Harris was described in early surveys as settlement in the 'bays' area. That which occurred in the Park district was probably similar, comprising little touns squeezed in around the many small embayments to be found around its coast. In 1818, the tack for the southern and eastern halves of Park was set to a sheep farmer and his brothers but it was not until 1836 that they started to clear touns, beginning with the block of land between Loch Erisort and the upper part of Loch Seaforth. The following year, they cleared the touns that had sprung up around the middle and lower parts of Loch Seaforth, on the Minch coast and along the southern and eastern sides of Loch Shell. Further clearances took place on the north side of Loch Shell in 1843, leaving all but Stiomrabhaigh wholly under sheep but even this was eventually cleared in 1857. By this point, the vast bulk of Park, the extreme north and north-west

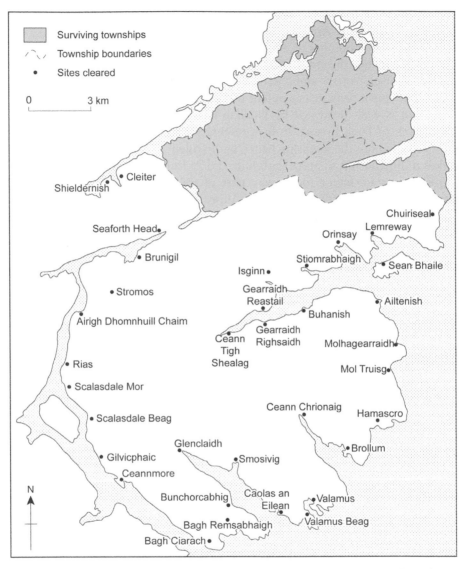

Figure 8.5 Touns, clearances and crofts in the Park district, Lewis, based on Caird, *Park*, Figs 1 and 2.

excepted, had been cleared. The Napier Commission was exercised over how many touns were actually cleared in this process, with estimates ranging from twenty-eight to forty-five. Caird's study suggests a best estimate of thirty-four.[84] By this time, too, the Mathesons had taken control of Lewis. Their management shifted the emphasis of change. Faced with runrig touns still surviving in the extreme north and north-west along the southern side of Loch

Erisort and on either side of Loch Quirn, many now crowded by decades of natural increase and those displaced by the clearances in Park, the estate carried out a reorganisation in 1850–52, abolishing their runrig and relaying them as crofts.[85] In fact, the estate even took steps to reverse part of the clearances, resettling Lemreway on the north side of Loch Shell with crofts, and laying out others on part of Shieldernish and Seaforthhead in the north-west of Park. The final makeover of Park's nineteenth-century landscape came in 1883, when the sheep tack covering most of the eastern and southern halves were reconverted into a deer forest. In an effort to comprehend these district-wide changes on Lewis, the Napier Commission conveniently produced a summary map of how its landscapes had changed (Figure 8.6).

Wester Ross and Sutherland

The extent of the Clearances in the North-west and Northern Highlands has already been discussed. While most were carried out without ceremony, they were not necessarily without obligation, a significant proportion of tenants being resettled in crofts on the coast. Combined with the reorganisation of those touns already on the coast into crofting townships, this resettlement produced a profound change in the landscape of many coastal areas.

Because of the estate-wide scale at which the Sutherland estate conceived change, Richards described it as an exercise in regional planning.[86] The clearances for sheep were only a half of this exercise. What turned it into a regional plan was not only the way in which new coastal crofting townships were created out of overcrowded coastal touns but also for absorbing those displaced from interior touns. Even by the 1790s, the estate had plans for converting the interior to sheep farms and moving those displaced to crofting townships on the coast. The scale of capital required for creating the infrastructure needed by such townships, however, forced them to wait until the marquis had inherited his family's wealth in 1803 and until the leases covering large parts of the estate began to expire in 1807.[87] Soon after, other key personalities arrived on the scene. These included William Young and Patrick Sellar who became involved with the estate as agents in 1809, at first in dealing with the planning of what they called the coastal 'villages', a matter in which Young had some experience, following his establishment of a new fishing village at Hopeman and new settlement on Achavandra Muir, both on his own estate in Moray.[88] Equally influential was the appointment in 1812 of James Loch as commissioner for all of Stafford's estates. By this time, Young and Sellar were already advising on where and in what form coastal townships could be developed. The earliest wave of clearances, however, those in 1806, did not involve any provision for resettlement but those begun in 1811 and in 1819 did, at least for those willing to accept the new arrangements. From what Loch later wrote, when it came to who was to be moved where, a simple catchment

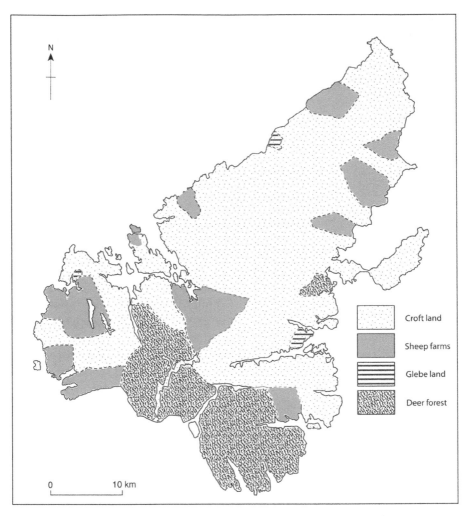

Figure 8.6 Map of land use on Lewis, 1884, based on BPP, *Evidence*, 1884, 25, Appendix.

principle applied, with people being moved to the crofting townships at the mouth of their respective straths, so that those from Strathnaver were settled in crofting townships lying between the mouth of the Naver and Bighouse; those from Kildonan in townships beside Helmsdale; and those from Strath Brora in townships in the vicinity of Brora.[89] The largest of the new coastal settlements were those at Brora and Helmsdale. These were intended to be diversified. At Brora, for instance, the estate promoted spinning and weaving, flax making, distilling, brewing, salt working, glass making, brick making, tile making and coal mining.[90] Elsewhere, the townships laid out along the northern coast, such as those between Strathy Point and Tongue, and the smaller number between Rispond and Durness, were founded on more modest expectations, being intended for crofter fishermen. We can only speculate whether those crofters in these northern townships, whose lots were positioned on nutrient-poor muir land and who had to invest a great deal of labour in land reclamation, let alone adapt themselves to sea fishing, were among those who Loch thought moved cheerfully to their new holdings. Fisher crofts were also used as the solution for the crofting townships created in Assynt. With the arrival of Young and Sellar, the idea of a smallholding economy in Assynt that faced seawards was taken up with renewed vigour, with a number of touns being reorganised into large crofting townships based on fisher crofts laid out along the coast between Drumbeg and Lochinver.

Just as understanding what happened on the Sutherland estate is not simply about positioning the estate within the broader pattern of trends but also about the personalities involved, the same is true for Gairloch. Dr John Mackenzie, who took over the management of the estate from his brother in 1841, had been greatly impressed by Flemish systems of farming based on the indoor feeding of cows, not just in winter but all year, and on the careful husbanding and use of the liquid manure produced, as well as dung, for fertilising arable.[91] To a degree, his thinking was really more in line with eighteenth- than nineteenth-century thinking on what to do with the smallholders present on his estate, seeing the question as how best to keep as many people as possible on the land and believing that adopting 'scientific agriculture' would achieve this.[92] During his brief tenure in charge of the estate, he devised a scheme for reordering touns on the estate into crofts that extended to no more than 4.5 acres (1.8 hectares), which would adopt his Flemish scheme, and would rely solely on the spade for cultivation.[93] Just as key resources of the farm were recycled to advantage, Mackenzie took the same view of his tenants, with 74 per cent of the 364 present *c.*1843 being given crofts within the same toun in the new order and the other 26 per cent being given crofts in another toun.[94] Even those living with parents or relatives were squeezed into the new order. The remapping of the Gairloch smallholdings was notable in another way. The job of surveying what was present and laying out the new crofts was tasked to Campbell Smith.[95] His approach fingerprinted Gairloch crofts in a way that makes them stand out for, in a good proportion of townships, he adopted a square format for crofts, though the demands of fitting squares on to one site, that of Big Sand, produced more of a diamond shape for them. Campbell Smith tried to lay out the new crofts over the pre-existing arable. Caird's before and after comparison of arable at Ormiscaig enables us to see the extent of arable in 1845, the moment recorded by Campbell Smith, with that of the new crofts abstracted by Caird from the 1875 OS 1:2500 map (Figure 8.7). The comparison shows just how much some crofts had to reclaim their arable afresh, even where crofts were laid out over former touns. The Ormiscaig crofts were not

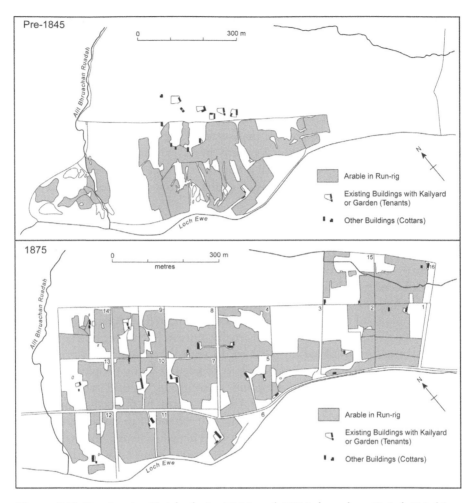

Figure 8.7 Ormiscaig, Gairloch, in 1845 and 1875, based on Caird, 'Making of the Gairloch Crofting Landscape', pp. 152–3.

alone in having a square-like format. Others townships around it, from South Erradale northwards to Mellon Charles, were laid out in a similar way. Once we reach Loch Broom, though, the format of small, ladder crofts once again becomes the norm, as at Badluarach and Scoraig (Figure 8.8).

Northern Isles

Both Shetland and Orkney acquired a broad base of consolidated smallholdings and crofts over the eighteenth and nineteenth centuries. There is this distinction between them, however. Far more crofts developed in Shetland than

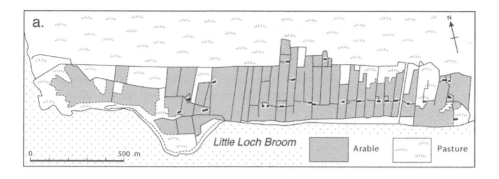

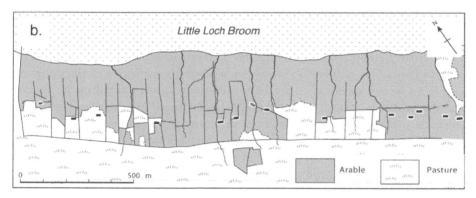

Figure 8.8 Crofts at Badluarach and Scoraig, Little Loch Broom. Based on First Edition OS 6-inch map.

in Orkney.[96] Admittedly, the distinction is lessened if we take into account the many small farms and smallholdings which emerged in Orkney that were owner occupied and, for that reason, were not classed as crofts under the terms of the Crofters' Holding (Scotland) Act, 1886.[97] Their presence stemmed from the fact that more Udal tenures persisted there, or were converted into feus, than in Shetland. Yet, while the smaller number of legally recognised crofts in Orkney became an important difference, we must not ignore the fact that, prior to the 1886 act and its narrower definition of what a croft comprised, smaller Udal tenures in Orkney were commonly described as crofts after they had been divided out of runrig.

Whether formed around heritors or tenants, the runrig organisation of arable appears to have been widespread across the Northern Isles by the mid- to late eighteenth century. Fenton, for instance, described 'most of the corn lands' in Orkney as in runrig by this point.[98] In part, the prevalence of runrig was due to the build-up of population over the eighteenth century. It was also due, however, to the way in which the rapid growth of supplementary forms of income, ranging from fishing and kelp burning to knitting, hat making and linen weaving, enabled smaller holdings to remain viable. Such forms of income provided a steady flow of cash rents, so landlords did not initially act to restrict subdivision. The strategy behind the landlords' thinking was summed up by the factor for the Buness estate on Unst, an area where fishing, especially haaf fishing, was a major source of income. Speaking of the years prior to 1820, he reported to the Napier Commission that 'there were few or no large farmers as the great object was to increase the number of fishermen, and thus the land was subdivided into small lots'.[99] For this reason, landowners allowed holdings to become more subdivided over the closing years of the eighteenth century in many coastal parts of Unst and Yell.[100]

Though some runrig touns in Orkney and Shetland were removed via a process of clearance, a number was removed through a commensurate division. In part, this was because of the presence of feuars and udallers whose rights could be addressed only through a commensurate or exact division of their interests. These were divisions carried out via a process in the court of session under the authority of the 1695 Act Anent Lands Lying Runrig. Where heritors held large blocks of land, such divisions could involve two interlinked divisions, one between heritors and one between their undertenants. Papa Stour provides such an example. The island was shared between Arthur Gifford and Sir Arthur Nicholson. Apart from the outsets of Hanavoe and North Banks, its 800 to 1,000 rigs of arable had been arranged into one runrig toun. Its division in 1851 involved the island's entire arable, some pasture within the head dyke and a small part of the common grazings.[101] Once complete, each now separated estate had the choice over how it arranged land among its tenants. Even where touns were held by a single landlord, the concern to sustain rental income from smallholders meant that the shares of the latter were carried over into new crofts via generalised divisions. This happened at Walls when it was divided in 1851,[102] and on the Buness estate on Unst when it was eventually divided in 1867.[103] If we treat planking as an attempt to simplify runrig by rearranging runrig holdings into large, croft-like blocks or planks, then its removal in some touns would have been even more elaborate.[104] At Birsay-be-South, a part of Birsay that was planked in 1748, such attempts at 'partial consolidation', as Thomson called it, proved to be little more than a temporary solution because its planks were subsequently subdivided into still smaller, albeit simpler, runrig systems.[105] Only in 1845 was what remained of runrig entirely removed.[106]

The changes that affected the scattald in the Northern Isles were among the most contested by landholders. At a time of rapidly rising population, the seemingly free space offered by the scattald, with its prospect of being regarded as heritable property if occupied continuously for forty years, was always going to be attractive to squatters. This was especially so in Shetland, where settlement tended to be patchy and more coastal than in Orkney, with large areas of hill ground beckoning those in search of a land-based subsistence alongside fishing. Smallholders in large numbers seized the opportunity. The scale of their opportunism is well shown by what happened at Sandstig and Aithsting on the Shetland mainland, where the creation of new outsets on the scattald had started in the late eighteenth century. By the 1840s, 104 holdings were reported to be in existence.[107] The heath land of Rousay, too, was affected by squatting before being formally divided.[108] As the nineteenth century unfolded, though, the context in which different interests competed for the scattald changed. Most had rights of use shared between different heritors. In a world in which the closer definition and more intensive use of all forms of property were becoming the norm, heritors pressed for the formal division of their skattalds so that they knew and controlled what belonged to them. Once divided, some merely used the opportunity to control the squatting process, ensuring that those settled were herded on to more convenient parts of their skattald and now paid a rent. In reality, of course, such controlled settlement of what lay beyond the head dyke served as a cheap form of land improvement. Often, it was land improvement on a large scale. On Assater, a block of skattald just inland from the head of Ura Firth, in the north-west corner of the mainland of Shetland, as many as five new townships were laid out among small crofters in the mid-nineteenth century.[109]

Controlling the settlement of skattald by smallholders was only part of the agenda that was starting to unfold by this point. Commercial stock farming, especially for sheep, had become important by this time. Once incorporated into policies of estate management, the interests of the smallholder became marginalised. We can see this in the way some touns or townships were now cleared or thinned to make way for sheep. In a reference to the mid-nineteenth century, it was said that 'of late years', as sheep spread, the better grounds on Unst, Yell and Fetlar, along with the vales of Dale, Laxfirth and Wiesdale were all put under sheep, a switch that involved 'numerous evictions'.[110] On Unst, ten smallholders were cleared from Balasta alone to make way for sheep. The same new pressures began to reshape landscapes in Rousay. Prior to the division of its skattald, parts of its heath-land areas had already been subject to squatting by smallholders. As landlords took a more aggressive approach to land management, the heath became a dumping ground for those displaced by forced clearances. When as many as forty families were cleared from Greendale and Westside to form a sheep pasture in the mid-nineteenth century, for example, some were resettled on the heath land at Sourrin.[111] Of course, a clearance was not the only way in which sheep farming spread. A commonly used strategy was to carve out sheep farms from the skattald. As ownership of the island slowly shifted into fewer hands, with General Frederick Burroughs becoming by far the dominant landlord, large areas of what had been skattald fell under his sole management. As we find in the southern and western Highlands during the early spread of sheep farming there, he was accused of responding to his inheritance by removing 7,500 acres (3,035 hectares) of rough grazing out of use as common land for his crofters so as to create a number of sheep farms, the largest being Westness.[112]

The closer interest now taken in the skattald by landlords was also a reflection of how ideas on improved farming had started to spread into the Northern Isles by the second quarter of the nineteenth century. The use of sown grasses and turnips, additives such as marling and liming, better techniques of drainage, including subsoil tile drainage, and low-cost government grants for draining all combined to drive local ideas on land improvement. Land outside, as well as below, the head dyke now acquired a potential value, as those in the van of land improvement set about lessening, if not removing, their differences. David Balfour's blanket imposition of a comprehensive grid of regular, mostly square-shaped 10-acre (4-hectare) fields across Shapinsay during the mid-nineteenth century provides the best example of this, with the basic grid for his system being laid out over heath land or skattald no less than across arable. Most of the field boundaries were shaped by a new system of drainage for which Balfour had received a substantial loan under the Land Drainage Act (Scotland) of 1847. The 10-acre (4-hectare) fields created then provided the basis for all landholding, whether let in groups of fields to middling farmers or set individually to smallholders.[113] While the regimented way in which Shapinsay's landscape became divided out and enclosed was notable even by Orkney standards, it was not exceptional. Block-like regular patterns also emerged on parts of Westray, Stronsay, Sanday and South Ronaldsay as well as in parts of the mainland.

Resource Exploitation and its Environmental Impacts, 1815–1914

Historically, the human exploitation of resources has had an environmental impact over large parts of the Highlands and Islands. A long-standing source of impact was the cutting of peat and turf (Figure 8.9).

Turf was being extensively cut for fuel as early as the Neolithic though, by the Late Iron Age, it was being used primarily for walling, roofing and insulation followed by a secondary used as compost. The use of peat, both for fuel and as a soil additive, meanwhile, would only have developed as climatic degradation led to the formation of deep deposits during the Bronze and Iron Ages. In practice, the cutting of peat and turf is not always distinguishable but, together, their exploitation had impacts that landlords tried to control. We see this from barony court records for the seventeenth and eighteenth centuries. Taking those concerning the cutting of peat first, we find courts issuing strict guidelines on the type of peat spade to be used, the need to cut peat faces in a straight line, the need to replace the grass cover, that tenants and cottars cast their peats in only one 'poole' per toun so that 'yair be not ane poole mair in ye toune', and not to cast their peat 'within corne landis' or within meadows or

Figure 8.9 Peat cutting, Lewis.

anywhere within the head dyke.[114] Estates even included guidance on how peat was to be cut in tacks. Their main concern was to minimise 'potts and holes' or what some referred to as the 'spoilling and hoilling of the moss' or 'ye peate'.[115] We know from sites such as Eilean Domhnuill on North Uist that the cutting of turf for fuel and walling was already having an environmental impact in the Neolithic. The extent to which its use was sustained over time for walling, roofing and compost and the scale of this use suggest that, as a practice, cutting turf resulted in a great deal of environmental damage. When seen through seventeenth- to eighteenth-century court records, however, the prime concern was to curb the cutting of green pasture or land within the head dyke.[116] This may always have been a problem but it is tempting to see such a restriction at this point as motivated by a heightened sense of how grazing should be valued.

Easily the most visible impact caused by the sustained harvesting of peat and turf is what Darling referred to as skinned land, that is, land from which the surface turf and peat have been mostly removed. His map of 'skinned land' in parts of north-west Lewis, like the 'skinned' half of Papa Stour in Shetland, captures how the sustained cutting of peat and turf could ultimately degrade land.[117] Though it is tempting to see the creation of such 'skinned land' as a fairly recent phenomenon, brought on in such areas by the build-up of population during the eighteenth and nineteenth centuries, the start of this kind of degradation is likely to have begun much earlier. The fact that most 'skinned land' lies close to former touns, as on Lewis, suggests that opportunity costs had a part to play, with turf and peat being harvested where it was most accessible. When used as a soil additive, communities were effectively sacrificing the soil in one part of the toun for the sake of enriching it in another. This is what happened on Papa Stour in Shetland, though submissions to the Napier Commission suggested that it was 'all scalped bare as much for the need for fuel as for soil additives.[118] The harvesting of soil and turf as a soil additive was widely practised in the Northern Isles. Though some refer to it as a harvesting of soil or moss earth, more explicit references make it clear that the turf capping was repeatedly cut and removed in the process,[119] which is why it was seen as a destructive practice, a 'scalping of the hills' that was damaging to pasture.[120]

Another cultural impact on the environment, one that came more to the fore over the nineteenth century, was the trenching of hill ground and low-lying wet muir or moss using a trenching spade or some form of draining plough. In some areas, such as in parts of Upper Tusdale and on the ground that slopes south-westwards from Loch Sleadale in Mingnish, Skye, we can find deep, irregular drainage trenches that contrast markedly with the more regular trenches being inscribed across waterlogged ground from the closing decades of the eighteenth century onwards. These may represent an older forms of open-ditch drainage, perhaps associated with specialised cattle farming. The trenching of more regular open drains probably spread in association with commercial sheep farming. When Sir John Sinclair penned his *General Reports* in 1814, the practice had already been used widely in Perthshire, Inverness-shire, Ross and Sutherland.[121] As the nineteenth century unfolded, large parts of northern Scotland, including large swathes of ground in the wet interior of Sutherland, as well as parts of the western seaboard and the Hebrides, were extensively trenched. In all cases, we can assume that they were trenched as part of the local investment in sheep farming, and intended as pasture improvement. The remarkable network of trenches we see cut across the isolated north-west corner of Rum, for example, can probably be dated to the brief period in the 1840s when the island formed a vast sheep farm. Just how quickly the practice followed in the steps of the first commercial sheep flocks is shown by Sutherland. Writing barely a decade after the first large-scale sheep farm had been established, Loch talked about how sheep drains had been trenched 'to the extent of many miles on every stock farm on the estate'.[122] In fact, it was commonplace throughout the nineteenth century for the large, specialist sheep farmers to employ trenchers during the summer months. Not all open sheep drains were subsequently maintained once cut but their sum effect was to leave hardly a corner of the region, even in areas that might otherwise be considered 'pristine' or untouched wilderness, striated by such drains. By the closing years of the eighteenth century, Highland farms had also started the task of trying to drain their arable and sown pasture more effectively. Early forms of subsurface arable drainage used ineffective techniques such as gravel-filled or bush-filled drains but drains around the side of arable could improve it. The extent to which we find references to the improvement of fields by draining suggests that whatever the limitations, there were benefits. It was not, however, until a system of cylindrical tile drains became available in the early nineteenth century that effective field drainage was possible. Even then, it was not until an act of 1847 enabled farmers and landowners to raise a loan for drainage from a state-provided fund that significant investments in tile drainage were made. Supported by the fund, tile drainage was put in place during the mid- to late decades of the nineteenth century across low-ground arable in areas such as Caithness[123] and easter Ross[124] as well as more selectively in some of the wider straths and glens. The 18,000 acres (7,285 hectares) reported to have been drained in Orkney after 1850, especially in the parishes of Westray, Burray, Sanday, Rousay and South Ronaldshay, were almost certainly based largely on tile drainage.[125] In some cases, estates attempted something more spectacular, such as the Duke of Sutherland's efforts at Shinness in the mid-1870s. This was an area of moss and wet muir lying between the lower part of the River Tirry and Loch Shin north of Lairg. Initially, the plan was to drain 1,175 acres (475.5 hectares) but the final total actually drained amounted to 1,829 acres (740 hectares) all divided into a series of farms (for example, Colaboll, West Shinness and Achnanearain) and small lots to which a portion of outrun or grazing was attached.[126] Within land improvement programmes, the clearing of stones from arable was just as important as drainage. Even in the southern Highlands, we find sufficient references to the clearance of stones for it to have still been a major part of land-improvement programmes during the early decades of the nineteenth century.[127] Efforts to clear stones, of course, had been part of land colonisation since the first farmers arrived, with many prehistoric communities simply heaping stones into clearance cairns as well gathering them into the stone dykes that bordered their fields. By the late eighteenth and early nineteenth centuries, removing the many clearance cairns that still survived was itself seen as an improvement, a task aided by the use of dynamite to blast larger stones into smaller blocks for easier removal.[128]

In many touns, especially in the southern and eastern Highlands, the burning of hill ground, especially where heather was present, was a long--established practice by farmers,[129] one initially designed to increase the proportion of younger, fresher growth, so that grazings were more palatable for stock. Of course, in time, landlords also had an interest, for ground-nesting game birds such as grouse not only used heather as cover but also as a source of feed during the plant's growing phase. In other words, for landlords interested in game no less than farmers interested in the grazing value of the muirs, burning was of value because it rejuvenated stands of heather, seasonally removing older, woodier growth.[130] Yet, far from being perfectly aligned, the interests of landlord and farmer contained a potential conflict for, from a landlord's point of view, burning needed to be carried out before game birds had started their annual breeding cycle. Landlords were so protective of their interests in this matter that we actually find a 1457 Act of Parliament stipulating the season for burning hill ground.[131] Despite this Act, barony courts can be found over the seventeenth to eighteenth centuries regularly having to fine tenants for burning out of season. In addition, there were regular restatements of the restriction over when and where burning could take place. As well as being banned wherever there were signs of tree growth, it was banned on green grass and, in one instance, anywhere 'outwith the hills and [in] sight of all Infeildis and manurit landis'.[132]

By the nineteenth century, the context in which burning took place had changed. To start with, the use of hill ground had altered, with specialised sheep production now widespread. While we might expect the thinned farming community of the nineteenth century to have been more responsive to estate regulations, there were clearly exceptions because at least one *NSA* report, that for Strathdon, noted that illicit burning, or burning out of season by farmers, was 'still too prevalent'.[133] Alongside the move towards more commercialised forms of stock production, the growth of leisure estates and the rising rent of shooting lets in the eastern and southern Highlands created pressure for more systematic regimes of muir management for game birds. Sheep and heather muirs can, and do, coexist. Heavy grazing by sheep can graze out heather but less intensive grazing can benefit a grouse muir, helping to prolong the heather at its 'building phase'.[134] In those areas where grouse muirs were developed more extensively, especially on the dry heaths of the eastern Grampians and around the southern and south-eastern edges of the Highlands, we can expect sheep-grazing strategies to have been adjusted accordingly as management priorities changed. The distinctive appearance of burnt heather muir, with its different patches at different stages of recovery, is very much a feature of these areas, but only with the introduction of more systematic forms of heather muir management during the nineteenth century are heather muirs likely to have taken on the highly patterned appearance that we see today (Figure 8.10).

Whatever the needs of shooting lets, the early spread of commercial sheep farming in the mid- to late eighteenth century had a more striking impact on heath muirs. As numbers surged, we get early reports of heather being grazed out. We see the problem through the *Old Statistical Account, 1791–1799*, especially in parish accounts for the southern and south-western Highlands, the area where sheep production made the greatest advances during the second half of the eighteenth century. In that for Arrochar (Dunbartonshire), close to where commercial sheep farming first developed in the Highlands, the reporter

Figure 8.10 Heather burning in the Grampians.

noted that, prior to the Clearances, the hills had been covered with heath but now, with the spread of sheep, 'the country has assumed a different aspect'.[135] At Balquhidder (Perthshire), the reporter tried to be more explicit, saying that the hills were previously 'partly covered with heather but mostly green, the heather of late years having been destroyed'.[136] The connection between sheep and the loss of heather recurs in the report for Kilmore and Kilbride (Argyll) where the hills were said to be mostly covered with heath 'excepting a few that are cropped by the sheep'[137] while, at Strachur and Stralachan (Argyll), the hills were said to be 'gradually growing green, since the sheep's stocks have been lately introduced; the heath is decaying fast'. In marked contrast to the approach taken by nineteenth-century shooting estates, those with an interest solely in sheep saw this greening of heath as beneficial. Thus, in the parish of Glenorchy and Inishail (Perthshire), it was reported that the 'hills and muirs, which, some years ago, were covered with heath and coarse herbage, are, since the introduction of large flocks of sheep into the country, gradually getting a richer sward and greener hue, and afford excellent pasture'.[138] At Inverchaolain (Argyll), the transition from heath to grass was seen as greatly increasing the value of the ground.[139] Some estates even set out to use sheep deliberately to change grazings, with new tenants in Lochaber being required by a new set of 1770 'to Shift and Remove their Sheal Houses Every third year into new Heathery ground' so as to 'meliorate' the ground.[140] The rapid expansion of sheep numbers across the northern Highlands during the early nineteenth century led to a similar grazing out of heath. Even by 1820, less than a decade after the formation of the largest farms in Sutherland, Loch was able to report that, with the spread of sheep, 'many thousand acres of the finest pasture now exist, where nothing but heath formerly grew'.[141]

Arguably, these are unlikely to have been Loch's last words on the matter for, by the 1850s, large-scale, mono-grazing by sheep had started to have another effect, one that, for many observers, amounted to an opposite effect, a deterioration in the quality of highland pasture.[142] The Napier Commission collected evidence on the issue for the 1880s and saw the decline as manifest in two ways: 'first, in the increasing prevalence of mosses and rushes on the green lands formerly tilled as arable or outfield by the crofting communities . . . secondly, over the general surface of the mountain grazings, which lie, and have always lain, in a state of nature'.[143] Others report green swards being replaced by poorer sward comprising bent (*Agrostis tenuis*), cross-leaved heath (*Erica tetralix*), Yorkshire fog (*Holcus lanatus*) and mosses.[144] So serious was the deterioration that many reported a drop in the stocking capacity of pastures. An 1880 survey of sheep farmers in Sutherland found that 'a great body of them' thought 'that there has been marked deterioration'.[145] Reports of pasture degradation were also claimed for Ross and Cromarty, Skye and Kintail.[146] Overall, what most concerned farmers was the fall in stocking capacities and lambing rates. In Sutherland, the 1880 survey estimated that many farms were carrying a third less sheep than twenty-five years earlier, a scale of reduction later confirmed by farm-level evidence gathered by the Napier Commission.[147]

Modern debate over these reports of hill pasture degradation and what caused it has shown little consensus. Some attribute it to the loss of nutrients and gradual soil impoverishment brought about by heavy stocking and cropping of sheep and wool over the nineteenth century.[148] The spread of specialist sheep production brought with it a fuller and more systematic use of hill grazings through closer management and better strategies of rotational grazing. The adoption of heavier breeds, such as the Cheviot, the greater attention given to wintering ground and winter fodder, the reduction in the age at which wedders were sold off Highland farms (four to five-and-a-half years in 1800 down to two-and-a-half years in 1880), coupled with the surge in numbers that occurred during the mid-nineteenth century (for example, in Sutherland, from 168,170 in 1853 to 240,096 in 1875) all combined to increase grazing pressures further as the century wore on.[149] Yet, despite the range of contemporary evidence in support of the part played by such grazing pressures, some have rejected it as an explanation. In a critical survey of the problem, Innes argued that modern analyses of how much nitrogen and phosphorus are extracted through the regular cropping of sheep and wool suggest that the quantities involved are more than offset by what is returned to the soil via dry atmospheric deposition alone, before we allow for that added by wet atmospheric deposition, so that pasture degradation is unlikely to have been caused by grazing.[150] More recent research has taken a different line, however, suggesting that whether the nutrients extracted by sheep are balanced by what is put back via dry atmospheric deposition is not clear cut, especially if one adds the nutrient loss caused by burning.[151] We must also allow for the possibility of local surpluses and deficits to the nutrient budget given the tendency for sheep to graze in some areas and to sleep and dung in others so that nutrients were effectively moved between them.[152] Yet, whatever role we attribute to the sheer number of stock now grazing the hills, it was only part of the story. The shift from the mixed grazing of the pre-Clearance toun to the mono-grazing of the specialist sheep farm was an additional factor. Admittedly, this shift was not immediate, for many farms in the southern and eastern Highlands maintained a fold of cattle down until the mid-nineteenth century,[153] but, once established, mono-grazing by sheep led to the selective grazing out of the more nutritious grasses so that, over time, those grasses which they find less nutritious or palatable, such as *Nardus* and *Molinia*, became dominant.[154]

The environmental impact brought about by the exploitation of woodland resources has already been touched on in earlier chapters. After *c.*1815, planting came more to the fore as the prime form of woodland management. Encouragement to plant trees around farmsteads and kailyards was given by an Act of Parliament as early as 1457 and then repeated at intervals across the first half of the sixteenth century.[155] By the seventeenth century, estate interest in planting was directed at planting scattered ornamental trees, tree-lined avenues and small plantations in the grounds that surrounded mansions or policies. The Earl of Argyll, for example, adorned the grounds around Inveraray by planting a mix of trees during the 1660s and 1670s, mostly Scots fir, beech, ash and plane.[156] At this time, what passed as a plantation was 'small, mixed and attractively arranged rather than extensive, uniform and utilitarian'.[157] We can get a rough idea of where early plantings were carried out from Roy's 1747–55 map of the Highlands. The regular alignments, avenues and clumps at Whitehouse, on the southern shore of West Loch Tarbert or the quite elaborate geometry of woodland around Balnagown in Inverness-shire are examples. Contemporary paintings, such as those for the area around Taymouth, Lord Breadalbane's seat, can also be of use as a record of planted woodland around policies by the eighteenth century, if we take the regularity of clumps and alignments as indicative of planting.[158] Even at this stage, we can assume that most efforts at planting woods made use of exotic as well as native species. In a report compiled in 1768, Archibald Menzies described how Mount Stewart on Bute had been surrounded with extensive 'plantations' that included exotic species such as New England pines, Newfoundland spruce, acacias, cedars and laurels, all clearly intended as ornamental planting.[159] When James Robertson passed Castle Grant on his 1771 trip, he talked about 'the large plantation of trees both indigenous and exotic'.[160] Elsewhere on their estates, landlords had started by this time actively to encourage tenants to plant fresh woodland through their tacks. In Assynt, for instance, tacks covering the period 1757 to 1766 stipulated that tenants had to plant trees. That for Farrbrask, Swordly and Kirktorny required the tenant to plant two thousand trees.[161] Once thinned, this would not have amounted to very much. Regulations issued for Lord Macdonald's estate on Skye are more forthcoming over the trees to be used, the estate's policy being that 'few things conduce more to beautify or Improve a country than planting', and recommending that every tenant be bound to plant around their dwelling or kailyard 'Firr, Ash, Sycamore, Beech, etc' provided by its nursery.[162] Instances of plantings on the MacLeod estate on Skye were more ambitious. A 1773 tack for the 3d land of Ardfreuch in Minginish required the tenant to plant 'at least' twenty thousand trees 'betwixt Firs & other kinds of Timber'.[163] The 1767 renewal of the tack for the 12d land of Talisker and the 5d lands of Meikle and Little Ardhousey was different, for it gave permission for MacLeod of Talisker to plant '40,000 firs and other timbers' rather than directing him to plant them.[164]

By the closing decades of the eighteenth century, some landlords started to carry out programmes of planting that fell more into the category of reforesting rather than ornamental planting. The Laird of Mackintosh was said to have established 'very considerable plantations of scots firs mixed with forest trees in Moy and Dalarossie, to which he is making additions annually'.[165] In the eastern Grampians, all the main landlords in Crathie parish had active phases of planting over the later decades of the eighteenth century. Farquharson of Invercauld was said to have planted fourteen million Scots firs and one million larch as well as other species, while the Earl of Fife and Gordon of Abergeldie pursued similar policies, with again, the main species being Scots fir.[166] For the times, schemes such as those implemented by Farquharson of Invercauld were exceptional. So, too, were the efforts by the fourth Duke of Atholl on his Atholl and Dunkeld Estates from the mid-1770s onwards, initiating what proved to be one of the largest afforestation projects. Starting in 1774, he reportedly planted over 15,000 acres (6,070 hectares) with twenty-seven million trees by the time of his death, 11,000 acres (4,452 hectares) in the Dunkeld area alone. Around a half of all the land planted was under larch, with the rest under Scots fir, spruce, birch and even a significant acreage (*c*.1,000 acres or 405 hectares) under oak.[167] The duke was especially single-minded about planting larch, seeing its quicker growth, ability to cope with poorer soils and the fertilising powers of its litter as making it the tree of choice for the Highlands.[168]

There were good reasons why large-scale schemes of planting were exceptional before *c*.1800. Sustainable woodland management required a different mindset, as well as significant capital investment, when it came to replacing timber that had been felled. With farming, one reaped the benefit of what one had sown in spring by the following autumn. Admittedly, building up a breeding flock or herd unfolded over a longer timescale. Neither was as long, however, compared to the time involved with woodland. The maxim was that you planted woodland for your grandchildren owing to what Dr Johnson referred to as 'the frightful long time between seed and timber'.[169] Anderson also thought that there was a learning curve involved in the shift to larger plantations.[170] By c.1800, some of the early attempts at reforestation, designed solely for their timber, had reached maturity, yielding valuable knowledge about what species succeeded best, under what ground conditions, and in relation to which markets, thereby removing part of the risk. The wars had also demonstrated the strategic importance of home-grown timber. This was reflected in plantations becoming a part of estate management over a wider area by the early nineteenth century. On Speyside and in Glen Urquhart, the Seafield estate planted 8,227 acres (3,329 hectares) between 1811 and 1845. The dominant species used by the estate was Scots fir but, generally, larch had become the main species for plantations by the 1830s and 1840s, being preferred because it grew more quickly and could be marketed more easily.[171] Even in native fir woods, closer management by foresters had an impact on the species present. Traditionally, such woods had always had a mix of species present but trees such as birch, alder and even oak were now thinned so that they became more fir dominant.[172]

While we can debate the differences of detail, such as how they cropped their arable, what sort of manures were used, their use of common grazings and so on, we do not have to stand too far back to see a degree of sameness about many traditional touns throughout the region in c.1700. As Walker said, 'the general inclination of the people is to have every thing on every farm'.[173] Only those tacksmen's touns which had taken a more commercial turn would have really stood out. The extensive restructuring which unfolded over the eighteenth to nineteenth centuries swept this degree of sameness aside and replaced it with a more disaggregated pattern of landholding and land uses: sheep farms, aggregated sheep farms, sheep farms with managed grouse muirs, small to medium-sized mixed farms, small to medium-sized cattle-rearing farms, deer forests, crofting townships and townships based on fisher crofts. Yet, while these more segmented land uses would have been the most visible difference between what we see c.1700 and what had emerged by the nineteenth century, they were not the only form of change. The growing tensions between landlords and the crowded communities of crofters, cottars and landless in the far west, north-west and north of the region, along with the growing power of external markets to change how land and its product were valued and the growing use made of the region by those outsiders who flowed into it in ever-increasing numbers, produced other no less significant changes in values, rights and attitudes.

In the final chapter, I want briefly to review these other forms of change and how they, too, contributed to how the region was altered.

Notes

1. BPP, *Report*, 1884, p. 2.
2. M. Bangor-Jones, 'Sheep farming in Sutherland in the eighteenth century', *Agricultural History Review*, 50 (2002), p. 186.
3. NLS, SP, 313/3476.
4. E. Richards, *Leviathan of Wealth*, (London, 1973), p. 9.
5. Ibid., pp. 7–10.
6. E. Richards, 'Structural change in a regional economy: Sutherland and the Industrial Revolution, 1780–1830', *Economic History Review*, 26 (1973), p. 68.
7. A. Mackenzie, *The History of the Highland Clearances* (Glasgow, 1883), p. 22.
8. J. Macdonald, 'On the agriculture of the county of Sutherland', *THASS*, 4th series, xii (1880), p. 63.
9. Loch, *Improvements*, Appendix 11, pp. 39–40.
10. Macdonald, 'On . . . Sutherland', p. 63.
11. E. Richards, *The Highland Clearances* (Edinburgh, 2008 ed.), pp. 177–95.
12. Macdonald, 'On . . . Sutherland', p. 24.
13. NLS, SP, 313/1060.
14. NLS, SP, 313/993.
15. Macculloch, *Western Highlands and Islands*, vol. ii, p. 453.
16. NLS, SP, 313/1000.
17. NLS, SP, 313/2151.
18. NLS, SP, 313/1000, 1818–19.
19. BPP, *Evidence*, 1884, vol. 24, p. 1732.
20. Gaskell, *Morvern Transformed*, p. 23. See also, T. M. Devine, *Clearance and Improvement: Land, Power and People in Scotland 1700–1900*, (Edinburgh, 2006), pp. 175–86.
21. Turnock, *Highland Development*, pp. 31–2; D. Turnock, 'Evolution of Farming Patterns in Lochaber', *TIBG*, 41, pp. 145–58.
22. Ibid., p. 25.
23. Dodgshon, *Chiefs to Landlords*, pp. 187–9.
24. Devine, *Great Highland Potato Famine* (Edinburgh, 1995), pp. 33–56.
25. CDC, LMP, GD221/5914; NAS, RH2/6/24.
26. *NSA*, 1834–45, xiv, pp. 357–8.
27. BPP, *Evidence*, 1884, 22, pp. 330–1.
28. Ibid., p. 334.
29. Ibid., p. 277.
30. Watson, 'The rise . . . sheep industry', pp. 16–18.
31. BPP, *Evidence*, 1884, 22, pp. 476–7; *Report*, 1884, Appendix A, p. 26.
32. BPP, *Evidence*, iv, 1884, p. 2676.

33. J. Gilbert, *Hunting and Hunting Reserves in Medieval Scotland* (Edinburgh, 1979), pp. 360–1.
34. Ibid., pp. 364–5.
35. Gregory and Skene (eds), *Collectanea*, p. 189.
36. HMC, *Fourth Report of the Royal Commission on Historical Manuscripts*, part 1, *Report and Appendix* (London, 1874), p. 533.
37. W. Orr, *Deer Forests, Landlords and Crofters. The Western Highlands in Victorian and Edwardian Times* (Edinburgh, 1982), p. 52.
38. NAS, GD112/9/5/8/26.
39. W. A. Gillies, 'Extracts from the Baron Court Books of Menzies', *Transactions of the Gaelic Society of Inverness*, xxxix–xl (1942–50), pp. 111–12. See also, NAS, GD50/136/1.
40. For example, NAS, GD92/50B/10.
41. NAS, GD112/59/31/11; ibid. GD112/59/31/27; GD305/1/163/105 and 143.
42. NAS, GD305/1/63/143.
43. NAS, GD44/51/354/1/7; Gaffney (ed.), *Strathavon*, 32–4.
44. Orr, *Deer Forests*, pp. 29–30.
45. Ibid., p. 28; Smith, 'Deserted farms and shielings', p. 452.
46. Ibid., p. 31.
47. Ibid., p. 31.
48. Ibid., p. 47.
49. Gray, *Highland Economy*, pp. 155–70.
50. Devine, *Great Highland Famine*, Chapters 1 and 2.
51. CDC, LMP, GD221/116.
52. NAS, GD221/4190/13.
53. Ibid., GD221/4208/1.
54. HULL, DDKG/99.
55. HULL, DDKG/103.
56. CDC, LMP, GD221/5904.
57. Ibid., GD221/116.
58. Ibid., GD221/5904.
59. 1851 survey data from HULL, DDKG/103.
60. BPP, *Evidence*, 1884, 22, pp.336–8, 347, 415.
61. Ibid., p. 346.
62. Ibid., p. 384.
63. Ibid., p. 284.
64. Ibid., pp. 202, 207.
65. For example, DC, MDP, 2/490/8; DC, 2/485/35/1.
66. DC, MDP, 2/485/55.
67. Ibid., 2/485/59.
68. MacLeod, *Book of Dunvegan*, ii, p. 118.
69. DC, MDP, 2/485/73/1; 2/709/1–4.
70. Ibid., 2/490/300; BPP, *Evidence*, 1884, 22, p. 207.

71. BPP, *Report*, 1884, Appendix A, p. 25; DC, MDP, 2/490/20/4; ibid., 2/709/1–4.
72. BPP, *Evidence*, 1884, 22, p. 80.
73. Ibid., p. 80.
74. NAS, GD403/2.
75. CDC, LMP, GD221/5914.
76. H. A. Moisley, 'North Uist in 1799', *SGM*, 77 (1961), pp. 89–92.
77. NAS, GD46/1/277.
78. BPP, *Report*, 1902, p. lxx.
79. Ibid., p. lxx.
80. BPP, *Evidence*, 1884, 23, p. 989.
81. NAS, GD46/1/278.
82. BPP, *Evidence*, 1884, 23, p. 989.
83. Ibid., 23, pp. 1133–4.
84. J. B. Caird, *Park. A Geographical Study of a Lewis Crofting District* (Nottingham, 1958), p. 3.
85. Ibid., p. 3.
86. Richards, 'Structural change', pp. 63, 68.
87. Richards, *Leviathan of Wealth*, p. 169.
88. Ibid., pp. 171, 173, 176.
89. Loch, *Improvements*, pp. 100–1.
90. Richards, 'Structural change', p. 70.
91. J. B. Caird, 'The Making of the Gairloch Crofting Landscape', in Baldwin (ed.), *Peoples and Settlement in North-West Ross* (Edinburgh, 1994), p. 143; N. Magillivray, 'Dr. John Mackenzie (1803–86): proponent of scientific agriculture and opponent of Highland emigration', *Journal of Scottish Studies*, 33 (2013), esp. pp. 90–2.
92. Ibid., pp. 81–100.
93. Caird, 'Gairloch', p.143.
94. Ibid., pp. 147–8.
95. Ibid., pp. 137–58.
96. H. A Moisley, 'The Highlands and Islands: A crofting region?', *TIBG*, 31 (1962), p. 85; Fenton, *Northern Isles*, p. 57.
97. Turnock, *Highland Development*, p. 107; D. Turnock, 'Small farms in the North of Scotland: an exploration in historical geography', *SGM*, 80 (1964), pp. 164–81.
98. Fenton, *Northern Isles*, p. 49.
99. BPP, *Evidence*, 1884, 23, p. 130.
100. Fenton, *Northern Isles*, p. 56; J. R. Coull, 'Walls: a Shetland crofting parish', *SGM*, 80 (1964), pp. 136–43.
101. Fenton, *Northern Isles*, pp. 71–88. See also, the map of Papa Stour's runrig on the cover of Crawford and Ballin Smith, *The Biggins*, 1999.
102. Coull, 'Walls', pp. 136–43.
103. BPP, *Evidence*, 1884, 23, p. 1300.

104. Ibid., pp. 42–3; Thomson, *Orkney*, pp. 332–4.

105. Ibid., p. 334.

106. BPP, *Evidence*, 1884, 23, p. 1527.

107. Fenton, *Northern Isles*, p. 56.

108. T. Farrell, 'On the agriculture of the Islands of Orkney', *THASS*, 4th series, vi (1874), p. 97; see also E. A. Cameron, 'The Scottish Highlands as a special policy area, 1886–1965', *Rural History*, 8 (1997), p. 199.

109. BPP, *Evidence*, 1884, 23, p. 1316.

110. Ibid., p. 1428.

111. Ibid., p. 1535.

112. Ibid., p. 1534.

113. Ibid., pp. 386–7.

114. For example,. NAS, GD112/17/4; ibid., GD112/1711.

115. NAS, GD44/25/2/76; NAS, GD170/431/27; ibid., GD170/431/41.

116. For example, NAS, GD50/159, 27 March 1622; NAS, GD80/384/11, 8 December 1740.

117. Darling, *West Highland*, pp. 275–6.

118. BPP, *Evidence*, 1884, 23, p. 1378.

119. *NSA*, 1834–45, xv, pp. 13, 66–7; BPP, *Evidence*, 1884, 23, p. 1214.

120. Ibid., p. 1207.

121. J. Sinclair, *General Report of the Agricultural State and Political Circumstances of Scotland* (Edinburgh, 1814), ii, p. 397.

122. Loch, *Improvements*, p. 146.

123. Macdonald, 'On the agriculture of the county of Caithness', *THASS*, 4th series, vii (1875), p. 228.

124. Macdonald, 'On the agriculture of Inverness-shire', *THASS*, 4th series, iv (1872), p. 31.

125. Farrell, 'Islands of Orkney', p. 73.

126. J. Macdonald, 'On ... Sutherland', pp. 28–39; see also Macdonald, 'On ... Caithness', pp. 202, 214–15.

127. NAS, GD112/12/1/2/12; NAS, GD112/16/134/14–15.

128. For example, NAS, GD112/12/1/2/36–7; NAS, GD112/14/3/2/32; NAS, GD112/12/1/2/13.

129. For signs of its prehistoric roots, see G. Whittington and J. McManus, 'Dark-age agricultural practices and environmental change', in Mills and Coles (eds), *Life on the Edge: Human Settlement and Marginality*, p. 117.

130. C. H. Gimingham, 'Heaths and moorland; an overview of ecological change', in D. B. A. Thompson, A. J. Hester and M. B. Usher (eds), *Heaths and Moorland: Cultural Landscapes* (Edinburgh, 1995), pp. 16–17.

131. APS, ii, p. 61.

132. NAS, GD112/17/4.

133. *NSA*, 1834–45, xii, p. 549; NAS, GD112/16/10/5/17.

134. Gimingham, 'Heaths and moorland', pp. 15–16.

135. *OSA*, 1791–99, iii, p. 430.
136. Ibid., vi, p. 99.
137. Ibid., xi, p. 122.
138. Ibid., viii, p. 338.
139. Ibid., iii, p. 163.
140. NAS, GD44/25/2/76. See also, NAS, GD50/2/3.
141. Loch, *Improvements*, p. 146.
142. Dixon, *Gairloch*, p. 137.
143. BPP, *Report*, 1884, p. 44.
144. Macdonald, 'On . . . Sutherland', pp. 83–4; J. Hunter, 'Sheep and deer: highland sheep farming, 1850–1900', *Northern Scotland*, 1 (1973), p. 203.
145. Macdonald, 'On . . . Sutherland', pp. 45, 52, 65, 83.
146. Hunter, 'Sheep and deer', pp. 202–3.
147. Macdonald, 'On . . . Sutherland', pp. 45, 52, 65, 83–5; A. S. Mather, 'The environmental impact of sheep farming in the Scottish Highlands during the nineteenth century', in T. C. Smout (ed.), *Scotland since Prehistory: Natural Change and Human Impact* (Aberdeen, 1993), pp. 79–88.
148. Ibid., pp. 79–88; R. A. Dodgshon and G. A. Olsson, 'Heather moorland in the Scottish Highlands: the history of a cultural landscape, 1600–1800', *JHG*, 32 (2006), pp. 21–37.
149. Watson, 'The rise . . . of the sheep industry', pp. 16–18; Macdonald, 'On . . . Sutherland', p. 65.
150. J. L Innes, 'Landuse changes in the Scottish Highlands during the nineteenth century: the role of pasture degeneration', *SGM*, 99 (1993), pp. 144–6.
151. Mather, 'The environmental impact of sheep farming', pp. 79–88; T. C. Smout, *Nature Contested. Environmental History in Scotland and Northern England since 1600* (Edinburgh, 2000), p. 126.
152. P. R. Latham, 'The deterioration of mountain pastures and suggestions for their Improvement', *THASS*, 4th series, xv (1883), pp. 111–12.
153. For example, NAS, GD112/16/14/7/3; GD112/16/14/7/29; GD112/16/14/8/13; GD112/10/2/4/45.
154. Darling, *West Highland Survey*, p. 112; Smout, *Nature Contested*, p. 126.
155. APS, ii, p. 51.
156. *OSA*, 1791–99, v, p. 296; *NSA*, 1834–45, vii, pp. 13–14.
157. Lindsay, 'The commercial use of woodland', p. 275.
158. Smout, 'Pre-improvement fields', pp. 47–55.
159. NAS, E729/9/1.
160. Henderson and Dickson (ed.), *Naturalist in the Highlands*, p. 166.
161. NLS, SP, 313/3160/6.
162. NAS, GD92/50B.
163. DC, MDP, 2/2/3.
164. Ibid., 2/27/2. See also, GD92/50B/11.
165. *OSA*, 1791–99, viii, p. 503.

166. *OSA*, 1791–99, xiv, pp. 337–8.

167. M. L. Anderson, *A History of Scottish Forestry* (Edinburgh, 1967), ii, p. 174. See also, Smout et al., *History of the Native Woodlands*, p. 277.

168. Jonsson, *Enlightenment's Frontier*, pp. 150–6.

169. Johnson, *Journey*, p. 139.

170. Anderson, *Scottish Forestry*, ii, p. 148.

171. For example, *NSA*, 1834–45, ix, pp. 564 and xiv, pp. 104, 495.

172. Smout et al., *History of the Native Woodlands*, p. 74.

173. Walker, *Economical History*, ii, p. 48.

THE YEARS OF CHANGE: AN OVERVIEW

The fundamental changes that swept across the region during the eighteenth and nineteenth centuries involved more than a rearrangement of what was on the ground. They involved changes in attitudes and in how people interpreted, or constructed, what existed there. By way of a conclusion, I want to step back from the details of change and briefly review some of these changes in perception, changes in how people without, as well as within, saw the region.

The Highlands and Islands, c.1700: a Closed Society?

A simplified reading of how the region was transformed over the eighteenth to nineteenth centuries would be to see it as moving from a relatively closed and inward-looking world to being one more open and more engaged with the world beyond. The reality was more complicated than this. Undoubtedly, the region lived far more within its own terms c.1700 than it did c.1914 and was perceived by the world beyond as doing so. The way in which the government objectified it as a problem over the century or so before the Forty-five, highlighting its endemic disorder and backwardness, was instrumental in shaping this impression. The uncompromising attitudes that fed into the Statutes of Iona (1609) come across from James VI's description of the region's mainland inhabitants as 'barbares' but who still had a modicum of civility, as opposed to the 'alluterlie barbares without any sorte or shew of civilitie' who lived in the Isles.[1] Such state attitudes, especially towards those of the far west, remained down to the Forty-five, with the Annexing Act (1752) reusing words like 'barbaric' to describe the region. Yet what was really at issue was the state's ultimate monopoly over power and violence and the way this jarred with how parts of the region were self-referential when it came to such matters, with the clan system fostering different attitudes towards how land should be acquired, granted and held, how rents should be raised and used and how disputes over land or marriages should be settled.[2] It was this self-referential character that did most to create the impression of a region that looked inwards on itself, at least from a political perspective.

Economically, too, there were some who saw the region as largely self-contained and self-sufficient, the embodiment of a natural economy, before change swept over it from the late eighteenth century onwards. Even Adam Smith portrayed the typical Highlander as someone self-sufficient in all their needs,[3] from food to cloth. The description of the farming toun by others as a 'petty commonwealth' that did not need to look outside itself made the same point. The reality was somewhat different. Many pre-Clearance Highlanders were already touched by the markets that fringed the region. Whether they engaged with them indirectly through their payment of rents in kind (grain, butter, cheese, stock, and so on) or directly through what was left as surplus, many would have had more than a rudimentary understanding of how they determined prices. The grain payments handed over as rent led to sizeable amounts being marketed from some parts of the region, some delivered to markets by the tenants themselves. For most ordinary and lesser tenants, however, what they had left after handing over rents in kind, and covering for their seed or their basic subsistence needs, would have been modest. The fact that most rents also had a cash component, coupled with the need to cover shortfalls in the farm economy, though, made the marketing of some produce necessary. From the early seventeenth century, the gradual growth of the cattle trade provided an extra solution, one that connected tenants in the remotest corners of the region to a wider, national market. Of course, in the most cases, it was about the main tenants of a toun supplying an animal every year or so in lieu of their cash rents and was not something that reshaped their toun economy. Only where tacksmen developed a more substantial, specialist engagement with cattle droving can we speak of the toun economy being reshaped by the trade. Yet these qualifications made, and even allowing for the indirectness and marginality in the way markets had started to penetrate the region, we cannot describe the toun economy of the region as a pure, natural economy when seen *c*.1700. Parts were already touched by the needs of a wider world, quite apart from how its own needs in covering for the deficiencies or imbalances of the toun economy forced it to engage with the wider world through the market.

The World Beyond and its Reach

Whatever its degree of closure or openness *c*.1700, the changes that gathered momentum over the eighteenth century shifted the balance markedly towards openness. Some of the government actions after the 1715 rebellion and, later, the Forty-five were undoubtedly instrumental in laying the foundations for this shift. As mentioned in Chapter 7, part of its response to 1715 and again to the Forty-five was to build new roads across the region. Admittedly, they were military roads, designed to facilitate the rapid deployment of troops from the barracks that were also built, like the one at Ruthven after the 1715 rebellion, or, on an even larger scale, that at Fort George after the Forty-five. It would be easy to dismiss these roads as having a simply a strategic value for control of the region, and little else, but – along with the estate roads that started to be built soon after – they facilitated the freer and safer movement of everyone, soldiers, drovers and travellers alike. What is especially striking is how quickly perceptions of whether the region was safe for investment shifted after the Forty-five. The spread of large scale commercial sheep farming, the most far-reaching economic change of the entire period, began within a decade of the 1745 uprising. Given how many of the tenants who took on tacks of early sheep farms were from outside the region, and the scale of capital that underpinned their decisions, we can be in no doubt that, after the Forty-five, there was a change not just in what was on the ground but also in people's perception of what the region could offer. Just as new roads ultimately helped to open up the region to greater interaction with the outside world, as well as with itself, so also can we see the spread of sheep farming as the change that probably did most to open up the region to the needs of the world beyond. The cattle trade had its specialist producers but they were comparatively few in number. The spread of commercial sheep farming was far more significant because farm specialisation and with it, the market orientation of farm output, now developed on a far larger and more extensive scale. The way Campbell reputedly paid a fourfold increase in rent for Glenovoe and the way rents generally were said to have risen by three or four times or more with the initial adoption of commercial sheep production, such as in Ross-shire,[4] encapsulates the impact of this new market orientation in a single measure. Furthermore, while the market opportunities had to be present in the first place, rent increases were widely seen as the way landlords drove change, forcing new, more market-responsive systems to be adopted.[5] Indeed, when reviewing the problem of the region in the 1880s, even a government minister suggested that it was the eagerness to maximise rents in pursuit of rising markets that had led to the marginalisation of crofters.[6] Nor was the emergence of specialised sheep farming the only way in which the demands of burgeoning national markets reached into the region. By 1800, communities in the remotest of areas had already become heavily involved in, and changed by, their trade in items such as kelp, whisky, cattle and fish.[7] Needless to say, there were hard lessons to be learnt from this growing dependence on wider markets. The extent to which the collapse of prices for sheep, wool and kelp stressed the region at the end of the Napoleonic Wars was not the first, nor was it the last of them.

The World Beyond and its Touch

The opening up of the region to the wider world had other consequences. What we can call its indigenous nature was seriously breached in a number of different ways. It was breached extensively, given the scale of new plantations, by what we see planted in the landscape, with the introduction of non-native or exotic tree species. From the moment landowners such as the Duke of Atholl began to reforest parts of his estate in the 1770s, as opposed to merely tree planting, to use Anderson's distinction between the spread of plantations rather than just the ornamental dressing of policy grounds,[8] the species favoured by the new foresters altered. Though Scots fir was widely used, it was gradually displaced by non-native or exotic species, especially the European larch and, to a lesser extent, Norway spruce. By the mid-nineteenth century, some foresters were also testing species such as Sitka spruce, Douglas fir, Corsican pine, though the wider use of the last two was more a feature of the twentieth rather than nineteenth century. The forester's use of exotic species for plantations was, of course, matched by those foresters-cum-gardeners who planted exotic trees, shrubs and flowers in the gardens that were laid out around castles and mansions, often to spectacular effect, such as at Inverewe. Even the sheep farms which spread across the region saw the native give way to the non-native or exotic with the Cheviot becoming by far the most dominant breed of sheep, in place of the Blackface, as numbers rose rapidly after 1800.

Yet arguably, the most significant way in which what was indigenous was breached was the way in which many of the tenants who took over sheep farms were from outside the region and by the way increasing number of landlords were also from outside. Traditional Highland landlords, struggling with a mix of poor rental incomes, rising expenditure and accumulating debt, sold out to a new breed of owner.[9] Devine estimated that by 1850, 60 per cent, or 1,139,707 acres (461,233 hectares), of the large estates in the Highlands and Hebrides, outside of Sutherland, had passed into new ownership, with most of the transfer occurring after 1800.[10] Some of these new owners were from other parts of the region, such as the Countess of Sutherland, who purchased the Reay Estate in 1829, or Sir John Matheson, who bought Lewis in 1844, but the majority were from outside. Furthermore, while some drew on old money, as when the Second Marquis of Salisbury, attracted by its suitability as a deer estate, bought the island of Rum in 1845,[11] or when, more notably, Queen Victoria purchased the Balmoral estate in 1852, many purchases involved new money made in trade or industry. Octavius Smith, the buyer of the Achranich and Ardtornish estates in Morvern, made his money in distilling.[12] Sir John Bullough took over Rum in 1888 on the back of profits from the making of machinery for the Lancashire cotton industry.[13]

These new owners from outside the region brought different perspectives and different values. This was especially true of those whose capital had been accumulated in trade and industry. Yet it would be wrong to assume that they invariably applied the kind of hard-nosed business acumen, by which they had made their money, to the management of their Highland estates, and that much of the tension, increasingly evident over the nineteenth century, was rooted in this uncompromising collision between such values and those of the Highlander. Yes, we can find instances of long-established communities being swept away by new owners without regard for whether they had 'earned' rights of livelihood but there were new owners, who had made their money in commerce or industry, who accepted they had responsibilities as well as rights regarding their smallholders. Whether they treated their sitting tenants and smallholders responsibly or not, what we find is that many brought access to a scale of capital that was beyond that which traditional Highland landowners could generate. The way in which the marriage of the Countess of Sutherland to Lord Stafford ultimately gave the Sutherland estate access to the vast mining and industrial wealth of the Leveson-Gower estate is the obvious example of how change now became harnessed to the fortunes of a different world.[14] We must also allow for this new breed of landlords drawing on multiple values, rather than a single value, in making investments, with different expectations applied to different parts of their estate. On some parts, every effort was made to maximise rental income through a policy of clearances, farm amalgamation, and the like. Meanwhile, other parts might be converted to a deer forest for which the landlord was prepared to absorb heavy costs by constructing hunting lodges, access roads, the employment of stalkers, ghillies and gamekeepers, and the building of cottages for them.

The World Beyond and its Gaze

Withers has claimed that the Highlands 'are both real . . . and they are a myth; a set of ideologically laden signs and images'.[15] Among the 'myth' makers were the increasing numbers of travellers who visited the region over the eighteenth and nineteenth centuries and who did so in many guises: observers, surveyors, field scientists, antiquarians, writers, artists, photographers, sportsmen, tourists, and so on. Few, if any, can be labelled as passive visitors, most being among the 'ideologically laden sign and image' makers, carrying with them as much baggage in their minds as they carried in hand and adding complex layers of meaning to how the world beyond read the region's landscapes. An extensive literature now exists on how they constructed the Highlands. My intention here is not to review this literature at any length but briefly to note the different ways in which this construction of the region found expression.

In any review of how visitors from outside constructed the Highlands, Sir Walter Scott's reworking of its myths and histories into works of fiction, a literary vision that reinvented Highland tradition and romanticised our image of it through novels such as *Rob Roy* or poems such as 'The Lord of the

Isles', can be seen it as one of the most powerful layers of meaning imposed on the Highland landscape. His work drew on a powerful mix of historical accounts, traditions, mythologies and straightforward invention, a *construction* in every sense. Establishing what was real or authentic, though, was not necessarily what mattered most with such fiction. Indeed, he was able to write about the region *in extenso* while paying scant attention to its harsher realities, with no reference to a land almost emptied by the Clearances or the communities living in its more remote corners who endured extreme deprivations of livelihood.[16] Ignoring some of the harsher realities was also a feature of those who flocked to the region in search of the picturesque and the sublime. Appreciation of the sublime and picturesque in landscape was first articulated by William Gilpin, the eighteenth-century writer and artist. As shown during a visit to the Highlands in 1776, his sense of the picturesque valued 'unadorned nature',[17] landscape shorn of its human touch. Gilpin's fellow Cumbrian, Dorothy Wordsworth, followed in his footsteps in 1803. Her sense of the picturesque also inclined towards 'unadorned nature'. Even her description of a house interior in the Trossachs naturalised what she saw: 'the smoke came in gusts, and spread along the walls and above our heads in the chimney, where the hens were roosting like light clouds in the sky'.[18] More telling is her description of Glencoe. There was no hint of the horror or threat that some early travellers saw in mountains but, as one might expect of someone from the Lake District, a pure appreciation of their grandeur; 'the grandest that I have ever seen'.[19] When Millais painted John Ruskin, it was not in a studio setting, but posed against the setting of Glenfinglas Falls, Perthshire, where Ruskin apparently was anxious that the gneiss be visible. Had a wider angle of setting been used, one could easily imagine it as a portrait by Landseer, an artist who was a friend of Millais. The choice was probably made with equal deliberation by Ruskin for he believed that beauty, the picturesque, increased 'in proportion to the increasingly mountainous character' of landscape.[20] No less powerful as a source of 'symbols and images' were those who now had privileged access to the fishing, or stalking and shooting of game in the region. The symbolism and representation of such sports acquired a central place in the region's iconogaphy. The routines of the shooting parties on the hills, the working of a beat on a salmon river, the representation of the stag first at bay on the hill and then again, after the kill, with the stalking party arranged in different poses, with or without the dead animal in the foreground, all became part of the region's projected image, by painters such as Landseer[21] and later, by early photographers.[22] The sporting estate stood for more than just imagery, however. It also exuded a complex set of related values. By their very nature as large, expansive areas of reserved space, the nineteenth-century deer forests help to foster the notion that the region was about untamed wilderness, landscape as nature intended.[23] In such a setting. it is not surprising that deer stalking became associated with masculinity, endurance, fitness and other such qualities.[24] With vast areas of presumed wilderness or natural landscape, it was also inevitable that the Highlands became valued as an observatory for field scientists. The botanist, James Robertson, was among the first to seize the opportunity, the prime objective of his 1767–71 tours being to study the plant life of the region. Such field studies were part of how, in Withers's words, Scotland 'came to know itself'.[25] It was a task performed as much by those within as without the region. The numerous parish reports of the *NSA*, 1839 to 1845, make it clear just how much local field knowledge was already culturally stored about the region's natural history. By the second half of the nineteenth century, the twin tasks of recording and interpretation were aided by the numerous societies established in Scotland devoted to field studies with some, like the Scottish Alpine Botanical Club, founded in Edinburgh in 1873, and the Northern Institution, founded in Inverness 1876, having a specific Highland focus in their remits.[26] Major debates, such as that which emerged over the region's geology in the form of the so-called Highlands Controversy, show that its natural history was just as open to the imposition of constructs as other areas of knowledge. As made clear in my Introduction, the recording and interpretation of the region's natural history had its corollary in the recording and interpretation of information about its cultural landscape. That the region has some of the finest prehistoric sites and landscapes in western Europe cannot be questioned but, from the eighteenth century onwards, we find more and more commentators arguing that what had survived down into the early modern period by way of the farming toun and its various traits was no less a part of this prehistoric inheritance. In fact, this presumption that the pre-Clearance or crofting landscape was an ancient landscape, one that had somehow survived across the millennia without change, has been one of the more enduring constructs imposed over the region in recent centuries but one which is no more than that, a construct.

The World Beyond and the Smallholder

For smallholders, crofters and cottars, the most significant way in which the world beyond touched the region was through the passing of the Crofters' Holding Act of 1886 and subsequent legislation such as the Land Settlement (Scotland) Bill of 1919.[27] In contrast to other changes, this actually had the effect of ring fencing crofts and their economy from the more disruptive effects of wider market change, enabling crofters to maintain their values and livelihood. In this sense, the 1886 Act really echoed eighteenth-century attempts at land reform. Its background was in the making throughout the nineteenth century, as the Clearances, displacement on to poorer ground, overcrowding and the removal of grazing land slowly eroded the viability of holdings for many crofters and left others landless. Increasingly, how they viewed solutions to their plight conflicted with how landlords perceived the problem. By the early 1880s, a sequence of events brought matters to a head. It began with poor harvests in 1881 and 1882 and was exacerbated by a storm that damaged the fishing boats by which many supplemented their livelihoods. On some estates, tenants began to engage in direct action to secure more adequate holdings and reasonable rents, announcing rent strikes and occupying extra land without agreement. Events moved quickly. A major confrontation between crofters and police at Breas, south of Portree on Skye in 1882 turned violent. Another, at Glendale on Skye soon after, also became violent, with the authorities this time calling on the army as well as the police to control the situation. In the minds of many crofters and their supporters by this point was the passing of the Irish Land Reform Act in 1881. Not only did it provide a template for how Highland land reform might be legislated for but it also demonstrated the power of direct action as a means by which the government could be persuaded to act. In an effort to provide both the organisational direction and a political voice for such aspirations, the Highland Land Law Reform Association was formed in 1883. Its headquarters was in London but it set about creating branches across the Highlands and Islands. Anxious to control what was fast becoming a volatile situation, the government responded, in 1883, by setting up the Napier Commission, or the Commission of Inquiry into the Conditions of the Crofters and Cottars in the Highlands and Islands. The commission gathered evidence directly from landlords, crofters and cottars alike across the Highlands.[28] We learn a great deal from this evidence but what stands out most from it is the extent to which the attitudes of landlords and their agents, on the one hand, and those of the crofter or cottars, on the other, had polarised, each reading what had happened – and what each other's rights and responsibilities were – in quite the opposite ways.

Initially, the government fumbled over its response to the commission's report but, faced with continuing unrest in the Highlands, it eventually moved to pass the Crofter's Act in 1886. As well as establishing the rights of crofters to compensation for improvements and to a fair rent, it also gave them security of tenure. These provisions applied only to smallholders in qualifying parishes in certain counties, in what became known as the seven crofting counties: Argyll, Inverness-shire, Ross and Cromarty, Sutherland, Caithness, Orkney and Shetland.[29] To qualify, a parish had to have, or to have had in the previous eighty years, smallholders possessing what was the standard croft – arable plus share of common grazings – with a rent of less than £30. As Cameron pointed out, qualification, and with it, the definition of the crofting counties, hinged on the presence of common grazings. This was critical because it provided the means by which the government was able to ring fence the impact of legislation to smallholdings in those parts of the Highlands where the 'crofting' problem existed, avoiding the problem of all smallholders in Scotland being affected by its provisions.[30] In fact, while the Act delivered part of what had been called for, the HLLRA, and crofters themselves, were dissatisfied, because it did not address all the core grievances, particularly the recovery of cleared land and the creation of new crofts. For this reason, social unrest continued into the late 1880s, partly encouraged by poor returns from fishing and the accompanying hardship.[31] Once the newly formed Crofters Commission began its work, however, enlarging the arable and adding more grazing land to those crofts that had been unduly clipped, and once it had considered early submissions over fair rent and had responded by reducing rents and by making significant reductions in the arrears that had accumulated on islands such as Skye as part of the protests, then both the HLLRA and crofters began to see its benefits.

As the scale of protests subsided, a still more critical change took place. Initially, the government had been frustrated at the continued crofter unrest after the 1886 Act but, during the early 1890s, its approach shifted. It no longer saw its prime focus as suppressing unrest and, instead, began to look at ways in which the Highland economy generally might be developed.[32] The geographical framing established by the Act helped it to underpin what, in effect, became a 'Highland policy area',[33] one that became the territorial focus for further initiatives and government agencies such as the Congested Districts Board. Created in 1897, and fashioned after an equivalent Irish board that had been established a few years earlier, the board was a significant step because it could, and did, create new crofts, albeit only a few in number. Locally, though, it made a difference, as with the laying out of sixty new crofts at Eoligarray, Barra. Generally, its ability to create new crofts was made easier by the fact that landlords were more open to releasing land at this point owing to a collapse in the profits of sheep farming but it was made more difficult by the board's still limited budget. This budget was extended when the board's responsibilities were passed to a 'Board of Agriculture' created within the department of agriculture in 1911. As in the case of earlier wars, the drafting of men from the Highlands into the army and other services for World War I helped to alleviate the land problem but, with the end of hostilities, the problem was brought back into sharp focus with returning soldiers now feeling that they had an even greater claim to part of the land they had fought for. Social unrest and land raids returned to the region. Aided by the Land Settlement (Scotland) Bill of 1919, the Board of Agriculture slowly responded, purchasing land and distributing it as crofts. In fact, the amount of land that the board purchased and redistributed over the five or six years after the Great War ended was *in cumulo* quite extensive. On Skye alone, for instance, it acquired 51,000 acres (20,639 hectares), adding extra grazing land to crofts and laying out over two hundred new ones.[34]

Though one of the criticisms of the 1886 Act is that it froze the croft as a unit of landholding, the wider crofting landscape did keep on 'evolving': the result was a great deal of intraregional variation.[35] When seen later, in the mid-twentieth century, the size of crofting townships, as with the earlier toun, was higher in the Hebrides than on the mainland. In terms of the size of crofts, though, the opposite was true, with larger crofts along the eastern side of the Highlands, where crofting was a less significant component within the landholding structure, and smaller crofts across the west and north-west of the region where, locally, crofts had become a dominant feature of the landholding structure. There were exceptions to this, such as on Tiree and in the Uists, where the new crofts created by the Board of Agriculture through the subdivision of farms led to quite high average croft sizes.[36] Though the Crofters Commission and then the board created 2,700 new crofts, enlarged 5,160 others and added 750,000 acres (303,521 hectares) of pasture to others from 1886 onwards,[37] analysis suggests that, overall, the number of crofts probably declined down to the mid-twentieth century. Estimates for Sutherland, for instance, suggest that, while the total number of crofts remained more or less stable between the mid-nineteenth century and the early years of the Crofters Commission, thereafter their number fell by as much as 20 per cent by 1963.[38] Much of this decline was through amalgamation, a trend that affected most crofting areas despite concern over its effects on the viability of communities, if not crofts. The viability of communities has also been affected by the growing level of absenteeism during the twentieth century.[39] As a trend, this has been more marked on the mainland than in the Hebrides, with the tenants of 25 per cent of all crofts in an area like Sutherland being classed as absent by 1960.[40] What stands out most, though, from the flurry of studies over the mid-twentieth century that took stock of the crofting problem is that, despite the efforts of the government agencies, the viability of crofting as a farm economy had not been secured by enlargements or resettlements and that it sat uncomfortably in the modern world.[41] Even Darling concluded that the system was 'anachronistic', when seen beside the values and demands of a wider world,[42] but he would have been the first to argue that its cultural viability depended on a different set of values and needs.

Farming societies, even in mountain areas, are rarely entirely open or entirely closed to the world beyond. As Robert Redfield, the American anthropologist theorised, all peasant societies – and not just those of mountain areas – are by nature part societies. He couched their degree of closure or openness in terms of their mix of belowness and aboveness, the former being what it provided from within itself and the latter being what was 'imposed' down on to it or drawn from the outside world.[43] If we had to place the Highlands and Islands of the early eighteenth century within such a scheme, it would clearly have had more belowness than aboveness. The changes described in Chapters 7 and 8 altered this balance and did so dramatically. Yes, there were enclaves, notably the new crofting townships, within which the pace of change as regards the adoption of market values and the orientation of the farm economy worked itself out slowly. Elsewhere, though, we find marked shifts taking place in the structural ordering of the land economy and its market orientation. Along with its greater political and economic integration into the world beyond its bounds, the way in which it was constructed in people's mind shifted. Old constructs about it being a cultural backwater remained but an increasing and diversifying flow of tenants, landlords, visitors, observers, sportsmen and tourists from outside added new constructs. Of course, even those constructs developed from within now formed a fractured perspective, as the increasing contrast between the perspective of the crofter and landlord make clear. We might also separate out the perspective held by the many who migrated out of the region over these centuries and who could lay claim to a hybridised or insider/outsider perspective, as could the many tenants and landlords who moved into the region from outside, their perspectives melding what was without with what was within. When we bring all these ways of seeing together, we do not find a single history but one that could be retold, symbolised and memorialised in different ways. It is in this respect that we need to see how what had emerged by 1914 was not just about the new structures of landholding or new land uses, the patterned realities of what existed on the ground, but, equally, about the complex web of meanings now draped over the landscape, part woven from how the region sees itself and part out of how the world beyond has used and imagined it.

Notes

1. Cited by Withers, *Geography, Science and National Identity*, p. 51.
2. R. A. Dodgshon, 'Pretense of blude and place of thair duelling: the nature of Highland clans, 1500–1745', in R. A. Houston and I. D. Whyte (eds), *Scottish Society, 1500–1800* (Cambridge, 1989), pp. 169–98.
3. A. Smith, *An Inquiry into the Nature and Causes of the Wealth of Nations* (London, 1930 ed.), i, p. 19.
4. *Farmers Magazine*, 1803, iv, p. 108.
5. Ibid., p. 108; T. Pennant, *Tour in Scotland and a Voyage to the Hebrides, 1772* (Chester, 1772), p. 228.
6. Cameron, 'The Scottish Highlands', p. 197.
7. Shaw, *Northern and Western Islands*, pp. 154–82.
8. Anderson, *Scottish Forestry*, ii, pp. 153–4.
9. Devine, *Clearance and Improvement*, pp. 136–42.
10. Ibid., pp. 109–111.
11. Orr, *Deer Forests*, p. 29.
12. Gaskell, *Morvern Transformed*, pp. 57–80.
13. G. Ritchie and M. Harman, *Argyll and the Western Isles* (Edinburgh, 1985), p. 48.
14. Richards, *Leviathan of Wealth*, pp. 283–97.
15. C. Withers, 'The historical creation of the Scottish Highlands', in I. Donnachie and C. Whatley (eds), *The Manufacture of Scottish History* (Edinburgh, 1992), p. 143.
16. Withers, ''The historical creation . . . Highlands', p. 153.
17. Smout, *Nature Contested*, p. 22

18. D. Wordsworth, *Recollections of a Tour Made in Scotland, AD 1803* (New York, 1874 ed.), 103.
19. Ibid., pp. 173–4.
20. D. A. Finnegan, 'Naturalising the Highlands: geographies of mountain fieldwork in late Victorian Scotland', *JHG*, 33 (2007), p. 795.
21. T. Pringle, 'The privation of history: Landseer, Victoria and the highland myth', in D. Cosgrove and S. Daniels(eds), *The Iconography of Landscape* (Cambridge, 1988).
22. For example, Gaskell, *Morvern Transformed*, plate 16; C. Withers, 'Picturing Highland landscapes: George Washington Wilson and the photography of the Scottish Highlands', *Landscape Research*, 19 (1994) pp. 68–79.
23. H. Lorimer, 'Guns, game and the grandee; the cultural politics of deerstalking in the Scottish Highlands', *Cultural Geographies*, 7 (2000), p. 404.
24. Ibid., p. 410.
25. C. W. J. Withers, 'How Scotland came to know itself: geography, national identity and the making of nation, 1680–1790', *JHG*, 21 (1995), pp. 371–97; C. W. J. Withers and D. A Finnegan, 'Natural history societies, fieldwork and local knowledge in nineteenth-century Scotland: towards an historical geography of civic science', *Cultural Geographies*, 10 (2003), p. 341.
26. Ibid., p. 341; Finnegan, 'Naturalising the Highlands', pp. 795–8.
27. See W. E. Gladstone's words in Cameron, 'The Scottish Highlands', p. 196.
28. BPP, *Evidence*, 1884, vols 22–5.
29. Cameron, 'The Scottish Highlands', p. 197.
30. Ibid., pp. 197–8.
31. For example, J. Hunter, *The Making of the Crofting Community* (Edinburgh, 1976), pp. 172–7
32. Ibid., pp. 178–9.
33. Cameron, 'The Scottish Highlands', pp. 195–215.
34. Hunter, *Crofting Community*, p. 205.
35. Ibid., p. 93.
36. Moisley, 'Crofting region', p. 93.
37. Hunter, *Crofting Community*, p. 206.
38. Wheeler, 'Landownership', p. 47.
39. Cameron, 'The Scottish Highlands', p. 203.
40. Moisley, 'Crofting region', p. 85.
41. Turnock, *Highland Development*, p. 107
42. Darling, *West Highland Survey*, p. 208.
43. Redfield, *The Little Community* (Chicago, 1956).

SELECT BIBLIOGRAPHY

Armit, I. (1991), 'The Atlantic Scottish Iron Age: five levels of chronology', *PSAS*, 123: pp. 181–214.

Armit, I. (1992), *The Later Prehistory of the Western Isles of Scotland*. BAR British Series, no. 221, Oxbow: Oxford.

Armit, I. (1996), *The Archaeology of Skye and the Western Isles*. Edinburgh University Press: Edinburgh.

Ashmore, P. J. (1996), *Neolithic and Bronze Age Scotland*. Batsford and Historic Scotland: London.

Atkinson, J. A., Banks, I. and MacGregor, G. (eds), (2000), *Townships to Farmsteads. Rural Settlement Studies in Scotland, England and Wales*. BAR British Series, no. 293, Oxbow: Oxford.

Atkinson, J. A. (2012), 'Settlement Form and Evolution in the Central Highlands of Scotland, ca. 1100–1900', *International Journal of Historical Archaeology*, 14: pp. 316–34.

Ballin Smith, B. (ed.) (1994), *Howe: Four Centuries of Orkney Prehistory Excavations 1978–1982*, PSAS Monographs, no. 9: Edinburgh.

Ballin Smith, B. and Banks, I. (eds) (2002), *In the Shadow of the Brochs. The Iron Age in Scotland*. Tempus: Stroud.

Bangor-Jones, M. (2002), 'Sheep farming in Sutherland in the eighteenth century', *Agricultural History Review*, 50: pp. 181–202.

Barber, J. (2003), *Bronze Age Farm Mounds and Iron Age Farm Mounds of the Outer Hebrides*, Scottish Archaeological Internet Reports 3 (www.sair.org.uk).

Bil, A. (1990), *The Shieling 1600–1840. The Case of the Central Scottish Highlands*. John Donald: Edinburgh.

BPP, *Evidence taken by Her Majesty's Commissioners of Inquiry into the Condition of Crofters and Cottars in the Highlands and Islands* (London, 1884). N.b., all text references based on Irish University Press BPP reprint, vols 22–5.

BPP, *Report of the Commissioners of Inquiry into the Condition of the Crofters and Cottars in the Highlands and Islands* (London, 1884).

Branigan, K. and Merrony, C. (2000), 'The Hebridean Blackhouse on the Isle of Barra', *Scottish Archaeological Journal*, 22: pp. 1–16.

Branigan, K. and Foster P., (2002), *Barra and the Bishop's Isles*. Tempus: Stroud.

Caird, J. B. (1958), *Park. A Geographical Study of a Lewis Crofting District*. Geographical Field Group: Nottingham.

Caird, J. B. (1994), 'The Making of the Gairloch Crofting Landscape', in J. R. Baldwin (ed.), *Peoples and Settlement in North-West Ross*. Society for Northern Studies, Edinburgh, pp. 137–58.

Cameron, E. A. (1997), 'The Scottish Highlands as a special policy area, 1886–1965', *Rural History*, 8: pp. 195–215.

Campbell, E. (1999), *Saints and Sea-kings: The First Kingdom of the Scots*. Birlinn and Historic Scotland: Edinburgh.

Campbell, E. (2001), 'Were the Scots Irish?', *Antiquity*, 75: pp. 285–92.

Cottam, M. B. and Small, A. (1974), 'The distribution of settlement in Southern Pictland', *Medieval Archaeology*, xviii: pp. 43–65.

Crawford, B. E. (1987), *Scandinavian Scotland*. Leicester University Press: Leicester.

Crawford, I. (1983), 'The present state of the settlement history in the West Highlands and Islands', in A. O'Connor and D. V. Clarke (eds), *From the Stone Age to the 'Forty-Five*. Edinburgh University Press: Edinburgh, pp. 350–67.

Crawford, I. and Switsur, I. (1977), 'Sandscaping and C14: the Udal, N. Uist', *Antiquity*, 51: pp. 124–36.

Cregeen, E. R. (1969), 'The tacksmen and their successors. A study of tenurial reorganization in Mull, Morvern and Tree in the early 18th century', *SS*, 13: pp. 93–114.

Davidson, D. A., Harkness, D. D. and Simpson, I. A. (1986), 'The Formation of Farm Mounds on the Island of Sanday, Orkney', *Geoarchaeology: An International Journal*, 1: pp. 45–60.

Davies, A. L. and Tipping, R. (2004), 'Sensing small-scale human activity in the palaeo-ecological record: fine spatial resolution pollen analyses from Glen Affric, northern Scotland', *The Holocene*, 14: pp. 233–45.

Dixon, T. N. (1982), 'A survey of crannogs in Loch Tay', *PSAS*, 112: pp. 17–38.

Dodgshon, R. A. (1977), 'Changes in Scottish township organization during the medieval and early modern periods', *Geografiska Annaler*, series B, 55: pp. 51–65.

Dodgshon, R. A. (1992), 'Farming practice in the Western Highlands and Islands before Crofting: a study in cultural inertia or opportunity costs?', *Rural History*, 3: pp. 173–89.

Dodgshon, R. A. (1993), 'West Highland and Hebridean settlement prior to crofting and the Clearances: a study in stability or change?', *PSAS*, 123: pp. 419–38.

Dodgshon, R. A. (1993), 'West Highland and Hebridean landscapes: have they a history without runrig?', *JHG*, 19: pp. 383–98.

Dodgshon, R. A. (1993), 'Strategies of farming in the western Highlands and Islands of Scotland prior to crofting and the clearances', *Economic History Review*, xlvi: pp. 679–701.

Dodgshon, R. A. (1998), 'The evolution of Highland townships during the medieval and early modern periods', *Landscape History*, 20: pp. 51–63.

Dodgshon, R. A. (1998), *From Chiefs to Landlords. Social and Economic Change in the Western Highlands and Islands, c.1493–1820*. Edinburgh University Press: Edinburgh.

Dodgshon, R. A. (2004), 'Coping with risk: subsistence crises in the Scottish Highlands and Islands', *Rural History*, 15: pp. 1–25.

Dodgshon, R. A. (2008), 'Bones, bows and byres: early dairying in the Scottish Highlands and Islands', in P. Rainbird (ed.), *Monuments in the Landscape*. Tempus: Stroud, pp. 165–76.

Dodgshon, R. A., (2012), 'The Clearances and the transformation of the Scottish Countryside', in T. M. Devine and J. Wormald (eds), *The Oxford Handbook of Modern Scottish History*. Oxford University Press: Oxford, pp. 130–58.

Dodgshon, R. A. and Olsson, G. A. (2006), 'Heather moorland in the Scottish Highlands: the history of a cultural landscape, 1600-1800', *JHG*, 32: pp. 21–37.

Driscoll, S. T. (1991), 'The archaeology of state formation', in W. S. Hanson and E. A. Slater (eds), *Scottish Archaeology: New Perspectives*. Aberdeen University Press: Aberdeen, pp. 81–111.

Edwards, K. J. and Mithen, S. (1995), 'The colonisation of the Hebridean islands of Western Scotland: evidence from the palynological and archaeological records', *World Archaeology*, 26: pp. 348–65.

Fairhurst, H. (1964), 'The surveys for the Sutherland Clearances of 1813–1820', *SS*, 8: pp. 1–18.

Fairhurst, H. (1968–9), 'The deserted settlement at Lix, West Perthshire', *PSAS*, 101: pp. 160–99.

Fenton, A. (1978), *The Northern Isles: Orkney and Shetland*. John Donald: Edinburgh.

Fleming, A. (2005), *St Kilda and the Wider World*. Windgather Press: Macclesfield.

Fojut, N. (1994), *A Guide to Prehistoric and Viking Shetland*. Shetland Times: Lerwick.

Gailey, R. A. (1963) 'Agrarian improvement and the development of enclosure in the south-west Highlands of Scotland', *SHR*, 134: pp. 105–25.

Gaskell, P. (1968), *Morvern Transformed. A Highland Parish in the Nineteenth Century*. Cambridge University Press: Cambridge.

Graham-Campbell, J. and Batey, C. E. (1998), *Vikings in Scotland*. Edinburgh University Press: Edinburgh.

Gray, M. (1957), *The Highland Economy, 1750–1850*. Oliver and Boyd: Edinburgh.

Guttmann, E. B. (2005), 'Midden cultivation in prehistoric Britain: arable crops in gardens', *World Archaeology*, 37: pp. 224–39.

Guttmann, E. B., Simpson, I. A., Davidson, D. A.and Dockrill, S. J. (2006), 'The management of arable land from prehistory to the present: case studies from the Northern Isles of Scotland', *Geoarchaeology: An International Journal*, 21: pp. 61–92.

Harden, J. and Lelong, O. (2011), *Winds of Change. The Living Landscapes of St Kilda*. Society of Antiquaries of Scotland: Edinburgh.

Harding, D. W. (2000), *The Hebridean Iron Age: Twenty Years On*, University of Edinburgh, Department of Archaeology, Occasional Papers no. 20: Edinburgh.

Hedges, J. W. (1987), *Bu, Gurness and the brochs of Orkney*, BAR British series, Part III. Oxbow: Oxford.

Henderson, J. (1998). 'Islets through time: the definition, dating and distribution of Scottish crannogs', *Oxford Journal of Archaeology*, 27: pp. 227–44.

Holden T. (with contributions by L. M. Baker) (2004), *The blackhouses of Arnol*. Historic Scotland Research Report: Edinburgh.

Hunter, J. (1973), 'Sheep and deer: highland sheep farming, 1850–1900', *SS*, pp. 202–3.

Hunter, J. (1976), *The Making of the Crofting Community*. John Donald: Edinburgh.

Jonsson, F. Albritton, (2013), *Enlightenment's Frontier. The Scottish Highlands and the Origins of Environmentalism*. Yale University Press: New Haven.

Lamont, W. D. (1981), '"House" and "Pennyland" in the Highlands and Isles', *SS*, 25: pp. 65–76.

Lorimer, H. (2000), 'Guns, games and the grandee: the cultural politics of deer-stalking in the Scottish Highlands', *Ecumene*, 7: pp. 403 31.

Macgregor, L. J. and Crawford, B. E. (eds) (1977), *Ouncelands and Pennylands*. University of St Andrews, St John's House Papers no. 3: St Andrews.

Macinnes, A. I. (1996), *Clanship, Commerce and the House of Stuart, 1603–1788*, Tuckwell Press: East Linton.

McKerral, A. (1947), 'The tacksman and his holding in the South-West Highlands', *SHR*, xxvi: pp. 13–14.

McKerral, A. (1950–51), 'The Lesser Land and Administrative Divisions in Celtic Scotland', *PSAS*, lxxxv: pp. 52–64.

McKichan, F. (2007), 'Lord Seaforth and Highland estate management in the First Phase of Clearance (1783–1815)', *SHR*, 86: pp. 50–68.

MacKie, E. W. (2008), 'The broch cultures of Atlantic Scotland: origins, high noon and decline. Part 1: Early Iron Age beginnings *c.*700–200 BC, *Oxford Journal of Archaeology*, 27: pp. 261–78.

MacKie, E. W. (2010), 'The broch cultures of Atlantic Scotland. Part 2: The Middle Iron Age: High noon and decline *c.*200 BC–AD 550', *Oxford Journal of Archaeology*, 29: pp. 89–117.

Mackillop, A. (2000), *'More Fruitful than the Soil'. Army, Empire and the Scottish Highlands, 1715–1815*. Tuckwell Press: East Linton.

Mather, A. S. (1970), 'Pre-1745 land use and conservation in a Highland glen: an example from Glen Strathfarrar, North Inverness-shire', *SGM*, 86: pp. 159–69.

Mather, A. S. (1993), 'The environmental impact of sheep farming in the Scottish Highlands during the nineteenth century', in T. C. Smout (ed.), *Scotland since Prehistory: Natural Change and Human Impact*. Scottish Cultural Press: Aberdeen, pp. 79–88.

Mills, C. M., Armit, I., Edwards, K. J., Grinter, P. and Mulder, Y. (2004), 'Neolithic land-use and environmental degradation: a study from the Western Isles of Scotland', *Antiquity*, 78: pp. 886–95.

Mithen, S. (2000), 'Mesolithic sedentism on Oronsay: chronological evidence from adjacent islands in the Southern Hebrides', *Antiquity*, 74: pp. 298–304.

Mithen, R., Pirie, A., Smith, S. and Wicks, K. (2007), 'The Mesolithic–Neolithic transition in western Scotland: a review and new evidence from Tiree', *Proceedings of the British Academy*, 144: pp. 511–41.

Moisley, H. A. et al. (1961), *Uig. A Hebridean Parish*, Parts 1 and 2. Geographical Field Group: Nottingham.

Moisley, H. A. (1962), 'The Highlands and Islands: A crofting region?', *TIBG*, 31: pp. 83–95.

Nicolaisen, W. F. H. (1976), *Scottish Place Names. Their Study and Significance*. Batsford: London.

Noble, G. (2006), *Neolithic Scotland: Timber, Stone, Earth and Fire*. Edinburgh University Press: Edinburgh.

Orr, W. (1982), *Deer Forests, Landlords and Crofters. The Western Highlands in Victorian and Edwardian Times*. John Donald: Edinburgh.

Parker Pearson, M. (ed.), (2012), *From Machair to Mountains*. Oxbow Books: Oxford.

RCAHMS, (1990), *North East Perthshire*. RCAHMS: Edinburgh.

RCAHMS (1993), *Strath of Kildonan. An Archaeological Survey*. RCAHMS: Edinburgh.

RCAHMS, (1993), *Afforestable Land Survey. Waternish*. RCAHMS: Edinburgh.

RCAHMS and Historic Scotland (2002), *But the Walls Remained*. RCAHMS: Edinburgh.

Richards, E. (1972), *The Leviathan of Wealth. The Sutherland Fortune in the Industrial Revolution*. Routledge: London.

Richards E. (1973), 'Structural change in a regional economy: Sutherland and the Industrial Revolution, 1780–1830', *Economic. History Review*, 26: pp. 63–76.

Richards, E., (2008), *The Highland Clearances*. Birlinn: Edinburgh.

Ritchie, A. (1976–77), 'Excavation of Pictish and Viking-age farmsteads at Buckquoy, Orkney', *PSAS*, 108: pp. 174–227.

Robertson, I. M. (1949), 'The head-dyke – A fundamental line in Scottish geography', *SGM*, 65: pp. 6–19.

Ross, A. (2006), 'The dabhach in Moray: a new look at an old tub', in A. Woolf (ed.), *Landscape and Environment in Dark Age Scotland*, St Andrews University, St John's House Papers no. 11: St Andrews.

Shaw, F. (1980), *The Northern and Western Islands of Scotland*. John Donald: Edinburgh.

Simpson, I. A. (1997), 'Relict properties of anthropogenic deep top soils as indicators of infield management in Marwick, West Mainland, Orkney', *Journal of Archaeological Science*, 24: pp. 365–80.

Smith, B. (2000), *Toons and Tenants. Settlement and Society in Shetland 1299–1899*. Shetland Times: Lerwick.

Smout, T. C. (ed.) (1997), *Scottish Woodland History*. Scottish Cultural Press: Dalkeith.

Smout, T. C.and Lambert, R. A. (1999), *Rothiemurchus. Nature and People on a Highland Estate 1500–2000*. Scottish Cultural Press: Dalkeith.

Smout, T. C., Macdonald, A. R. and Watson, F. (2005), *A History of the Native Woodlands of Scotland, 1500–1920*. Edinburgh University Press: Edinburgh.

Stewart, J. (1990), *The Settlements of Western Perthshire. Land and Society North of the Highland Line 1480–1851*. Pentland Press: Edinburgh.

Storrie, M. (1965) 'Landholdings and settlement evolution in west highland Scotland', *Geografiska Annaler*, series b, 43: pp. 138–61.

Storrie, M. (1981), *Islay: Biography of an Island*. Oa Press: Port Ellen.

Thomson,W. P. L. (1970), 'Funzie, Fetlar: a Shetland runrig township in the nineteenth century', *SGM*, 86: pp. 170–85.

Thomson,W. P. L. (2001), *The New History of Orkney*. Mercat Press: Edinburgh.

Tipping, R. (1994), 'The form and fate of Scotland's woodlands', *PSAS*, 124: pp. 1–54.

Tipping, R., Ashmore, P., Davies, A. L., Haggart, B. A., Moir, A., Newton, A., Sands, R., Skinner, T. and Tisdall, E. (2008), 'Prehistoric *Pinus* woodland dynamics in an upland landscape in northern Scotland: the roles of climate change and human impact', *Vegetation History and Archaeobotany*, 17: pp. 251–67.

Turner, V. (1998), *The Shaping of Shetland*. Shetland Times, Lerwick.

Turnock, D. (1964), 'Small farms in the North of Scotland: an exploration in historical geography', *SGM*, 80: pp. 164–81.

Turnock, D. (1967), 'Evolution of Farming Patterns in Lochaber', *TIBG*, 41: pp. 145–58.

Turnock, D. (1969), 'North Morar – The Improving Movement on a West Highland Estate', *SGM*, 85: pp. 17–30.

Turnock, D. (1970), *Patterns of Highland Development*. Macmillan: London.

Turnock, D. (1982). *Historical Geography of Scotland Since 1707*. Cambridge University Press: Cambridge.

Turnock, D. et al. (1977) *The Lochaber Area. A Case Study of Changing Land Use in the West Highlands of Scotland*. Geographical Field Group, Regional Studies no. 20: Nottingham.

Watt, D. (2006), '"The laberinth of thir dfficulties": the influence of debt on the Highland elite. *c.*1550–1700', *SHR*, 85: pp. 28–51.

Whittington, G. (1974–75), 'Placenames and the settlement pattern of dark-age Scotland', *PSAS*, 106: pp. 99–110.

Withers, C. (1988), *Gaelic Scotland: The Transformation of a Culture Region*. London: Routledge.

Withers, C. (1992), 'The historical creation of the Scottish Highlands', in I. Donnachie and C. Whatley (eds), *The Manufacture of Scottish History*. Polygon: Edinburgh, pp. 143–56.

Youngson, A. J. (1973), *After the Forty-Five: The Economic Impact on the Scottish Highlands*. Edinburgh University Press: Edinburgh.

INDEX

Aberdeenshire, 44, 48, 65
Abernethy, 18, 143, 196
A'Cheardach Mhor, South Uist, 26
Achindown, lordship of, 107
Achmelvich, 124
Achnacarnan, Assynt, 207
Achnacross, 147
Achnashellach, 143
Achranich and Ardtornish estates, 273
Acts of Parliament
    Act of Union(1707), 216
    anent lands lying runrig (1695),
        209–10, 252
    Annexing Act (1752), 2, 193, 270
    Crofters' Holding (Scotland) Act
        (1886), 251, 276
    distilling acts (1785 and 1823), 200
    Irish cattle imports (1686), 197, 216
    Irish Land Reform Act (1881), 277
    'putting . . . of men out of their tacks'
        (1546), 210
    season for muirburn (1457), 258
    tree planting (1457), 261
    'warning of tenants' (1652), 210
    winter herding (1686), 109
Adam, R. J., 123
alder, 145, 263
Alyth Burn, hut circles, 15
An Sithean, Islay, hut circles and field
    systems, 21, 33
Anderson, J., 194, 273
Angus, 45, 48, 218
annexed estates see forfeited estates

Appin of Dull, 87
Applecross, 64
arable, 25, 85, 105–6, 108–9, 119,
    122–3, 127, 129, 132, 183
    ecology, 105, 120–1, 124, 128, 136–8,
        167, 180
    ploughed land, 129–30
    spaded, 128–30
Ardgour, 143–4, 196
Ardluing, 213
Ardnamurchan, 51, 67, 73, 91–2, 100,
    143, 168
Ardvoile, 94–5
Argyll, 51, 53, 147, 217
Argyll, Duke of, 86, 124, 145, 193, 204,
    207, 212, 213–14, 220, 237
Argyll estate, 143, 147, 220
Arisaig, 146, 156
Armit, I., 15, 24–6, 30, 32, 36, 162, 164
Arnol, 161–4, 243–5
Arnprior, 193
Aros, 102
Arran, 16, 51
Assater, Shetland, 253
assessments, 64–73, 96, 101, 105, 181–2
    land, 64, 70–1, 105
    person, 70–1, 106
Assynt, 64, 66, 123–5, 134, 165, 169,
    207–8, 216, 230, 249, 262
Atholl estate, 262
Atholl, Duke of, 262, 273
Atkinson and Marshall, sheep farmers,
    229

Atkinson, J. A., 160
Auchuirn, 88

Bäcklund, J., 71
Baddidarach, 207
Badenoch, 121, 140, 160, 233
Badluarach, 250–1
Bal-, 49–50
Balallan, 245
Balchladich, 207
Balemore, 242
Baleshare, North Uist, 21, 33, 29, 63, 85, 241–2
Balewilline, Tiree, 126
Balfour, D., 130, 132, 180, 254
Ballemore, 94–5
Ballin Smith, B., 97
Ballone, 241
Balmoral estate, 273
Balnagown, 210, 219, 261
Balnasuim, Perthshire, 109
Balquhidder, 216
Balranald, South Uist, 66, 242
Bangor-Jones, M., 71, 73
Bannerman, J. W., 51
Barcaldine estate, 66, 199
bark peeling, 144–5, 196, 198
barley, 2, 21, 33–5, 69, 201–2, 215
Barnhouse, Orkney, 13, 15, 17
barony court acts, 108, 110, 144–6, 183–4, 258
Barra, 31, 161–2, 164, 166, 169, 178, 233
Barrisdale, 157, 201, 212
Barrow, G. W. S., 48, 64
basket houses *see* creel houses
'bays' area, Harris, 128, 245
Beinn na Mhic Aongheis, South Uist, 164
Belly parish, 200
Ben Lawers project, 160
Benderloch, 158
bere, 33, 67, 91–2, 120, 122, 124, 134, 136, 200–1
Bernera, 85
Berneray, 129
Bhaltos Peninsula, Lewis, 30
Bigelow, G., 33, 59, 142
Bighouse, 248
Bil, A., 110–11
birch, 121, 143
Birsay, 209, 252
Blackadder, J., 133, 232, 238–9
blackface, 229, 273

blackhouses, 3, 161–5, 245
Board of Trustees for the Manufactures, 202
bolls of sowing, 64, 67, 109, 125, 119
Book of Deer, 48, 64
Boreraig, 232, 239
Borghastan, 243
Bornais, South Uist, 61, 164
Borrafiach, Waternish, 169–70, 180
Borve, Harris, 128
Boswell, J., 2, 157
Botuarie Beg, 141, 145
bowmen, 142
Brae of Mar, 110
Braes, Skye, 277
Bragar, Lewis, 94, 96, 164–5, 243–4
Brand, A., 132
Branigan, K., 31, 162
Breadalbane estate, 108–9, 111, 119, 134, 141–4, 146, 158, 197–200, 212–13, 217–18
Breadalbane, Lord, 143–4, 216
Bresikor, Tusdale, 171
brewing, brewseats (inc. alehouses), 92, 200, 210
British Linen Bank, 202–3
British Society for Extending the Fisheries, 206–7, 212
brochs, 24–5, 29, 46, 53
Bronze Age arable, 21
Bronze Age, farm mounds, 20
Bronze Age, settlement, 16, 19–23, 47
Brora, 248
Broun, D., 71
Buay, Orkney, 103
Buchan Plateau, 48
Buckquoy, Orkney, 29, 45
Buness estate, Unst, 252
Bunessan, 207
Burray, 257
Burroughs, General Frederick, 253–4
busses, 132, 205
Bute, 48, 51
butter, 33, 60, 91, 110, 132, 138, 141–2, 182, 271
byres, 35, 47, 58–60, 62, 91, 134, 140–1, 146, 157–9, 161–3, 165–6, 244

Caird, J., 245–7, 249–50
Caithness, 18, 47, 229, 257
calf, slaughter, 33, 60, 141–2
Callanish, Lewis, 17
Callander, 202, 212

Cameron, E. A., 277
Campbell of Glenure, 144
Campbell of Knockbuy, 215–16
Campbell, A., 3
Campbell, Alexander, 124
Campbell, E, 31
Campbell, J., 217
Campbell, K., cottar, 230
Cape Wrath, 207
Carie, Rannoch, 122
Carmicheal, A., 3
cas chrom, 2, 124, 138, 167–8
cas dhireach, 124, 138, 167–8
casting of turf, 254–5
Castle Grant, 262
cattle droving, 215–17, 271
cattle, 121, 123, 131, 138, 215–17, 233, 271
cattle, specialist producers, 216, 220
cave sites, 8
Cavers, G. and Hudson, G., 36, 176
Central Highlands, 22–3, 51, 86, 102, 119–23, 139, 200–1
changehouses, 201
Chapman, G., 243
charcoal, 10, 12, 196–8
cheese, 33, 60, 69, 91, 110, 138, 141–2, 182, 271
Cheviot breed, 228–9, 260, 273
claims of kindness, 210
Clanranald estate, 203
Clarke, Charles, sheep farmer, 230
Clashgour, 111
Clashmore, 207
clearance cairns, 13, 18–19, 173
clearances for sheep, 4, 118, 120, 142, 154, 176, 192, 210, 217–20, 227–33, 240, 253–4
clearances, touns cleared in Farr and Strathy district, 229
Clettravel wheelhouse, 33
clientage, 53, 70, 88
climatic degradation, 18, 29, 35, 82–3
Cluny estate, 87, 122
Cluny Mains, 138
Clushfern, 83
Clydesiders, 205–6
Clyne, 228
Coigach, 193, 196, 199, 212
Coile a Ghasgain, Skye, hut circle, 15
Coll, 31, 51–2, 85, 125, 155–6, 169, 195
Colonsay, 9–10, 66, 201
Commission of Inquiry into the

Condition of the Crofters and Cottars in the Highlands and Islands, 277
Congested Districts Board (1897), 278
coppicing, 145, 199
Corpach, 219
cottars, 120, 124, 127, 132, 220–1
Coupar abbey, 101
couples, dairying system, 141–2
Cowal, 51
cows, 109, 138, 141–2
crannogs, 23, 53
crannogs, timber use, 53
Crathie parish, 262
Crawford, B. E., 56, 61, 97
Crawford, I., 164
creaga, 161–2, 164
creel houses, 145, 156–7, 159
Creich, Mull, 207
Crieff, 202, 215
croft land, 156
croft layouts, 243–4
croft size, 212–13
crofter fishing townships, 192, 208, 212–13, 249, 252
crofters, direct action, 276–7
crofters, grievances, 276–8
crofting, definition, 251, 277
crofting counties, 277
crofting townships, 230–1, 235–54
crofting townships, concept of, 212
crofts, 199, 211, 232, 248
crofts, absenteeism, 279
crofts, decline in total number, 279
crofts, pre-clearance, 90, 132, 185
crofts, removal of hill grazings, 240
crofts, re-organisation, 240
crofts, sub-division, 237–9, 244–5
Cromartie, Earl of, 206
cropping, 107–8, 124, 133–4, 134–6
crown feudalism, 71, 89–90
Cùl a'Bhaile hut circle, Jura, 15
Curle, A. O., 47
Curwen, L., 3
cutting of peat, 254–5

Da Biggins, 56–7, 96, 130
dairying, 45, 60, 91, 141; *see also* couples, calf slaughter
Dalarossie, 262
Dalriada, *Dál Riata*, 50, 52, 68
Darling, F., 255, 279
Davies, A., 12

davoch, *dabhach*, 64–6, 68, 71, 93, 100–2, 106–7, 122–3
Dawson, A. G., 83
Deeside, 196
demographic change, 80–2
Deskie, 98
deterioration of sheep pastures, 260–1
Dibidale, 241
Disher and Toyer barony court, 108, 146, 158–9
Disher, 217
Dockrill, S., 34
Dola, Lairg, 19, 180
domestic industry, 202–3
Dornoch, 228
Dr Webster, 82
Drimnin enclave, Morvern, 220
Drimore, South Uist, 60–1
Driscoll, S. T., 50
droving, 92, 216–17, 232, 271
Druim nan Dearcag, Loch Olabhat, 164–5
Drumturn Burn, Perthshire, hut circles and field systems, 14, 33
Drying barns, kilns, 157, 173
Drymochoind, 129
Drynoch, 240
Duirinish parish, 232
Dun Bharabhat, cellular huts, Lewis, 29, 46
Dun Carloway broch, 25–7
Dun Cuier, Barra, 46
Dun Vulan, South Uist, 33
Dunbartonshire, 217
Dunbeath Water, Sutherland, 19
Duncan, H., 52
Dundas, Sir Laurence, 209
Dunkeld, 262
duns, 24, 31, 51–3, 180
dutch fishing, 132
dutchas, 89
dykes, 124, 134–5

Easson, A. R., 68, 70
Easter Raitts, Badenoch, 140, 160
Easter Ross, 257
Eastern Highlands, 22, 107, 196
economic self-sufficiency, 271
Eilean Domhnuill, North Uist, 11–12, 16, 255
enclosures, 125, 155–6, 168–72, 174, 176, 179–80, 214–15
Ensay, 129–30

environmental degradation, 11, 17–18, 255–6
Eoligarray, Barra, 278
Europie, Lewis, 137
Evans, Estyn, 2, 4, 177

Fairhurst, H., 19–20, 154, 160
family settlements, 99
famine, 81, 84
farm enlargement, 208
farm mound, Bronze Age, 20–1
farming toun, development, 177–84
    how defined?, 93
    'proto' forms, 177–8
Farquharson, John, 119–20, 136
Fenton, A., 167, 251
Fetlar, 97, 130–1, 253
feu tenures, 133
feudalised tenures, 89–90
Fife, 48
Fife, Earl of, 262
Finlarig, 145–6
fir woods, 121, 142
fishing, 9, 132, 192, 205–8, 212–13, 230, 249, 252
flock composition, 233
flock size, 229
flooding, 12, 16, 86–8
fodder crops, 211, 33
Fojut, N., 21
folds and flakes, wattle construction, 135, 141
foot plough, 127, 137–8, 167–8, 232; *see also cas chrom, cas dhireach*
Forbes of Culloden, Duncan, 136, 195
forfeited estates, 2, 120, 123, 193, 206, 219, 270
Forfeited Estates, Board of Commissioners, 2, 120, 193, 206, 211
forts, defended enclosures, 23
forts, timber laced, vitrified, 23
Four-mail lands, Tiree, 213–14
fuel, turf, 12
Funzie, 130–2, 182

Gailey, R. A., 96, 215
Gairloch, 98, 101, 142, 169, 172, 176, 207, 249
Garnett, A., 206
garths, 180, 130, 183
Gaskell, P., 220
Geddes, A., 96, 155, 182

Geikie, A., 3
George Gunn, sheep farmer, 230
Gilpin, W., 27
Glassary, 215
Glen Achary, 219
Glen Affric, 12
Glen Bracadale, 127
Glen Brittle, 232
Glen Calder, 110
Glen Camgory, 123
Glen Campsie, 170
Glen Cannich, 143
Glen Corievou. 110
Glen Dessary, 219
Glen Etive, 196, 198
Glen Isla, 23
Glen Livet, 100, 121–2, 200
Glen Lochay, 218–19
Glen Lyon, 87, 111, 135, 141
Glen Morriston, 202
Glen Nevis, 143
Glen Quoich, 219
Glen Shee, 23
Glen Strathfarrar, 110, 143
Glen Urquhart, 263
Glenarigolach, 172–3
Glenavon, 235
Glencoe, 275
Glendale, 240, 277
Glenelg, 48–9, 64, 143
Glenelgchaig, 88
Glenevoe, Loch Lomond, 217, 272
Glenfalloch, 217
Glenfinglas Falls, Perthshire, 275
Glengarry, 110, 219, 231
Glenmore, 122, 143, 196
Glenmoriston, 157
Glenorchy, 119, 143, 196, 198, 218
goats, 109, 121, 145
Golspie, 228
Gordon estate, 65, 110, 145, 159,
    210–11, 235
Gordon of Abergeldie, 262
Gordon of Straloch, 65, 100, 107
Gortien, 31
graddaning, 2
grain marketing, 92
Grampians, 210
grass-arable systems, 136
Gravis, Orkney, 108
Gray, M., 217
grazing, regulation, 109
Great Glen, 48, 111, 137, 143, 193

Great Wood of Caledon, 5, 142
Greaulin, 239
Greninish, 241
Grimbister, 99
Grimsay, 204, 242–3
Gruting, Shetland, 13, 15
Gurness, 25–9, 46
Gurness, Bronze and Iron Age huts, 25–9

haaf fishing, 252
hained grass, 107, 109, 110, 139
Hamilton, J. R. C., 57
'hamlet farms', 155
Harris, 67, 85, 92, 96, 104, 127–30, 137,
    144, 168, 243
Harris, J., 22
Harris, Rum, 165
harvest failure, 84
harvest stubble, 107, 109, 134, 168
hat making, 252
haugh land, 86, 121
hay, 120, 139–40, 186
Hay, sir George, 196
hazel, 143, 145
head dyke, 69, 105, 106, 120, 138, 155,
    169, 172, 177, 180, 183, 186, 219
Hebrides, 2–3, 9, 12, 22–3, 29, 31, 44,
    48, 68, 73, 81, 83, 89–90, 102–3,
    107, 118, 124–30, 135, 140, 199
Hebrides, removal of runrig, 238
Hedges, J. W., 25
Helmsdale, 248
henge monuments, 16
herring, 206–7
'Highland controversy', 3
Highland Land Reform Association,
    277
Highland myths, 274–5
'Highland policy area', 277
Highland symbols and images, 275
Highlands, changes in landownership,
    193, 231
Holm, 209
Holme parish, Orkney, 96
Holocene, 18
Home, John, 123–4, 134, 207
Home's survey, 207
Hopeman, 247
horses, 33, 109, 121, 167
house construction, stone, 147, 156,
    158–9
    cycles of construction and destruction,
    157–8, 160

turf, 12, 16, 156–9, 160, 164, 173, 254, 260
  wattle, 145–6, 156–8
house custom, *bes-tige*, 51, 70
'house system', 51, 68, 70, 97
housing change, 161–7
Houstry Burn, 19
Howe, 36
Howe, Iron Age settlement, Orkney, 24, 29, 45
Huisnish, 128
Hunter, J. R., 36
hunter-gatherers, 8–9, 11
hunting forests, 110–11, 128–9, 173, 234–5, 237, 245, 248, 274
  keeping of greyhounds, 234
  livestock grazing, 235
  nineteenth-century recovery, 235
Huntly, Earl of, 111
hut circles, 19–23

Ibidale, 127
Iceland, volcanic eruptions, 18, 82
Illeray, 85–6, 165–6, 241–2
impact of commercial sheep farming on heather muirs, 258–60
Improvers, 208–15
infield, 2, 4, 35, 105–7, 120–1, 123–4, 134, 139, 156, 177, 180, 187
infield, 'proto', 35, 105
infield-outfield, 2, 4, 107–8, 112, 133–4, 177
Inveraven parish, 200
Invercauld, 262
Invermally, 123
Invernally, 110
Iona, Statutes of (1609), 91, 195, 270
Iron Age settlement, 19–22, 24, 29, 45, 47, 177
Iron Age, arable, 32
Iron Age, field systems, 33, 47
iron smelting, 197
Islay, 9–10, 53, 66, 73, 96, 119, 169
Isle Martin, 206

Jackson, K. H., 3
Jacobite rebellion (1715), 193, 271–2
Jacobite rebellion (1745), 120, 193, 270–2
Jarlshof, Shetland, 28, 58–60
Johnson, S., 2, 157, 167, 263
Jonsson, F. A., 2, 194, 211
Jura, 9–10, 51, 92, 197, 201, 203

kailyards, 93, 123, 155, 165, 170
Kebister, Shetland, 21
kelp, kelp making, 203–5, 207, 230–1, 238, 252
Kenknock, 219
Kenmore, Loch Fyne, 207
Kil-, 50–1
Kilbaan parish, 84
Kilcolmkill perish, 84
Kildonan parish, 29
Killigray, 129–30
Killin, 200
Kilmaluag, 127
Kilmorack, 21, 81, 140, 157
Kilmuir, 127
Kilpheder, South Uist, 24, 61
Kilphedir hut circles, Sutherland, 14–15
Kiltyrie, Perthshire, 160
kin groups, 51, 70
kin-based tenures, 88–9
Kincardineshire, 218
King William's lean years, 1690s, 81, 132
Kingarloch, 144
Kings cottagers, 212
Kingsburgh, Trotternish, 171, 180
Kingussie, 121
Kinloch Rannoch, 212
Kinloch, Rum, 9
Kintyre, 51–2, 83–4, 91, 96–7, 69, 100, 134, 141, 202, 216
Kirkhill and Pharnaway, 211
Kirkibost, 241–2
Knapdale, 215
knitting, 252
Knock, Mull, 197
Knockintorran, 242
Knockline, 242
Knoydart, 134, 156, 193, 199, 206, 231

labour services, 49, 53
ladle or horizontal mill, 199
Laggan, 121
Laggenfern, 123
Laidchroisk, 110
Lairg, 160, 228, 299
Lamb, H., 82
Lamont, W. D., 51, 68, 70
land allocation, 108, 124, 133
land drainage, 156–7, 254, 256–7
land improvement, 254, 257, 257

Land Settlement (Scotland) Bill (1919), 276–7
landlord debt, 192–3, 273
Landseer, E. H., 275
Langland, G., 155, 213
lasts, 130
Laxobigging, 130–1, 182
lazy beds, 124, 128, 136–7, 243
Ledour, 141
leisure estates, 234–6, 275–6
Lerwick, 205
Leveson-Gower estates, 228, 274
levy, military, 68, 70
Lewis, 5, 29, 54, 97, 155, 165, 173, 178, 201, 216, 243, 248, 256
Lianach, Perthshire, 101, 161
Lindsay, J. M., 199
linen, 202, 205, 252
Lismore, 144, 199
Little Formestoun, 133
Little Ice Age, 34, 81–2, 84–7
livestock, 22, 33, 120, 198
livestock, indoor wintering, 35, 60, 140–1
livestock, outdoor wintering, 140, 144, 198
livestock regulations, 109–10
livestock soums, 67, 69, 109, 141
Lix, 4, 160, 162
Loch Arkaig, 123, 196
Loch Assynt, 123
Loch Awe, 23, 52
Loch Badanloch, 20
Loch Beag, 127
Loch Broom, 20
Loch Creran, 199
Loch Eil, 123, 220
Loch Eishort, 232
Loch Ericht, 138
Loch Erisort, 247
Loch Fyne, 206, 215
Loch Leven, 83
Loch Linnhe, 219
Loch Maree, 143, 197
Loch Naver, 228
Loch Ordais, 94
Loch Quirn, 247
Loch Quoich, 231
Loch Sleadale, 256
Loch Sunart, 144
Loch Tarbert, 215
Loch Tay, 23, 94, 134, 199

Loch, J., 141–2, 157–8, 228, 249, 256, 260
Lochaber, 88, 111, 143, 145, 231, 260
Lochbay, 206–7
Lochbroom, 99
Lochbuie estate, 159, 197–8
Lochdochart, 217
Lochiel estate, 123, 157, 219
Lochtayside, 119–20, 136, 139, 143, 146, 183, 214, 216, 218
longhouse forms, 47–8, 161–5
Lordship of Huntly, 200
Lorgill, 127, 240–1
Lovat estate, 159, 211
Lower Kinchrackin, Breadalbane, 198
Luing, 83, 213
Luskentyre, 128
Luss parish, 199

McArthur, John, 119
McCormick, F., 142
Macculloch, J. R., 168, 229
Macdonald, Lord, 202, 232
Macdonald estate, 127, 195, 201, 238, 241, 262
MacDougall, S., 119
Mackay, Donald and William, sheep farmers, 229
McKenzie, Kenneth, sheep farmer, 230
McKerral, A., 68, 71, 73
MacKie, A., 24
Mackillop, A., 211–12
Mackintosh estate, 144, 262
Macklin, M., 10
MacLeod estate, 127, 129, 155, 155, 159, 170, 240, 262
MacLeod of Dunvegan, 232
machair, 18, 20, 29–30, 32–3, 61, 63, 84–6, 128–9, 136, 167, 184
Machrie Moor, Arran, hut circles and field system, 16
Mains of Clunie, 121
malies, 67–8, 103
malt, malthouses, malting, 73, 92, 185, 200–1
Mamlorne forest, 111, 135, 140
Manish, Harris, 128
manure, 21, 61, 62, 108, 139, 186
  animal dung, 34–5, 61, 105, 107–8, 136, 140, 249, 261
  ash, 21, 34, 61–2, 143
  ferns, 123, 134
  'manurit' land, 108, 134

peat, 21, 62–3, 136, 256
  seaweed, 132, 134, 136, 187, 203
  soil, 109, 136
  thatch, 62–3, 107, 134, 157
  turf, 21, 34, 62–3, 134, 136, 157, 254, 256
marketing, 91, 193–4, 220–1, 271
Marshall, W., 136
Martin, M., 142
Marwick, H., 55–7
Mather, A., 110
Matheson family, 243, 246, 273
Maunder Minimum (1645–1702), 81–3
meadow, 74, 86, 88, 105–7, 110, 120–1, 138–9, 183, 219
medieval climatic optimum, 82
Megalithic tombs, 17
meithing and marching of land, 103–5, 108, 182
Mellon Charles, 250
Melness, 229
Menzies, A., 141, 261
Menzies estate, 144
merklands, 66–9, 72–3, 109, 141
Mesolithic–Neolithic transition, 11
Mey estate, Caithness, 203
midden sites, 8, 10, 13, 21, 29
milk cows, 138, 141
milk production, 17, 33, 47, 60, 141–2
Millais, J. E., 275
mills, 128–9
Mingulay, 31, 169, 233
Mithen, S., 9
Moisley, A., 242
Monaltrie, 193
Moniack, Inverness-shire, 99
Monro, M., 108
Morar, 231
Moray, 48, 65–6, 72, 247
Morton and Culley, sheep farmers, 229
Morvern, 67, 144, 216, 220, 231
Morvern, changes in landownership, 231
Mounth, The, 48
Moy, 262
muir, 121–2
muir or heather burning, 144–5, 257–8
Mull, 51, 67, 83–4, 102, 142, 159, 169
multiple tenancy, 101–3, 124, 133, 155, 182
Murchison, R., 3
Murlagan barony, 120
Murray, Sir Alexander Murray of Stanhope, 119

Napier Commission, 105, 194, 230, 232–3, 235, 245, 252, 256, 260, 277–8
Napoleonic Wars (1793–1815), 199
Naver, 248
Neolithic, early farming, 11–13
Neolithic, hut circles, 13, 16
Neolithic fields, 16
Nesting, Shetland, 109, 131
Netherlorn, 51, 67, 83, 96, 119, 134–6, 141, 201, 212–13, 218
New Extent, 65, 69, 106
New Statistical Account, 200, 232, 276
New Tarbert, 202
Nicolaisen, W. F. H., 49–50, 54–5
Nieke, M. R., 31, 52
Nisibost, 128
Nithsdale, 217
Noodmore, 121
Norse, assessment, 69, 71, 96, 184–5
Norse, nature of colonisation, 46, 54–7
Norse housing, 61, 166
Norse housing, longhouse forms, 46, 57–9, 164, 166
Norse place names, 54–6
Norse settlement, 36, 54–7, 61–2
Norse settlement, chronology, 33, 57–60
North Dell, Lewis, 21
North Uist, 67, 162, 203, 232, 241
Northern Highlands, 4, 13, 17, 22–5, 31, 33–4, 43–4, 48, 53–7, 68–9, 86, 89, 96–7, 105, 108, 118, 176
Northern Institution, 276
Northern Isles, 130, 184, 205
Northern Isles, crofts and smallholdings, 250–54
Norway spruce, 273

oats, 22, 33–5, 67, 109–10, 120, 122–4, 134–6, 138
Oban, 8, 10
Old Extent, 65, 69
Old Scatness, 34, 36
Old Statistical Account, 88, 215, 220
Onich, 219
Orkney, 5, 11, 55, 57, 63, 69, 71, 89, 99, 137, 155, 210, 257
Ormiscaig, 249–50
ornamental tree planting, 261
Oronsay, 8–11
Ortoun, Elgin, 104
ounceland, 66, 68–9, 71; *see also* tirunga

outfield, 2, 4, 106–7, 120–4, 134–5, 170–1, 187
outsets, intakes, 96, 107, 172
over-population, 193, 203–4, 214
oxgates, 64

Pabbay, 85, 92
Paiblesgarry, 242
Papa Stour, 57, 63, 97, 130, 252, 255–6
Papa Westray, 62
Park district, 168, 245–7
Parker Pearson, M., 29, 35–6, 62, 166, 177–8, 184
peat spade, 254
peat, 5, 21, 29
pennyland, double, 67
pennylands, 51, 66–70, 85, 93, 96, 101–2, 19, 181, 210
Perthshire, 22–3, 33, 48, 180, 217
Perthshire, field systems, 33
Pictish, proto-, 43, 45–6
Pictish hut circles, cellular, 'figure of eight', 44–6
Pictish settlement, 44–50, 180
Pictish standing stones, 44–5
pigs, 33
pine forest, 122, 143, 196
Pit- place names, 44, 48–50
Pitcarmick-type dwellings, 44, 46–7, 180
plaggen soils, 63
planking, 209, 252
planting of larch, 262–3
planting of Scots fir, 262–3
ploughlands, ploughgates, ploughs, 64–6, 100–2, 106
Pont, Timothy, 83
Pool, Orkney, 36, 62–3
population growth, 193, 252
potato, 82, 92, 204–5, 221, 238, 232
potato, failure of harvests, 231, 238

quarterlands, 66
querns, 22, 34, 199–200

Ramasaig, 94, 97, 127, 240–1
Rannoch, 82, 88, 110, 143–4, 218
Reay estate, 228, 273
Redfield, R., 279
Reef, The, 86
renders, 51, 65
rents, 83, 90–1, 96, 123
rents, cash conversion, 216
rents in kind, 90–2

Richards, E., 247
Ridderach, 172
Righcopag, 172–5, 177, 180, 182
Ring of Brodgar, Lewis, 17
Ritchie, A., 43, 45
River Beauly, 18
River Brorar, 228
River Connin, 87
River Dee, 48
River Don, 48
River Helmsdale, 20, 228
River Spey, 48, 87
River Tay, 48
roads, 193, 272, 274
Roag, 241
Robertson, I. M. S. L., 183
Robertson, J., 135, 156–7, 206, 215, 262, 276
Rogart parish, 198, 228–9
Rønneseth, G., 184–5, 105
roof couples, 143, 146, 158–9
Rosal, Strathnaver, 4, 154
Ross, A., 65, 72, 181
Ross, Sir John Lockhart, 219
Ross-shire, 272
Rothiemurchus, 18, 121–2, 143, 196
roundhouses, 24–5, 31–2
Rousay, 17, 253, 257
Roy, General, 119, 142, 215, 261
Rum, 9, 256, 273
runrig, 2, 101–4, 108, 127, 130–3, 178, 184, 220, 251
  commensurate divisions, 209–11
  generalised divisions, 209, 213–14
  proprietary, 130, 209–10, 252
  removal, 103–4, 208–14, 251–2
  reorganisation into 'whole township' runrig, 185
Ruskin, J., 275
rye, 33

Saddell abbey, 51
Sahlins, M., 70
St Kilda, 63
  Gleann Mór field system, 16
  Tobar Childa field system, 16
St Ola, 209
sale of highland estates, 273
Salisbury, Marquis of, 273
Sallachan, 144
Sallachul, 83, 111
sand blows and erosion, 20, 29, 63, 84–6
Sanday, 62, 254, 257

Sandvick, 60
Sandwick, 209
Scarinish, 202–3, 207–8
Scarista, Meikle and Little, 85, 104, 128
Scarpa, 129
scattald, 252–3
Schulting, R. J. and Richards, M. P., 11
scientific study of region, 276
Scolpaig, 241
Scoraig, 250–1
Scord of Brouster, Shetland, 13–14
Scots fir, 273
Scott, Sir Walter, 274
Scotti, 50–3
Scottish Alpine Botanical Clud, 276
scriddan, 88
Seafield estate, 263
Seaforth estate, removal of runrig, 243–4
seaweed, restrictions on use, 167
Seil, 83
Sellar, P., 229, 247, 249
Senchus Fer nAlban, 51
Settlement, colonisation, 128, 132
    continuity/discontinuity, 4, 6, 20, 29,
        35–6, 45, 47, 50, 53–4, 178, 180,
        185
    expansion, 111, 132, 245
    forms, 93–6, 154
    mounds, 20, 61–2
    shifts, 100
Shapinsay, 209, 254
Sharples, N., 29, 62
Shaw, F., 96, 155
Shawfield estate, Islay, 66
'sheelings', Assynt, 124, 134, 175
sheep, 33, 109, 121, 138, 140–1
sheep, clearances, 217–18, 227–33
sheep farm, re-structuring, 233
sheep farm, scale, 229–30
sheep farms, consolidation, 233
sheep numbers, increase, 229, 233, 272
sheep pens, 176
shell food, 8–11
shepherds, 219–20, 228–9
Shetland, 11, 69, 97, 105, 142
Shetland, spread of sheep, 253–4
shielings, 22, 33, 55, 110–11, 128, 138,
        173, 217, 220, 231, 235
shieling, Ben Lawers, 138, 173
shieling, Monachyle Glen, 174
shieling, Sleadale, Skye 174
shielings, Duirinish, 139
Shinness, 257

silverweed, 136
Sinclair, Sir John, 256
single tenancies, 102, 127, 133, 210, 241
skali, hall house, 57, 60
Skara Brae, Orkney, 13, 16
skat, 55, 57, 69
skattald, 55–6, 69, 253–4
    colonisation by smallholdings, 253
    division of, 253–4
Skene Manuscript, 73
Skene, W. F., 68
skinned land, 5, 255–6
Skye, 12, 31, 44, 67, 126–7, 135, 155,
        159, 199, 201, 231–2, 238
Sleat, 127
sliabh, Slew-, 50
sliochd, 195
Slisgarrow barony, 120–1
smallholdings, 132, 137, 194, 211
Smith, Adam, 194
Smith, Annette, 193
Smith, Campbell, 249–50
Smith, Octavius, 273
Smout, T. C. and Watson, F., 196, 198–9
Snizort parish, 127
soil, settlement mounds, Orkney, 62
soils, 5, 12, 21, 29, 32, 34, 48, 62–3, 84,
        120, 128–9, 138, 167, 262
Sollas, North Uist, wheelhouse, 28–9
souming *see* livestock soums
South Ronaldsay, 62, 254, 257
South Uist, 178, 185, 201
South Uist, settlement shifts, 184
Southern Highlands, 139
Southern Uplands, 217–18
South-west Highlands and Islands, 102,
        215
South-west Highlands, 31, 50, 52–3,
        68–9, 71, 89
sowing rates, 136
spaded land, 128, 137–8, 167
Speyside, 138, 196, 263
splitting, place names, 98–9, 101, 104
sporting estates *see* leisure estates
spouty land, 87–8
Stafford, Lord later Marquis of, 228
stake and rice (ryce), 145, 156
Stanydale, Shetland, 13, 15
Stewart, 174
Stiennes, A., 71
stocking capacities, 260
stocking levels, 119
stockings, 205

Stones of Stenness, Orkney, 17
Storer Clouston, J., 99, 185
storminess, 83, 85
Strachur and Stralachlan parish, 88
strategies of change (1700–1914), 194
Strath Brora, 19
Strath Fleet, Sutherland 19
Strath Frithe, 228
Strath of Kildalton, 22
Strath of Kildonan, Sutherland, 19
Strath Oykell, 219
Strath Rusdale, 219
Strath Seilge, 228
Strath Skinsdale, 228
Strathardle, 23
Strathavon, 98, 104, 121, 121–2, 211
Strathcarron, 219
Strathdon parish, 258
Strathearn, 111
Strathfillan, 119, 217
Strathglass, 143
Strathnasalge, 172, 176
Strathnaver, 19, 169, 248
Strathyre, 3
Stromness, 209
Stronsay, 203, 254
Struan estate, 120, 212, 139, 172
Stylegar, F.-A., 105, 183, 185
subletting, 102
subsistence, 92
subsistence shocks, 81–2
Suisnish, 232, 239
Sunart, 69, 73, 91, 98, 138, 143
sun-divisions, 103–4
Sutherland, 18, 44, 47, 198, 227, 229–30
Sutherland, Elizabeth, countess of, 228,
    273–4
Sutherland, field systems, 33
Sutherland estate, 228

tacks, 101, 103, 133, 147
tacksmen, 85, 90, 102, 124, 127, 132–3,
    195, 209–10, 216, 220, 230, 232,
    239–43, 263, 271
tacksmen's houses, 195
Talisker, 262
Taminlienen, Glenlivet, 100–1
Tanera, 206
tanning 196
tanning bark, 144–5, 147, 196–9
Taransay, 129
tathing, tathfolds, 107, 109, 123, 134,
    136, 140, 146, 171, 187

Taylor, D. B., 19–20
Taymouth, 214, 261
tenant numbers, 155
tenures, verbal agreements, 103
thanage, 48–50
thirling, 199
Thomas, F. W. L., 161, 164
Thomson, W. P. L., 57, 71, 97, 99, 105,
    130, 132, 182, 185, 252
tile drainage, 248, 254, 257
timber felling, 12, 144, 147, 196
timber use, charges, 146–7
Tipping, R., 12
Tiree, 31, 51, 66–7, 72, 82–7, 92, 99,
    102–3, 109, 124–5, 127, 135–7,
    140, 147, 155, 202–4, 213–14
tirunga, 66–7, 72, 93, 102
Tobermory, 206–7
Tofts Ness, Sanday, Orkney, 33–4, 36
Tongue parish, 229
Totardor, 127
touns
    abandonment, 82–4
    definition, 154
    expansion, 186–7
    occupancy, duration of agreements,
        133
    splitting, 57, 97–9, 100–1, 104, 181,
        186
Toyer and Disher, 108
tree planting by tenants, 261–2
tree planting, exotic species, 261–2
trenching, 256–7
Trotternish, 127, 169, 239
Trumpan-more and -beg, 127
turf, 5, 86, 158, 254–6
Turnbull, J., 20–2, 67, 86, 124, 126
Turnock, D., 231
Tusdale, Minginish, 165, 170, 232, 256
Tyndrum, 217

Udal, The, North Uist, 4, 20, 46, 61,
    164
Udal tenure, 88–9, 97, 130, 133, 251
Uhlig, H., 177
Uists, 29, 30–2, 85, 108, 125, 136, 137,
    161, 203–4, 233
Ulinish, 127
Ullapool, 206, 212
Ulva, 169
Unapool, 207
Unish, 155
Unst, 253

Vallay, North Uist, 31, 241
Vatersay, 31
Vatten, 241
Victoria, Queen, 273
Viking phase, chronology, 57–60

Wade, General, 193
wadsets, wadsetters, 195, 210
wags, 44, 47
Wainwright, F. T., 54
Walker, Dr John, 2, 85, 124–5, 155, 194,
    201–2, 211, 263
Waternish, 72, 155, 169
Watson, F., 143
wattle, 121, 123, 135, 143
weaving, 202, 248, 252
weeds, 136
Wester Ross, 143, 197
Western Highlands, 6, 49, 200
Western Isles, 4, 17, 21, 25, 31, 43,
    53–5, 57, 61, 63, 102
Westray, 62, 66, 254, 257
wheat, 33
wheelhouses, 24, 28, 31, 33, 46

whisky distilling, 92, 200–2, 248, 273
Whittington, G., 48–9
Wickham-Jones, C., 9
Williams, G., 71–2, 181
winter feed, 107, 215
wintering and wintering ground, 105–6,
    110, 123, 140, 144–5, 183, 233,
    215–16, 229
Withers, C., 274, 276
woodland, 5, 119, 142, 196–9
    clearance, 5, 11–12
    damage by livestock, 121, 140,
        143–5
    enclosure, 198
    wood keepers, rangers, officers, 143,
        147
    wood leave, 144, 147
wool prices, 231, 233
Wordsworth, D., 275

yarromanna, 103
Yell, 253
Young, W., 207–8, 247, 249
Youngson, A. J., 200